A First Course in Computational Physics and Object-Oriented Programming with C++

Because of its rich object-oriented features, C++ is rapidly becoming the programming language of choice for science and engineering applications. This text leads beginning and intermediate programmers step by step through the difficult aspects of scientific coding, providing a comprehensive survey of object-oriented methods.

Numerous aspects of modern programming practice are covered, including object-oriented analysis and design tools, numerical analysis, scientific graphics, software engineering, performance issues, and legacy software reuse. Examples and problems are drawn from an extensive range of scientific and engineering applications. An emphasis on the fundamental logical principles of the language, combined with short, focused illustrations and discussions, helps promote rapid learning. The book also includes a CD-ROM with a full set of free programming and scientific graphics tools that facilitate individual learning and reduce the time required to supervise code development in a classroom setting.

This unique text will be invaluable both to students taking a first or second course in computational science and as a reference text for scientific programmers.

DAVID YEVICK is a leading scientist in the application of numerical modeling to optical communication systems, notably guided electric field propagation in waveguides and fibers, optical processes in semiconductors, and most recently optical communication systems. Over the last 25 years, he has collaborated with numerous industrial and government research establishments, including spending four years with the IBM Center of Advanced Studies, Canada, on practical applications of the VisualAge for C++ toolset.

Dr. Yevick has extensive teaching experience in both electrical engineering and physics. Since 1999, he has been a full professor of physics at the University of Waterloo, Ontario having previously held positions in the Electrical and Computer Engineering Departments of Queen's University at Kingston and the Pennsylvania State University, in the Solid State Physics Department of Lunds Universitet and at the Institutet för Optisk Forskning, Stockholm. He has authored or co-authored over 120 refereed journal articles. Dr. Yevick is a fellow of the American Physical Society, the Institute of Electrical and Electronics Engineers, and the Optical Society of America and is a registered Professional Engineer, Ontario.

A First Course in Computational Physics and Object-Oriented Programming with C++

David Yevick

PUBLISHED BY THE PRESS SYNDICATE OF THE UNIVERSITY OF CAMBRIDGE
The Pitt Building, Trumpington Street, Cambridge, United Kingdom

CAMBRIDGE UNIVERSITY PRESS
The Edinburgh Building, Cambridge CB2 2RU, UK
40 West 20th Street, New York, NY 10011–4211, USA
477 Williamstown Road, Port Melbourne, VIC 3207, Australia
Ruiz de Alarcón 13, 28014 Madrid, Spain
Dock House, The Waterfront, Cape Town 8001, South Africa

http://www.cambridge.org

First published 2005
Reprinted 2006

Printed in the United Kingdom at the University Press, Cambridge

Typeface Times New Roman 10/13 pt. and Univers *System* LaTeX 2_ε [TB]

A catalog record for this book is available from the British Library

Library of Congress Cataloging in Publication data

ISBN 0 521 82778 7 hardback

The publisher has used its best endeavors to ensure that the URLs for external websites referred to in this book are correct and active at the time of going to press. However, the publisher has no responsibility for the websites and can make no guarantee that a site will remain live or that the content is or will remain appropriate.

To my parents, George and Miriam, who were the first to teach me physics and mathematics, my wife Susan for her unending support and to Ariela, Hannah and Aaron who I hope one day will find this book useful

Contents

Part I

C++ programming basics

Chapter 1
Introduction

This textbook is the result of seven years of experience with teaching scientific programming in both science and engineering departments. The book has a single, clearly defined goal – to convey as broad an understanding of the entire scientific computing field as possible within a single term course. Accordingly, while the C++ programming language is explained concisely, its conceptual foundation is emphasized. Once this framework is understood, the complex language syntax can be far more easily retained. Further, all features of modern programming of relevance to scientific programming are surveyed with emphasis on strategies for simplifying coding. Free software tools are included that minimize the technical hurdles of coding and running programs.

1.1 Objective

As stated above, this textbook presents a broad introduction to modern scientific programming. This includes numerical analysis, object-oriented programming, scientific graphics, software engineering, and the modeling of advanced physical systems. Consequently, knowledge of the material will provide sufficient background to enable the reader to analyze and solve nearly all normally encountered scientific programming tasks.

1.2 Presentation

This text is concise, focusing on essential concepts. Examples are intentionally short and free of extraneous features. To promote retention, the book repeats key topics in cycles of gradually increasing difficulty. Further, since the process of learning computer language shares many similarities with that of acquiring a spoken language, the most important sample program segments are highlighted in gray. *Memorizing these greatly decreases the time required to achieve proficiency in C++ programming.*

1.3 Why C++

The relevance of C++ to scientific computing is somewhat controversial. On the one hand, an extensive amount of legacy procedural code exists in FORTRAN and C, while recent programming languages, such as Java and C#, possess formal advantages. Programs can often as well be written far more rapidly in dedicated scientific and symbolic languages, such as MATLAB®, Maple®, and Mathematica®. The relative advantages and disadvantages of C++ will therefore be considered briefly in the following paragraphs.

Object-oriented and procedural languages: A procedural language represents the transformation that a physical or abstract entity performs on a set of data by a high-level computing module, termed alternately a procedure, subroutine, or function. The module can be reused with the same or different input data when the transformation is repeated. A program effects an ordered flow of data into and out of a succession of these modules. The underlying simplicity of this paradigm yields programming languages that are easily learned and applied. The first and most successful of these from a scientific perspective was FORTRAN. Over decades, highly efficient function FORTRAN packages have been written that implement extensive sets of, for example, mathematical, scientific, and user-interface algorithms. However, the syntax of FORTRAN, while easily acquired, permits numerous unsafe constructs that lead invariably to subtle coding errors.

In contrast, the fundamental high-level unit in modern programming languages is an object. An object is a simplified model or *abstraction* of a particular entity. To illustrate, consider for definiteness a voltage meter. The meter has many attributes – in the extreme case the position and velocity of each atom – but only a few of these are typically of interest. These relevant attributes, which could include both user-accessible, **public**, data and behaviors, such as the voltage reading and the meter's response to depressing the power-on switch, and inaccessible, **private**, characteristics, such as the currents through individual circuit elements, compose the abstraction of the object. Similarly, in a C++ program, **public** properties can be accessed throughout the code, while **private** members are only accessible to functions that exist within the object itself, limiting the extent of inadvertent errors.

A framework for creating objects that share similar properties is provided by a class. Functions and data are segregated according to their enclosing class. That is, two functions with the same name but belonging to different classes can be defined without incurring errors arising from name collisions. A class can extend the functionality of a second class by *inheriting* its non-**private** properties. Objects of the new, derived, class can use or redefine the properties of the original class without having to recopy code. Further, refinements to the original class automatically propagate to the new class.

Features such as objects and classes introduce complex programming syntax. Accordingly, an object-oriented language requires far more time to learn than a procedural language. However, the enhanced feature set of C++ simplifies many programming tasks and naturally structures a program into logically independent units. Object-oriented development is therefore often said to be required by programs with over 250,000 code lines.

Since C++ constituted the first widespread object-oriented language and additionally incorporated the earlier C procedural programming language, numerous scientific programming packages have been composed in or translated into C++. Further FORTRAN programs can be, with some effort, accessed from within a C++ program, as detailed in Appendix D.

The additional functionality of C++ enables manipulations that, while rarely used, can introduce unexpected dependences among variables. The language cannot therefore normally optimize code as effectively as FORTRAN. Advanced C++ language features can, however, often be employed to circumvent these difficulties, leading to execution speeds that even surpass those of equivalent FORTRAN programs, cf. Chapter 21.

C++ and other object-oriented languages: A second criticism of C++ as a scientific programming language relates to certain structural limitations relative to more recently introduced object-oriented languages, in particular Java and C#. Java in particular provides a broader standard feature set than C++. Classes that, for example, handle graphics or internet communications are built into the language and are therefore supposedly the same across all Java implementations (although, in reality, version and machine dependencies exist). Java and C# additionally include newer, more flexible procedures for structuring code.

Unfortunately, since the Java and C++ languages are primarily oriented toward the much larger corporate business market, many of their underlying design choices are unfavorable to scientific programmers. Most significantly, C and C++ provide nearly the same range of control of hardware resources through high-level language constructs that is available through the native machine instruction set. This range includes addressing and modifying the contents of individual memory locations and allocating and subsequently releasing the memory available to a program during execution. Since many programmers find such facilities error prone, Java and C# instead perform much of this functionality automatically. Unfortunately, this can lead to unpredictably long execution times and precludes simple corrective remedies. Additionally, the built-in facilities in Java and C#, while useful, do not necessarily address the far more complex requirements of large scientific programs, and are not fully consistent over different software and hardware platforms (C# in particular is bound to the Windows® operating system). Perhaps most significantly, complex mathematical or scientific program libraries generally take years to adapt to new programming languages. Consequently, except for specialized uses such as receiving and viewing program results and data over the internet, C++ will likely remain the foremost object-oriented scientific programming language for years to come.

Specialized programming environments: While the extensive C++ feature set benefits large scientific programming projects, for small proof-of-principle computations the C++ syntax is often cumbersome. Specialized C++ scientific applications require external mathematical and graphical libraries that can be unreliable and platform dependent. In contrast, specialized scientific languages, such as MATLAB®, or symbolic manipulation languages, such as MAPLE® or Mathematica®, provide a simple, easily learned user interface to a unified built-in array of easily called and highly optimized numerical, scientific, and graphical libraries. Programs written in MATLAB can be transformed into C++ code through an add-on product, while C++ and FORTRAN code can be called by a MATLAB program after some effort. Further, a given version of a dedicated package such as MATLAB originates from a single commercial source and therefore functions nearly identically across all supported platforms (assuming the same MATLAB version number). On the negative side, the absence of high-level features, such as classes, type-checking, and user-controlled memory management, can lead to structural confusion, programming errors and runtime inefficiency for larger problems. As well, the software is unavailable at many sites because of its substantial cost.

A brief introduction to MATLAB is presented in Appendix A. A working knowledge of the language can be quickly acquired from this appendix in conjunction with the extensive MATLAB help system.

1.4 C++ standards

As requirements evolve, programming languages undergo periodic revision by a standards organization. For a new language such as Java, revisions can be major and involve central, frequently employed features. Language elements can become deprecated (unsupported) and in certain cases cease to function. In contrast, revisions to the mature C++ language are relatively minor and do not affect core functionality. New C++ standards are often implemented slowly, especially by large software manufactures that have substantial investments in competing solutions. Therefore, programs built on a new standard may only function on a limited set of compilers. Features that must be optimized by the compiler writer for specific hardware platforms or operating systems are deliberately omitted from the language standard. The most recent C++ standard as of the time of writing of this book is termed ANSI/ISO/IEC 14882: 1998. This standard was fixed in 1988 and is now well supported by most compilers.

The programs in this book are generic with one or two exceptions and should function on any recent C++ compiler. Where language features introduced in the 1988 standard are employed, alternative programming methods are typically identified in the text.

1.5 Summary

The organization of this book is as follows: After providing instructions for installing a free C++ programming environment, the basic structure and syntax of the C++ language and of computer hardware and software architecture are summarized in the initial chapters. Chapters 6 and 7 then provide a survey of the principles, structure, and syntax of modern object-oriented programming based on the industry-standard Rational Rose® software engineering package.

Subsequent chapters examine basic C++ constructs, such as control constructs, arrays, input and output, numerical methods, and functions that are required for elementary numerical programming. This is followed by an introduction to numerical analysis including such topics as differentiation, integration, root-finding methods, error estimation, and numerical interpolation.

The last part of the text surveys complex aspects of C++ scientific programming. These include language features, such as pointers, references, dynamic memory allocation, templates, and inheritance. Program optimization and numerical techniques for Monte-Carlo methods and partial differential equation solvers, together with a series of appendices that discuss the MATLAB languages, alternative free compilers, code profiling, calling FORTRAN programs from C++, and, finally, C++ coding standards conclude the text.

The programming exercises at the back of each chapter comprise both standard examination questions and fundamental scientific programming examples from different areas of physics, chemistry, biology, and engineering. A central feature of these examples is that object-oriented skeleton code is generally provided. In this manner, the student naturally absorbs object-oriented design, that is the translation of physical problems into a proper object-oriented framework, which is inevitably the most challenging topic for the beginning programmer. As well, by providing difficult code lines mistakes are avoided that would normally require many hours to resolve. As the book progresses, an increasing amount of detail is omitted from the examples. Solving these problems will rapidly lead to proficiency in object-oriented scientific programming.

1.6 How to use this text

To employ this text to the greatest possible advantage, the reader is encouraged to follow the steps below for the material in each chapter:

(1) Skim through the text.

(2) Reread the chapter, programming and running all the sample programs in the text, the more complex of which are provided in the **textbookprograms** directory (folder) on the enclosed CD-ROM. Attempt to extend these programs and note the results.

(3) *Memorize the programs or program sections marked in gray in the text.*

(4) Complete the exercises in the first part of each exercise set. These are representative examination problems.

(5) Finally complete the more challenging second part programming projects, which illustrate real-world scientific programming. Within this context, review again the material in the chapter.

1.7 Additional study aids

There are many free C++ resources on the Web. Some useful resources, a free tutorial, and links of various difficulty can be found at http://www.research.att.com/~bs/C++.html, the site of Bjarne Strostrup, the originator of C++, and at http://www.cplusplus.com/.

While this textbook makes every attempt to be self-contained, numerous highly readable books on the C++ language intended for general programming audiences are cited in the References and further reading at the end of this textbook, which also contains a listing of more specialized books, grouped according to subject matter.

1.8 Additional and alternative software packages

The Dev-C++ software and Windows operating system chosen for this textbook provide a high level of functionality yet require little student time and instructor intervention to use and maintain. An identical Dev-C++ package exists for Linux, although only the Windows installation procedure is discussed here because of installation differences among Linux platforms.

For readers that prefer command line compilers (which compile more rapidly and are ultimately simpler to use), a command line Windows version of the g++ compiler together with the Borland C++ command line compiler is included on separate directories of the accompanying CD-ROM. Directions for installing and using these products are given in Appendices B and C. As noted in the second of these appendices, once Dev-C++ is installed, its compiler can also be easily accessed within a command line environment.

While numerous comprehensive freeware and commercial C and C++ numerical libraries can be found through a simple Web search, such routines are typically designed for a restricted set of hardware and software platforms. Therefore, for smaller programs that do not require meticulously optimized code, a well-documented elementary source code library often provides the most efficient option. Perhaps the best of these is that which accompanies *Numerical Recipes for C++*, however extensive sets of numerical programs can be found in numerous other textbooks. Additionally, the Linux gsl software library offers a full set of scientific routines for the Linux C (gcc) compiler that can also be called from within C++ (g++) routines. Instructions on installing this code for the Dev-C++ and g++ compilers included on the CD are given in the next chapter.

Other particularly useful libraries include those that address the significantly (20–1000%) slower performance of C++ relative to FORTRAN through the implementation of advanced C++ features for certain mathematical operations, cf. Chapter 21. Commercial C++ mathematical and scientific routine packages include the standard IMSL and NAG libraries. One of the oldest and most stable source code libraries for scientific applications, the CERN library, is also transitioning from FORTRAN and C to C++; however, due to its size, installation is non-trivial and the number of supported compilers is restricted.

Chapter 2

Installing and running the Dev-C++ programming environment

We begin by presenting a step-by-step procedure for installing the freeware Dev-C++ development environment and DISLIN graphics software contained on the accompanying CD-ROM. The programs in this book can also be compiled and run somewhat more efficiently through the stand-alone command line (MS-DOS window) Borland or mingw (g++) compilers. This software is contained in separate directories in the CD-ROM and discussed in Appendices B and C, respectively.

Dev-C++ is a simple integrated set of development tools based on the Linux g++ compiler. It is available both on the native Linux platform and as a Windows port. The platform contains, among other programs, a FORTRAN 77 compiler (g77), a debugger (gdb), and a profiler (gprof).

If you wish to install the latest versions of the software, as opposed to the version on the CD-ROM, the Dev-C++ homepage, which contains updated packages, is http://www.bloodshed.net/devcpp.html.

Links to all free compilers, including the command line versions of Borland C++ and Microsoft Visual C++, are at the time of writing found at http://www.thefreecountry.com/compilers/cpp.shtml. This site currently lists all free C++ compilers.

To install Dev-C++ from the enclosed CD-ROM simply navigate to and double click on the file devcpp4990setup.exe in the Dev-C++ directory. We assume below that you select to install this package into the directory X:\Dev-Cpp, where X: is the drive letter (partition).

2.1 Compiling and running a first program

(a) *Entering a program*: To edit a new program, double click on the Dev-C++ icon and in the program window depress the button marked "New" on the button bar. This activates an editor window and places the cursor at the beginning position of the window. Using the tab key for indentation, type in the lines (the **Sleep(4000)** function which requires the **#include <windows.h>** statement holds the output window open for 4000 milliseconds or 4 seconds when the program is run)

```
// Hello world v. 1.0
// Aug. 11 2000
// (your name)
// This program tests the C++ environment

#include <iostream.h>
#include <windows.h>

main () {
  cout << "Hello World" << endl;
  Sleep(4000);
}
```

Note that any text on a line to the right of two forward slashes is treated as a comment and is not read as part of the C++ program. You may later want to try another variant of the program, which instead pauses the program indefinitely until any key is depressed

```
#include <iostream.h>
#include <conio.h>

main () {
  cout << "Hello World"  << endl;
  cout << "Press any key to continue" << endl;
  getch();
}
```

The two lines that start with **#include** *together with the* **Sleep( )** *or* **getch( )** *function will not be repeated in most of the remaining programs in this textbook, although they must generally be present in every program that is run from within Dev-C++ (otherwise only the* **#include <iostream.h>** *line is required).*

In entering programs, *names are frequently misspelled or improperly capitalized*, leading to unexpected consequences, especially when characters that look nearly alike are substituted for one another. In particular 0 and O and 1 and l are very difficult to distinguish at low screen or printer resolution and can lead to subtle errors, such as when $1 < 10$ ($1.0 < 10$) is used in place of $l < 10$ (the letter $l < 10$). *As well, the semicolon at the end of a statement is often omitted* and two successive lines are read by the compiler as a single statement. This often produces cryptic error messages with incorrect line numbering. Further, *fine details of the program syntax*, such as the use of double quotation marks rather than single quotation marks, braces instead of parentheses above and the lack of a semicolon after an **#include** statement, must be noted with great care, as these initially present major challenges.

(b) *Saving the file and running the program*: Save the file by depressing the fourth icon from the left on the upper button bar, which generates the float-over text "Save" when the cursor is positioned over it. Type in a name for the file (do *not* add a **.cpp** extension to your file name here – this is done automatically). Then click on the third icon from the left on the lower button bar with float-over text "Compile and Run." A compilation progress window should appear after which the program executes.

(c) *Opening a preexisting file*: To open a .cpp file in Dev-C++ click on the second icon from the left on the upper button bar with the associated float-over text "Open Project or File" and select the desired icon from the secondary file window. Once a file is saved, selecting the compile and run icon will automatically resave it.

2.2 Using the Dev-C++ debugger

Often the most efficient procedure for locating errors that arise when a program is executed is to inspect the values of the variables in the program during execution. The most straightforward procedure introduces additional code lines that write these values to the terminal or to a file. Alternately, variables can be inspected through a debugger. In this section, we introduce an error into our Hello World program in order to demonstrate both procedures.

Return to the program you created in the section entitled "compiling and running a first program." Incorporate both a programming error and debugging lines into the program as follows

```
#include <iostream.h>

main () {
   int i = 0;
   cout << "The value of i is " << i << endl;
   i = 5;
   cout << "The value of i is " << i << endl;
   int j = 0;
   cout << "The value of j is " << j << endl;
   int k = i / j;
   cout << k << endl;
}
```

Save this file by clicking on the fourth icon on the top button bar that resembles a floppy disk and has the "Save" float-over text.

To turn debugging on or off, select "Tools → Compiler Options" from the menu bar followed by the Settings tab (the arrow indicates that the "Tools" menu option should be selected followed by the "Compiler Options" submenu item). If you now click on "linker" in the left windowpane, "Generate Debugging Information" appears as a text label in the right-hand windowpane. Clicking on the word "Yes" or "No" to the right of this text generates a drop-down text box from which you can select the desired behavior. Profiling, discussed in Appendix C, can be activated or deactivated through a similar set of operations, after instead selecting "Code Profiling" in the left-hand windowpane. Notice as well for future reference the menu option "Optimization" in the left-hand windowpane. Increasing the compiler optimization level will generally decrease the execution time of the program, but should preferably only be applied at the end of the development process, as discussed in more detail in Chapter 21.

Now recompile the program using the first icon in the lower toolbar. Inside the editor window click in the gray area just to the left of the line **int i = 0;**. A check symbol should appear and the line should turn red; this is termed a *breakpoint*.

Click again on the Debug icon. A set of debugging menu items will appear at the bottom of the editor window. Locate the windowpane to the left of the main editor window and right click inside this area. A pop-up menu with the selection "Add Watch" will appear. Click on this icon. (If this fails simply click on the "Add Watch" icon in the debugging toolbar. A secondary window requesting a variable name appears. Type in **i** and depress the OK push button. An icon labeled **i** appears in the left-hand windowpane. Depress the "Run to Cursor" icon in the debug toolbar. The active line will advance to the first breakpoint. Now identify the upper left-hand icon in the debug toolbar labeled "Next Step" and select this icon repeatedly. The active line, marked in blue, will move through the program at the same time that the value of the variable **i** is seen to change. When the position of the error due to the invalid division by zero is reached, the active line cannot be further updated. Of course, in this example the error can also be obtained from the data displayed on the terminal through the output (**cout**) lines.

For later reference, the value of an array member such as **A[0]** will not update automatically in the Watch window and you will have to remove the watch and add it again to see the new value (however employing instead the array name **A** in the watch window redisplays all array elements after each update). Further, the Step Into icon is employed to enter a function, that is to change the scope of the debugger to that of a function. That is, to step into function bodies, the Step Into icon should be selected once the active line is located at the position of the function call.

2.3 Installing DISLIN and gsl

DISLIN is a high-performance professional scientific graphics package that is available for numerous C, C++ and FORTRAN programming environments. Updates to the DISLIN package are obtained at: http://www.linmpi.mpg.de/dislin/. *The DISLIN version for the mingw compiler must be employed for Dev-C++ installations.* This site has as well non-free versions that can be used with commercial compilers. Additionally, gsl is a scientific program library, found in the /gsl directory of the CD-ROM.

The installation steps for these products are:

(a) Return to the CD-ROM and double click on the folder labeled dislinmingw. Find the icon labeled setup and double click on this icon to install the product. Be sure to set, when prompted for a directory name, the same drive letter X: that you used for Dev-C++, while retaining the rest of the default directory name. The software will then be installed in the directory X:\dislin (if you are using both the mingw and the Borland C++ version of dislin, however, you will of course have to install one of these into a directory with a different name).

    To install gsl, double click on the icon for the gsl-14_008 program in the \gsl directory of the CD-ROM; however in this case, specify when prompted to select the destination directory X:\Dev-Cpp where you installed the Dev-C++ program – do not install into the mingw subdirectory of this directory!

(b) Alternately, for installations from the website you can edit the file mgclink.bat in your X:\dislin\win directory directly. Using e.g. the notepad editor, open this file and find the lines

```
:COMP
@set_ext=c
@set_int=%_dislin%
@if %_opt1%==-cpp set_ext=cpp
```

Replace =**c** by =**cpp** in the second line and **cpp** by **c** (twice) in the last line and save the file.

You can now follow the same sequence of steps as in Appendix B to run the program except for replacing bcclink with mgclink.

(c) From the start menu, select All Programs → Bloodshed Dev-C++ → DevC++ to start the program. Select Tools → Compiler Options from the menu bar and select the first check box entitled "Add the following commands when calling compiler" Click on the checkbox so that a check appears and in the associated text entry field enter.

```
"X:\dislin\dismgc.a" -luser32 -lgdi32
```

For future use, add

```
"X:\dislin\dismgc.a" -luser32 -lgdi32
```

into the lower text entry field marked "Add these commands to the linker command line" and enable the associated check box as well.

If you installed gsl, add the additional flags **-lgsl -lm** to the end of each of these two statements. *However you must remove these flags before running the built-in debugger, which will not function when gsl is included.* You can test your installation by running the following three-line test program:

```
#include <iostream.h>
#include <gsl/gsl_sf_bessel.h>
main() {cout << gsl_sf_bessel_J0(5.0);}        // Output:-0.177597
```

Now select the tab entitled Directories and then the subtab "C++ Includes." At the bottom of the list of directories enter

```
X:\dislin
```

Finally, depress the OK pushbutton.

2.4 A first graphics program

You will now write a sample graphics program in order to become familiar with the code lines required by all DISLIN programs. Type the following program into the editor window after pushing the "New" pushbutton on the right-hand side of the lower toolbar

```
#include <iostream.h>           // Required by DISLIN!
#include "dislin.h"             // Includes the plotting package
main () {
   int numberOfPoints = 2;
   float x[2] = {0, 1};
   float y[2] = {0, 2};
   qplot(x, y, numberOfPoints);
}
```

It is extremely important to observe that the command ***#include <iostream.h>*** *must be situated before* **#include "dislin.h"** *for DISLIN to function properly on a C++ file.* (The line **#include <dislin.h>** can be substituted for **#include "dislin.h"**.) Now select the third, "Compile and Run," icon from the left on the bottom button bar. Type in a suitable name for the file (again do not enter the .cpp extension). A graph of the two points should appear.

You can generate a TIFF, Adobe PDF, or postscript file in place of the screen plot by placing one of the lines

```
metafl("TIFF");
metafl("PDF");
```

or

```
metafl("POST");
```

into the **main()** program before the line containing **qplot**. A file named **dislin.xxx**, where **xxx** is respectively **tif**, **pdf**, or **eps**, is then produced in your directory when the program is executed (other graphics formats are also possible). If an .eps file called, for example, **dislin.eps** already exists a new .eps file will instead be called **dislin_1.eps**.

A procedure for printing the contents of any window, which can also be applied to a graphics window, is to depress the left mouse button inside the graph window and subsequently press the Print Screen key while holding down the ALT key. (Using the CTRL key instead of the ALT key instead captures the contents of the entire screen.) You can then any open an application program that accepts graphics such as Paint (Start→Programs→Accessories→Paint) or an appropriate word processor and select Edit→Paste from its menu bar to insert a bitmap of the captured window. The bitmap can then be printed through the application program's print functions.

2.5 The help system

Dev-C++ contains an abridged help system that can be accessed from the menu bar. The first section of the help menu contains information on Dev-C++ and describes debugging and compiling in detail. The second section summarizes the C language.

2.6 Linux alternatives

Dev-C++ and gsl are native to Linux and are in fact derivatives of the standard Linux g++ compiler. DISLIN can also be downloaded for Linux from the Absoft website. Linux generally includes several programming editors, the most versatile of which is EMACS (freeware Windows versions of this editor are also available). This editor has a large command set and an extensive set of shortcut keys. Since most commonly encountered editing operations can therefore be entered directly from the keyboard, eliminating hand movement between the keyboard and a mouse, the efficiency and accuracy of text entry during programming are greatly enhanced.

2.7 Assignment

This assignment requires that you install and run Dev-C++ and DISLIN on your computer following the sequence of steps given in this chapter.

Hand in the following:

(1) A printout of the modified Hello World program of Section 2.1.

(2) A printout of the program window while the debugger is running showing the contents of the variable **j** immediately before the line **int k = i / j;** is executed (use ALT–Print Screen) as indicated above to capture the window.

(3) A screenshot of the graph produced in Section 2.4.

Chapter 3
Introduction to computer and software architecture

Scientific programming is comprised of four basic elements: analyzing a physical problem, developing a numerical algorithm for solving the problem, designing a program that implements this solution within a clear and understandable framework, and, finally, determining the accuracy and the limits of validity of the numerical solution. To introduce our discussion, however, we first outline the basic concepts of computational methods and of software and hardware architecture.

3.1 Computational methods

While most problems encountered in university study are analytic in nature and have relatively simple closed-form solutions, typical real-world applications require numerical analysis. That is, although the equations that describe the physical system are known, analytic solutions are generally only available for highly symmetric geometries and a limited number of degrees of freedom. Further, even if these criteria are satisfied, simply formulated analytic solutions may be unstable, as in the case of a two-section pendulum with a nonlinear restoring force or water flowing through a cylindrical tube at high velocity. Such systems undergo a complicated motion whose form depends critically on the initial conditions. Therefore, to describe the time evolution meaningfully, numerical calculations should be performed for numerous initial conditions and statistical properties derived from the ensemble of results. As another example, to model the evolution of a stock price, time is discretized and a small random increment is added at each time step to the price. Repeating this calculation multiple times yields a probability distribution function for the future share price, from which the value of stock options can be determined.

Extending this approach, a numerical model of continuous quantities or systems replaces the global problem by a series of simplified coupled local problems, each of which describes a restricted spatial or temporal domain. Such a procedure maps continuous operators such as derivatives and integrals into a discrete representation in terms of differences and sums. That is, the response of a building to a complicated perturbing force, such as an earthquake, cannot be determined analytically. However, the forces on each brick of the building vary only slightly over the surface of the brick. The response of each discrete, rigid, brick to these applied forces is therefore easily evaluated. Coupling the forces on each brick to those on its neighboring bricks and to the forces applied at the boundary between the house and the ground generates a large linear equation system. Solving this system numerically yields the desired information. Similarly, weather can be predicted by modeling air motion in a small volume and coupling this behavior to that of adjacent volumes. In oil and gas exploration, the earth's crust is again approximated by a set of local elements that take the place of the bricks in our house example. The acoustic wave produced by an explosive charge at a point on the surface is evolved through these elements through a series of small time steps. The predicted fields at the positions of a set of sensors are then compared with the results of actual measurements. Finally, the assumed properties of the earth are iteratively adjusted until the difference between the calculated and measured results is minimized.

3.2 Hardware architecture

While most details of computer architecture have little relevance to ordinary scientific programming, a basic understanding of the interaction between software and hardware is often required for program optimization. The two most important issues are memory access and the machine instruction set, which are discussed below.

Consider first for simplicity a pocket calculator. This device is based on an integrated circuit called the Central Processing Unit (CPU) composed primarily of voltage-controlled transistor switches that implement logic functions. Pressing calculator buttons sends basic arithmetic and data handling instructions to the CPU. Data such as numbers that are entered from the keypad or the results of intermediate calculations are stored either in a series of integrated capacitors on the CPU chip or possibly in an external dedicated memory chip. The CPU also sends output data to a LCD display.

A computer differs from a calculator principally through the number of instructions in the CPU instruction set and the addition of numerous components that perform specialized functions. For example, arithmetic functions may be processed in a dedicated area of the CPU circuitry called the Arithmetic Logic Unit (ALU) or by a separate chip. Input or output can be directed to many devices such as a graphics card, printer, or hard disk, each of which may have additional processing circuitry. However, in all cases each target memory location or device must have a unique memory address. The CPU transfers data to or from a specific location by preceding the data by the corresponding address. This digital signal is placed as a packet on to a series of wires called the system bus. Each external device is attached to the bus wires. Dedicated circuits on each device intercept only the packets that contain the device address.

Two aspects of computer architecture particularly affect computational speed. A program can be optimized by rewriting mathematical or data operations so that they directly map to the CPU instruction set. This increase in efficiency is even more pronounced if the CPU manipulates several data units simultaneously.

Secondly, program execution is critically affected by the speed at which memory is accessed. Many types of memory are present in a computer with vastly different storage and retrieval times. From the fastest to the slowest, these can include:

(1) Memory registers – Memory circuits analogous to memory registers in a handheld calculator are located within the CPU. The latency time (additional clock cycles) required by the CPU to obtain or store data in these circuits is minimal.
(2) Cache memory – A block of memory, typically of the order of 1 MB in size, either inside the CPU or next to the CPU. The CPU accesses cache memory very rapidly through a dedicated memory bus. Cache memory is often organized into different *levels*, of which the smallest and fastest, L1 or level 1 cache, is physically closest to the memory. Successive levels, labeled L2, L3, . . . , are larger and slower.
(3) Random Access Memory (RAM) – Commercially available PC memory chips, placed manually into slots on the motherboard, that access the CPU through the memory bus.
(4) Read Only Memory (ROM) – A specialized, flash, memory chip that does not lose its information when the computer is shut down. The chip contains the initial program required to activate the basic functions of the computer upon startup (boot). As flash memory can be reprogrammed by applying voltages to the memory elements for a certain time, the ROM program can be altered by booting from an update floppy disk.
(5) Distributed memory – Memory locations on remote networked machines, accessed by modern operating systems through internet addresses. While normally slow, specially designed parallel computers feature hardware that hastens communication between separate processors.

A modern "virtual" operating system can allocate or reallocate the memory assigned to a running program to effectively arbitrary physical storage locations. Therefore, once all fast cache and RAM memory is in use by the programs running on the computer, blocks of data in memory that are not needed immediately by the operating system are reallocated to free space on the hard disk or other physical storage devices. This operation, which is called a page fault, is time-consuming, and if repeated frequently slows execution by orders of magnitude.

Dramatic increases in computational efficiency can often be achieved by restructuring numerical algorithms so that the sections that perform most of the computation require less memory. Once the size of the next faster memory structure is greater than these memory requirements, execution speed increases substantially, cf. Section 10.8.

3.3 Software architecture

While the hardware of a computer defines its ultimate performance limits, the degree to which these limits can be attained depends on software properties. Programming languages balance software performance against usability and have accordingly evolved from direct manipulation of the CPU and memory to high-level statements that closely resemble the objects to be modeled. Modern compilers then efficiently map these constructs on to the underlying CPU instruction set (although code that is very frequently used or that interacts directly with external devices such as sensors or measurement instruments may still be separately coded through low-level instructions that are not translated by the compiler). Below we trace the steps involved in this process.

Machine language: Machine language consists of a set of binary, system-specific instructions that map directly to CPU commands. That is, instructions are sent to the CPU chip as streams of binary data pulses on appropriate pins (external contacts). Each processor type employs a different set of binary data to represent an operation. For example, in certain Intel architectures, the bit pattern, also called an op code, that instructs the processor to copy the contents of one memory register to a second register is 1000100111. The subsequent two sets of three bits then specify the register from which the data are to be moved, for example 001, and the register to move these data to, such as 011. The complete machine-level instruction for the operation is therefore

1000100111001011

Machine language instructions were once entered into computers by raising or lowering a set of console switches but are now read from a mass storage device such as a disk file. That the .exe file created by Dev-C++ is machine language code can be surmised by returning to the program directory of the first assignment and opening the file *hello.exe* in the Notepad editor. This displays an apparently random string of characters that are generated whenever sections of the bit stream coincide with the binary representation of a particular character.

Assembly language: Machine language obviously exploits the full capability of a computer but lacks practicality as a programming medium. The first programming languages therefore mapped each machine instruction on to an easily remembered three-letter code words. These words are translated into a binary representation of the program called an object file through a computer program called an assembler. A linker then combines separate object files that may originate either directly from assembly language or from different high-level languages into a single object file or executable program. The name of an object file normally terminates with the .o or .obj extension, while an executable file has an .exe extension in Windows and an .out or no extension in UNIX and Linux.

To generate and examine an example of assembler code, we specialize to the Borland C++ compiler, which generates transparent object code from a C++ source file. The actual text representation that contains the assembler code can be generated through a set of compilation options named switches. To illustrate, consider the file below, named **assemble.cpp**

```
#include <iostream.h>

main() {
  int i = 3;
  int j = 4;
  cout << i + j << endl;
}
```

If this program is compiled by Borland C++ from the command window as follows

```
bcc32 -S -y assemble
```

where the **–y** flag adds line number information to the compiled program, while **-S** indicates that assembly language code instead of executable (.exe) code will be written, we obtain the file **assemble.asm.** A small part of this file contains

```
      main () {
;
;           int i = 3;
;
      ?debug L 3
      mov    ebx,3
;
;          int j = 4;
;
      ?debug L 4

; EBX = i
     mov     esi,4
;
;          cout << i + j << endl;
;
      ?debug L 5
; EBX = i, ESI = j
    add     esi,ebx
;
;   }
;
```

The above code can be easily deciphered as all lines beginning with a semicolon are comment lines that show the source code for each set of assembly instructions, while the lines labeled **?debug** print out the source line number. Thus the line **int i = 3;** results in moving the constant 3 into a memory register labeled **ebx**, while **int j = 4;** similarly places 4 into the memory register **esi**. Finally the statement **add esi,ebx** adds the contents of the two registers.

High-level languages: While assembler language is in many respects similar to entering commands into a calculator keypad, and is therefore far easier to learn than machine language, ideally a programming language should approximate spoken language. From this top–down view, a program models the evolution of physical or abstract objects in response to changes in their environment. A program should describe the properties of these objects and model their change in state resulting from the flow of interactions between them. As noted in the previous chapter, the first attempts at constructing such "high-level" languages were procedural languages. In these languages, the primary independent programming unit is a self-contained set of commands variously termed procedures, functions, or subroutines. An example of a procedure is the **sin( )** function. This function can be called repeatedly from within a program and can further be incorporated and thus reused by other programs. A program is ideally constructed from an unbroken sequence of procedures that transform a set of input data into output data.

In contrast, the basic premise of object-oriented programming languages is that the fundamental program unit should represent as closely as possible a physical object. For example, in the case of the sin function, the object that implements the function might be a handheld calculator. In this calculator, the user might enter a value such as 0.5 into the calculator's internal memory register, depress the sin button and finally display the new value of the memory register on the calculator screen. This leads to the C++ implementation

```
Calculator C1;
C1.inputValue(0.5);
C1.depressSineButton();
C1.displayValue();
```

where **C1** is the name given to the calculator object. The corresponding procedural code

```
double calculatorValue = 0.5;
double outputValue = sin(calculatorValue);
cout << outputValue;
```

does not require the introduction of a calculator object and is therefore far shorter. However, in a large program, combining functions with the data they operate on into an object leads to a natural and easily manipulated collection of self-contained program units that can be transported between programs or programmers. Further, object-oriented constructs form a natural basis for even higher-level programming idioms, such as graphical programming. As an example, in one graphical programming environment, IBM VisualAge®, objects such as the calculator **C1** are represented by an icon. Right clicking on the icon reveals a list of its attributes, which in this case could be the value, **calculatorValue**, held in the calculator's internal memory register and the **inputValue( )**, **depressSineButton( )** and **displayValue( )** functions. These can be connected by drawing arrows on the screen to other objects or to a graphical user interface constructed by dragging and dropping parts such as text boxes, entry fields, and pushbuttons from a menu on to a specialized window called a canvas.

3.4 The operating system and application software

When a computer is turned on, the first program that executes after the low-level ROM boot sequence completes is the operating system, such as Windows®, Linux, or UNIX. The operating system provides an interface between running programs and computer hardware, such as memory, the CPU, and external devices, such as the hard disk, terminal, and keyboard. External software programs that run under the control of the operating system are termed application software. A running application obtains hardware resources from the operating system by calling operating system functions. A modern multitasking operating system services these requests in a manner consistent with the relative priorities of other currently running programs. Lockups resulting from contention between two programs for the same resource can therefore be avoided.

In our Borland C++ example, a listing of all system calls initiated by the **assembly.cpp** program can be obtained if the program is compiled using the –M switch, that is **bcc32 –M assembly**. This produces a large file, **assembly.map**, that lists the system calls; some that pertain to memory allocation are

```
_virt_reserve
c200_0
_virt_alloc
_virt_commit
_virt_decommit
_virt_release
c202_0
__CRTL_MEM_GetBorMemPtrs
__CRTL_MEM_CheckBorMem
_malloc
__org_malloc
c208_0
__org_free
_free
_realloc
__org_realloc
```

3.5 Assignments

(1) Enter the program **assemble.cpp** into your Dev-C++ program. Select "Tools → Compiler Options" from the menu bar and then enter –S –y into the text entry field marked "Add the following commands when calling compiler" and depress the OK pushbutton (*be sure to undo this change after finishing this and the next problem!*). Now select the Compile icon on the button bar. You will generate code entitled **test.exe**; however, this code actually contains the assembly language version of the program.

Open the file in the Notepad editor and search for the text **LM2**; this corresponds to the second line of the program. Describe how this assembly language file maps to the Borland C++ code presented in Section 3.3.

(2) Perform the analysis of the preceding problem on the code

```
#include <iostream.h>
main(){
  int i = 10;
  int j = 20;
  int k = i * (j + 2);
  int m = k * k;
  cout << m << endl;
}
```

Analyze the core section of the resulting assembly code in the same fashion as in Section 3.3 and draw a diagram showing the contents of each of the key memory registers as each program line is executed.

(3) Consider the following physical situation: two radios are turned on (the station value and volume are initially unspecified). The dial of one radio is then set to 99.5 and the dial of the second to 104.2 and the volume of the first radio is set to 40 dB and the second to 50 dB (volume settings). How would you describe this in a procedural language? How would this description change in an object-oriented language? Use the calculator example in the text as a guideline.

Chapter 4
Fundamental concepts

Before undertaking a detailed discussion of the C++ language, we summarize the conceptual foundations of the language. While syntactically complex, C++ is tightly organized around a small number of basic principles to the extent that, even if a C++ expression is poorly understood, its effect can nearly always be inferred from these principles. Accordingly, acquiring the material in this chapter greatly hastens learning of C++ syntax.

4.1 Overview of program structure

Programs in C++ are organized in a hierarchical fashion. At the lowest level are tokens, which are the letters and symbols that can be processed by the compiler. Appropriate groups of symbols are then recognized as words, which include constants and identifiers (variable names), and are atomic (single-element) expressions. Operators act on expressions to form new expressions. Terminating a valid expression with a semicolon yields a statement, which is tantamount to a sentence. One or more statements can be grouped into a compound statement, known as a program block, corresponding roughly to a paragraph. A program block can encompass intermediate results that are isolated from the remainder of the program. Control structures are statements that govern program evolution in time by determining the order in which groups of statements or blocks are executed, while functions further isolate and modularize compound statements by restricting their interaction with external program variables. Finally, at the highest level, classes and objects structure groups of variables and functions into generalized arrays or data structures. These are collectively isolated from the remainder of the program and, like a chapter, describe a single topic, namely the properties of an object or group of objects. We survey each level of the programming hierarchy in this and the subsequent two chapters.

4.2 Tokens, names, and keywords

The C++ compiler, which is the part of the C++ program that constructs object code from the program text file (source code), processes source code lines in order of appearance and text within a line from left to right. A program is read as a sequence of characters called *tokens* separated by non-printing *whitespace* characters, which include tabs, carriage returns, and spaces. Except for comments, that are not processed by the compiler, the valid tokens are a–z, A–Z, 0–9 and the punctuation characters. Small and large letters represent different tokens so that **myVariable** and **myvariable** are unrelated names. A word is a sequence of tokens terminated by whitespace. A *reserved keyword* is a word, such as **int** or **if**, for which the complier has a special interpretation.

4.3 Expressions and statements

A segment of code in C++ that is terminated by a semicolon is called a *statement*. (A line that begins with a pound sign, #, is not terminated by a semicolon and is termed a *preprocessor directive* as it is processed before any statements are read by the compiler.) A statement is a control point in the sense that the compiler only processes a statement after the actions of all previous statements are complete. A statement is composed of one or more valid *expressions* separated by *operators*, such as $+$, $-$, $*$, or /. As an example, **13;** is a valid statement as is **a;** (although neither of these produce any visible effect and they would therefore normally be optimized away by the compiler); however these statements become expressions in the compound statement **a = a + 13;**.

Multiple whitespace characters are treated by the compiler as a single space. Thus, a line can contain several statements, while statements may span any number of lines or tab characters as in

```
int j = 10; int J   =   20; int
    k = 30;
```

C++ is a block structured language, in that statements are generally combined into an individual program unit, known as a compound statement or block, by enclosing them in braces. A block can be employed in place of any statement.

4.4 Constants, variables, and identifiers

Quantities that appear in a C++ program are classified as constants or variables. A constant is a fixed value that can only be placed in one or more memory location when a program is executed, as it does not have a dedicated storage location. A variable on the other hand corresponds to a labeled physical storage location that is reserved by the operating system at runtime (for example, when a program executes). This storage can hold a value that may or may not change with time.

The name of a variable, which is generally known as an *identifier*, has several restrictions – it cannot start with a number or contain punctuation marks (or spaces) except for the underscore character (_). Thus 3T4 and t!3 are not valid identifiers. Additionally, since the names of many variables that are used by the operating system begin with an underscore, most programmers avoid such identifiers.

4.5 Declarations, definitions, and scope

We have defined a variable as a labeled physical storage location or equivalently a labeled segment of memory that can hold a value. The amount of assigned memory space required to store the value of a variable and the interpretation of this value, however, are fixed by the variable *type*. For example, since the number of alphanumeric characters is small, a single letter can be represented in a few bits, while an array of floating-point values obviously demands a considerable amount of storage memory. Further, these bits must be interpreted when required as representing a character and not a numeric value. As a result, in C++ the compiler must be informed of the type of every variable (or of any other construct that requires the allocation of storage memory) before the variable can be used. That is, we can write

```
main(){
  int i = 1;
  cout << i << endl;
}
```

since the keyword **int** in the second line indicates that the variable **i** stores integer values. However, removing the **int** keyword yields a compile-time error.

A statement that indicates to the compiler the type and extent in memory of a variable (a *declaration*) can generally be separated from a statement that enables this space to be physically allocated (a *definition*). That is, declaring a variable informs the compiler of the variable's type, for example integer (**int**), real (**float** or **double**), or character (**char**), but does not lead to memory space being reserved for the variable at runtime. The distinction between declarations and definitions carries the following implications:

(1) Since a definition implies that methods will be initialized at runtime to reserve memory for a variable, it cannot be repeated inside a program unit (called a block), otherwise memory would be allocated for the object in two storage locations. In C++ the value stored in any individual memory location can be manipulated directly through its address. Hence two definitions for the same variable would incorrectly imply that a single variable could simultaneously have two different values. Hence

```
int i;
int i;              // Error!
```

yields an error during compilation (a compile-time error), since each of these statements reserves memory for the same variable **i**.

(2) Often, however, variables must be declared in advance of their actual definition statements so that their types can be resolved in succeeding program constructs, in particular when program units contain variables that appear as well in other units. Here a declaration does not activate any instructions for memory allocation, and can be repeated arbitrarily many times as long as each occurrence conveys the same information. For example, the type of a variable **i** defined in an external source file that will be linked with a given file can be declared to the compiler so that it can resolve future references to the variable. Such a declaration is written

```
extern int i;
extern int i;           // Valid
```

Since this statement only indicates the type of the variable **i** but does not reserve memory at runtime, it can be repeated an arbitrary number of times.

The *scope* of a variable refers to the block (or possibly file in a multifile program) in which it is defined. A variable with *global scope* is defined in the global space outside all blocks, including function blocks, appearing in the program. However, for future reference, other forms of scope exist. *Namespace scope* is nested in the global scope through user-declared namespace definitions. *Class scope* applies to variables declared within a class body. A variable may or may not always be *visible*; that is, accessible to program operations within its scope.

4.6 rvalues and lvalues

The C++ programming language generalizes the concept of a variable through the introduction of the term *lvalue*. This refers to any construct that is associated with directly accessible memory storage and can therefore be assigned a value. Such quantities can appear on either the left- or right-hand side of an = sign, which is termed the assignment operator since it places the value of the quantity to its right into the memory location associated with the variable to its left. As this is *not* the standard meaning of equals, mathematically incorrect statements such as **i** = **i** + **2;** yield meaningful results. (To make this distinction clearer, many languages reserve a special symbol such as := or <– for assignment.) In contrast an *rvalue*, typified by a numerical constant, does not correspond to a program-accessible memory location and, therefore, must always appear on the right-hand side of the assignment operator; that is, the statement **3** = **i**; is clearly invalid.

4.7 Block structure

A modern computer language must be structured, which enables the programmer to isolate individual program segments. Specifically, if a variable **i** is defined in one program segment and a second **i** is defined in a different segment, the two variables should be independent and, therefore, not overwrite each other. In C++, structure is implemented through *blocks*, which are code segments that appear within braces, {}. A variable defined in a program outside a block and before the start of the block is accessible (*visible*) from within the block – unless a variable of the same name is *defined* within the block. In this latter case, once the new variable is defined, it displaces the external variable of the same name until the block terminates and the new variable is destroyed. This process is called variable hiding. Thus a block is like a one-way valve that allows previously existing, that is defined, outside objects to be accessed from within the block – unless a similar object is created within the block – but does not allow objects inside the block to be examined from outside the block (except in certain specialized cases). A simple example serves to illustrate these concepts

```
int i = 3;

main(){
  int j = 4;                // i is 3 and j is 4
  {
    i = 10;
    int k = 5;              // i is 10, j is 4 and k is 5
    double j = 6.0;         // i is 10, j is 6.0 and k is 5
  }
  int m = 3;                // i is 10, j is 4, m is 3
                            // and k no longer exists
}
```

Note that **i = 10;** is not a definition, therefore it does not allocate a new memory location (create a new variable) for the name **i** within the innermost block.

A variable defined outside all inner blocks in the program, such as the first occurrence of **i** in the program above, is called a *global variable*. Such a variable possesses the special property that it can be accessed from anywhere in the program after it is defined, including an inner block within which a second variable of the same name has been previously defined. In the latter case, however, it is necessary to employ the syntax **::i**, where **::** is termed the *scope resolution operator* in order to access the global **i** and not the newly defined **i** variable.

To summarize, suppose an outer region R contains an inner block B. Then:

- Variables previously defined in R are accessible in B.
- Variables defined in B are destroyed (disappear) when B is exited.
- If a variable is defined in B with the same name as a non-global variable in R, the variable in R becomes inaccessible in B. Note that the new variable can even have a different type than the variable it displaces – the two variables are completely unrelated.

4.8 The *const* keyword

The value of a variable that is defined by including the keyword **const** in its type specifier (for example, by writing **const int** or **int const**) cannot be changed through its name (although it can be hidden by a non-const variable of the same name in the manner discussed in the previous section). Thus the second line below is illegal

```
const int i = 3;
i = 4;                                  // Compilation error
```

Clearly, a **const** variable must be assigned a value when defined, otherwise it will point to a random value that cannot subsequently be altered in the program.

4.9 Operators – precedence and associativity

Built-in constructs (primarily symbols) reserved by C++, that are dedicated to transforming or combining expressions into new, meaningful, expressions, are termed operators. A unary operator acts on a single expression, such as the – sign in –5, while a binary operator joins two expressions as in 3 – 4. In the latter case, one argument of the operator is to its *left* (the 3) and a second to its *right* (the 4).

Recall now that, in arithmetic, certain operations are performed before others. Thus in the expression 2 * 3 + 4, multiplication is performed before addition. This standard rule for applying operators can, however, be overridden through the inclusion of parentheses, such as in 2 * (3 + 4). In this modified expression, the operations inside the parentheses are performed before the operations outside the parentheses are applied. More precisely, the parentheses operator, which evaluates to the expression contained within the parentheses, is always applied before any arithmetic operators. The ordering of operations is termed *precedence* so that the parenthesis operation has a higher level of precedence than the * and / operators, which in turn have a higher precedence level than the + and – operators. The full set of C++ operators spans numerous different precedence levels that are shown for future reference in Table 4.1.

While the order of operations of different precedence levels is determined, an additional rule is often required to establish the result of a series of operations with the same precedence level. For example, consider the expression 5/4*3 in which the multiplication and division operators have equal precedence. The two possible ways of evaluating the expression, namely (5/4)*3 and 5/(4*3), lead to different results. Which of these two possibilities is implemented is determined by the *associativity* of the operators. Normally in C++, as in this example, operators are left-associative, indicating that the operators are applied from left to right so that 5/4*3 corresponds to the expression (5/4)*3. The most common right-associative operator is the assignment operator. The right-associativity insures that **i = j = 5** is a valid expression that first sets the value of **j** to 5 and subsequently assigns the result of this operation, namely 5, to **i**.

Whenever associativity is required to resolve an expression, unintentional errors frequently arise. In the above example, often the expression 5/(4*3) is intended

Table 4.1 *C++ operator associativity and precedence*

Operator name	Operators	Associativity
Primary scope resolution	::	left to right
Primary	() [ ] . ->	left to right
Unary	++ -- - + ! ~ & * size of new delete (typename) (C cast)	right to left
C++ Cast	(typename)	left to right
C++ Pointer to Member	.* ->*	left to right
Multiplicative	* / %	left to right
Additive	+ -	left to right
Bitwise Shift	<< >>	left to right
Relational	< > <= >=	left to right
Equality	== !=	left to right
Bitwise Logical AND	&	left to right
Bitwise Exclusive OR	^ or *	left to right
Bitwise Inclusive OR	|	left to right
Logical AND	&&	left to right
Logical OR	||	left to right
Conditional	? :	right to left
Assignment	= += -= *= /= <<= >>= %= &= ^= |=	right to left
Comma	,	left to right

but the parentheses are mistakenly omitted. Accordingly *parentheses should be employed in all such situations.*

4.10 Formatting conventions

While the variable names and statement indentation in C++ are only subject to the restrictions of Section 1.2, a consistent and logical set of conventions should always be observed. This text largely follows IBM practice, however conventions vary significantly among software development organizations. The following subset of the more comprehensive list of conventions in Appendix E is sufficient at an introductory stage:

(1) Variable and function names must be descriptive and are therefore generally formed from two or more words. All words except the first should be capitalized as in **numberOfGridPoints**.

(2) Spaces should be placed to the right and left of binary operators, but not unary operators, for example **i = j + −1.0**.
(3) Spaces should be inserted after commas, after the opening delimiters (and { and before the closing delimiters } and) as in **myFunction(int aI, int aJ)**;
(4) Each successive block should be indented by one further tab stop, cf. the example in Section 4.7 above.
(5) Blank lines should be used liberally between different program units and between lines of code that perform related functions.
(6) Function arguments should begin with a small **a**.
(7) The names of classes, structures, and objects should be capitalized.
(8) Internal class variables should begin with a small i.
(9) Boolean variables should begin with **is;** boolean functions should begin with the word **enable**.
(10) All letters of global constants should be capitalized.

4.11 Comments

Reconstructing the meaning of almost any non-trivial program without adequate documentation after a few months of absence from the code generally proves difficult, even for the program author. As a result documentation should always be provided, both in the form of program comments and in a separate document.

Two methods exist for incorporating comments into a C++ program. The first procedure encloses the comment region between the delimiters /* and */. This is useful for removing large segments of code for testing purposes but possesses a major drawback. Namely, if two non-adjacent segments of code are commented in this manner and the line containing the first end delimiter */ is removed by mistake, all code from the first occurrence of /* to the remaining delimiter */ is commented out, which can generate unusual errors. Accordingly, C++ (but not C) implements a second comment syntax in which all text on a code line appearing to the right of the two symbols // is ignored by the compiler. Note that in this case every comment line must be individually marked with the comment identifier.

As a rule, a program should begin with "prolog" lines specifying the title (which should reflect the purpose of the program), revision number, author, revision date, and program objective. The explanation of each section of code should be stated in comments immediately before it appears in the program. Finally, the interpretation of each significant line of code should be placed either directly above or to the right of the line. Comments can also be used to store test data values such that uncommenting lines activates test cases for debugging purposes if the program stops functioning during program development. A short program that illustrates the various uses of comments is given below

```
// Comment.cpp
// Revision 1.0
// April 30, 2004
// Illustrates the use of comments

/* A C comment
extends through
several lines here. */
// This function will demonstrate correct and incorrect program
// operation
main () {
      int i = 5;
// Uncomment this test value to test the division by zero case
//    i = 0;
      if ( i != 0 ) cout << 10 / i;        // Integer division
}
```

Since the abridged programs in this text are already documented by preceding discussion, the above conventions are largely ignored in subsequent chapters.

4.12 Assignments

Find the errors in the following programs:

(1)
```
main () {
    int i = 4;
    3 = i;
}
```

(2)
```
main () {
    j = 3;
    j = 4;
}
```

(3)
```
main () {
    int j = 3;
    int j = 4;
}
```

(4)
```
main () {
/*    int j = 3;
//    int j = 4; */             // Test cases
      int k = 5;
/*    int j = 5; */
      k = 6;
}
```

(5)
```
main () {
    int my
    Function;
    myFunction = 3;
}
```

(6)
```
main () {
  int j = 3
  j = 4;
}
```

(7)
```
main () {
  int myConstant = 5;
  myconstant = 6;
}
```

(8)
```
main () {
  {
  int j = 10;
  }
  j = 11;
}
```

(9) Which of the following are valid names for variables?

aNumberOfPoints!

3Points

p2Number

my Number

abignumber

_aResult

a_Result

aResult;

@result

What is the output of the following programs?

(10)
```
#include <iostream.h>
int i = 3;

main () {
  int i = 4;
  cout << i << endl;           // prints out i
}
```

(11)
```
#include <iostream.h>
int i = 3;

main () {
  i = 4;
  cout << i << endl;           // prints out i
}
```

(12)
```
#include <iostream.h>
int i = 3;

main () {
  i = 4;
  {
```

```
    int i = 5;
    }
    cout << i << endl;          // prints out i
  }
```

(13)
```
#include <iostream.h>
int i = 3;

main () {
  {
  i = 5;
  }
  cout << i << endl;          // prints out i
}
```

(14)
```
#include <iostream.h>
int i = 3;

main () {
  int i;
  {
  i = 5;
  }
  cout << ::i << endl;          // prints out i
}
```

Which of the following programs will run and, if so, what will the output be?

(15)
```
#include <iostream.h>
main () {
  int i, j;
  i = j + 1 = 3;
  cout << i << endl;
}
```

(16)
```
#include <iostream.h>
main () {
  int
  myFunction;
  cout << (myFunction = 3) << endl;
}
```

(17)
```
#include <iostream.h>
main () {
  int l;
  l = 3;
  cout << l << endl;
}
```

(18)
```
#include <iostream.h>
main () {
  int i, j;
  i = j = 4;
  cout << i + 1 << endl;
}
```

(19)
```
#include <iostream.h>
main () {
  int i, j;
  i = 12, j = 4;
  cout << (j + i / j * i) << endl;
}
```

Rewrite the following program using proper formatting rules; run it, possibly with different test values, and comment on it fully after you have determined what it does

(20)
```
#include <iostream.h>
void print(int i) {cout<<i;};
int square(int i) {return i*i;}
main () { int i=5; int j=10;
{int k=square(i)*square(j);
print(k);}
}
```

Chapter 5
Writing a first program

In the following three chapters, we introduce basic C++ program structure and syntax. This chapter summarizes elementary operations in the context of procedural programming. An introductory discussion of object-oriented programming follows in Chapters 6 and 7. The material in these chapters and in particular the assignments should be studied carefully, as they address many of the most significant challenges encountered by beginning programmers.

5.1 The *main( )* function

The entry point into a C++ program is provided by the **main( )** function; that is, the first statement executed when a program is run is always the first statement of the **main( )** function. Accordingly, every C++ program must contain exactly one occurrence of the construct

```
main() {
...statement lines
}
```

which can appear anywhere in the source file (program). The line **main( ) {** can be replaced by **int main( ) {** (or in the unlikely event that parameters are passed from the program from the command line by **int main(int argc, char* argv[ ]{**) in which case **return 0;** should immediately precede the closing brace.

5.2 Namespaces

A relatively new feature of C++ is the ability to group program elements into separate namespaces, such that, for example, a function or variable in a namespace **A** is referred to from outside the namespace by appending a prefix **A::** before its name. The resulting ability to segregate code into non-interacting code segments enhances library reuse. To illustrate

```
#include <iostream.h>

namespace A {
     int i = 1;
}
```

```
namespace B {
      int i = 2;
}
namespace C {
      main() {
            int i = 3;
            cout << A::i << " " << B::i << " " << i << endl;
                                                  // Output: 1 2 3
      }
}
```

If routines or libraries from numerous sources are employed in a program, this facility prevents the inadvertent inclusion of two similarly named program elements. However, when the programmer is certain that no such collisions will occur, the requirement that program elements in different namespaces be referred to with the corresponding prefix can be circumvented through the **using** directive

```
namespace A {
      int i = 1;
}
namespace B {
      int j = 2;
}
using namespace A;
using namespace B{
      main() {
            cout << i << " " << j << " " << endl; // Output: 1 2
      }
}
```

The block (braces) following **using namespace B** can be replaced by a single semicolon as in the case of the **A** namespace, since the namespaces **A** and **B** are both required from the **using** directives to the end of the program. A **using** directive must be introduced at global scope; that is, outside the bodies of all functions.

5.3 *#include* Statements

While a full implementation of the C++ language contains a wide variety of specialized routines for tasks, such as mathematical operations, physical device access, and multimedia playback, a typical program requires only a very small fraction of this capability. Therefore, to increase compiler speed and minimize memory usage, only the most commonly employed operations are enabled in the base C++ compiler. Additional features are activated by incorporating appropriate "header" files into the program at any point before the features are employed. This is accomplished by introducing a line containing **#include** followed by the name of the appropriate library package into the program. Recall that lines beginning with a pound sign (#) are not terminated by a semicolon. Some of the most important include files are

```
#include <iostream.h>        // Activates terminal and keyboard input
                             // and output
#include <fstream.h>         // Activates input and output from the
                             // hard disk
#include <math.h>            // Activates mathematics functions such
                             // as sqrt, sin
```

Throughout this book, we normally suppress the line **#include** <iostream.h> *which should begin each program unless otherwise noted.*

Newer compilers employ an alternate, nearly equivalent method for including libraries using the syntax

```
#include <iostream>
#include <math>
...                        // Other #include statements
using namespace std;
...                        // Remainder of program
```

since program elements defined in all header files named without a **.h** suffix are located in the **std** namespace. While this procedure avoids potential name conflicts, we employ the older header files in this text to enhance portability. Recall from the previous section that, if the **using** directive is omitted in the above program, each element of the include files, such as **cout**, **cin**, **exp( )**, etc., must be individually prefixed with **std::** to indicate their membership in the **std** namespace.

5.4 Input and output streams

Following the occurrence of the **#include <iostream.h>** statement, the so-called standard input and output streams **cin** and **cout** (pronounced see-in and see-out) are enabled. These streams implement intelligent buffers between the program and selected input and output devices. In particular, the **cin** stream reads character or numeric input from the standard input device (the keyboard). The values of these characters or numbers are extracted (piped) from the stream and placed into a program variable through the extraction operator >>. Similarly, the values of program variables can be piped into the standard output stream **cout** through the insertion operator <<. Since **cout** is attached to the standard output device (the terminal), the values are subsequently printed on the screen (the standard output device). Accordingly, the code

```
int i, j;
cin >> i >> j;
```

reads first the value of **i** and then the value of **j** from the input stream. Since C++ normally ignores whitespace characters, this means that two values entered from the keyboard that are separated by any whitespace character, such as a space, tab, or carriage return (but not a comma!), are stored in the variables designated by **i** and **j**. Similarly

```
cout << i << '\t' << j << '\n' << flush;
```

or equivalently

```
cout << i << '\t' << j << endl;
```

inserts (pipes) into the output stream the value of **i**, a tab ('\t' is called a tab character), the value of **j** and a newline character. The **endl** statement combines the newline character and the **flush** statement, where the **flush** statement immediately sends the contents of the stream buffer to the terminal. A **cout** statement that does not end with **flush** or **endl** could lead to a delay in displaying the output, which complicates debugging. For reference, some specialized characters are

```
'\b'        \\ a backspace
'\n'        \\ newline
'\t'        \\ horizontal tab
'\v'        \\ vertical tab
'\b'        \\ backspace
'\r'        \\ carriage return
'\f'        \\ formfeed
'\a'        \\ alert (bell)
'\?'        \\ question mark
'\"         \\ single quote
'\"'        \\ double quote
'\\'        \\ The \ character
```

5.5 File streams

The input and output streams **cin** and **cout** offer an interface of convenient user-accessible operators and functions, such as the insertion and extraction operators, that separate the manipulation of data in memory buffers from the low-level commands required to drive the keyboard and terminal. Extending this paradigm, if data are read or written to a different device destination, the same interface to the memory buffer can be preserved with only minor modifications. For a mass storage device, such as a floppy or hard disk, data are stored in files. File handling functions analogous to **cout** and **cin** are enabled by adding the line **#include <fstream.h>** to the beginning of the program (in many compilers, such as Borland C++, but not in g++, the **fstream** header file as well contains all the operations, such as **cout** and **cin**, of the **iostream** header file). However, to read from or write to a file, its name must be specified, in contrast to the keyboard and terminal, which are recognized as "standard" devices and therefore do not possess distinct names. This is done through a statement

```
fstream fs("input.dat");
```

which associates the disk file **input.dat** in the directory from which the program is being run with a stream, here arbitrarily called **fs,** that can be used for both input and output.

To generate a stream that can only be used for input (corresponding to **cin**) **fstream** should be replaced by **ifstream**, similarly an output file stream is created by instead substituting **ofstream**. The input file **input.dat** is created by entering the exact keystrokes including whitespace characters into a text or program editor that would be entered into the program from the keyboard if **cin** were instead employed.

For example, suppose **input.dat** contains a single line

```
2          5
```

Then the result of the program

```
#include <iostream.h>
#include <fstream.h>

ifstream fin ("input.dat");                // Contents input.dat: 2 5
ofstream fout ("output.dat");

main () {
        int i, j;
        fin >> i >> j;
        fout << i * j << endl;             // Contents output.dat: 10
}
```

is a new file **output.dat** containing the single line

```
10
```

The names **input.dat**, **output.dat**, **fin**, and **fou** can be chosen arbitrarily.

5.6 Constant and variable types

As mentioned earlier, the type of a variable (for example, **int**, **float**, **char**) determines both the amount of memory storage associated with the variable and the manner in which the stored bit pattern at the variable's memory location is interpreted at runtime. Thus a given sequence of bits in memory may be interpreted on the one hand as the letter A if typed as a **char** and on the other hand as the integer 64 if typed as an **int**. In this section, we examine the properties of the most frequently used variable types.

A byte refers to a standard unit of memory consisting of a set of eight binary memory locations. Each location is termed a bit and can hold a zero or a one. Consequently, a single byte can possesses 2^8 or 256 different values. In a 32 bit machine, memory is accessed through a bus of 32 wires that therefore naturally access a maximum of $2^{32} = 4{,}294{,}967{,}296$ different memory locations and send or retrieve four-byte packets. An **int** variable is accordingly normally stored in four bytes and therefore has $2^{32} = 4{,}294{,}967{,}296$ possible values that are mapped to integers from $-2{,}147{,}483{,}647$ to 2,147,483,648. This representation of an integer is exact; however, because of the details of the mapping, incrementing the largest allowable **int** by one generates the smallest allowable integer.

The number of bytes reserved for variable types, such as **int**, is machine and compiler dependent. In earlier versions of C++ created for 16-bit machines, an **int** occupies only two bytes, limiting the number of possible **int** values to only 65,536. (The size of any variable **i** in bytes can be determined at runtime from the value of the expression **sizeof(i)**.) The size that an **int** occupies in memory can generally be halved or doubled by adding the keyword **short** or **long** to the definition respectively. Further, the additional keyword **unsigned** indicates that the value of the **int** is positive, doubling the range of valid positive integer values.

An integer constant is written without a decimal point, while, for example, an **unsigned long** integer constant corresponding to **18** is written **18ul** or **18UL** (a specific **short** integer constant suffix does not exist).

To display the bit pattern that corresponds to a given integer in hexadecimal format, write, for example, **int i = 20; cout << hex << i << endl;**, which yields the output 14. In the hexadecimal system, each group of four bits has as output one of the numbers 0, 1, 2, ..., 9, A, B, ..., H, where H corresponds to the decimal value 15 or the binary value 1111. Thus the decimal value 16 maps to 10 in hexadecimal notation and 20 to 14.

Floating-point numbers are stored in scientific notation with an accuracy of approximately 7 digits for a **float** and 14 digits for a **double**. The first bits in the variable's memory store the mantissa (the 7 or 14 significant digits) and the last few bits store the exponent. Such a representation is inexact (unlike the representation of an **int**) but can span a wide range of magnitudes, up to about $10^{\pm 38}$ for a **float** and $10^{\pm 308}$ for a **double** (the exact values can be found by including the header file **float.h** and then inserting the line **cout << FLT_MAX** << " " << **DBL_MAX << endl;** into the program).

The two principal floating-point types are **float** and **double**. In most 32-bit C++ compilers both types reserve eight bytes of storage space, but the **double** keyword is generally preferred as **float** variables are often automatically converted to and stored as **double** values by setting the least-significant bits to zero, leading to increased computation time. A **long double** is a floating-point type with significantly greater precision and range than a **double**. A **double** constant is distinguished through use of the decimal point and can also be written in exponential notation by appending the suffix **e** or **E** followed by the exponent. That is, **3e−1** is the same **double** constant as **3.0e−1** or **0.3**. Constants of type **float** and **long** are generated by substituting **f** or **l** in place of **e** (except that the decimal point is required to distinguish a **long double** from a **long int** constant).

A third fundamental built-in (compiler-defined) data type is **char** which stores one byte of data that is interpreted at runtime as the code for a single character (letter). That is, writing

```
char aC = 'a';
```

yields the output **a** when the following line is executed

```
cout << aC << endl;
```

The character constant '**a**' is stored in a single byte of data. This should not be confused with the character *string* "**a**". A string is a two-element array of characters consisting of the character **a** followed by a null character, which is a byte with all its component bits equal to zero. A string can be of any length, since functions of string arguments stop processing the string when the null character is detected. Special properties of a string are that two successive strings are automatically concatenated; further, if a string is broken into two lines, a continuation character given by a backslash (\ ) must be placed at the end of the first line. To illustrate

```
cout << "t" "e \
st" << endl;
```

yields the output **test**.

Standard 1-byte **char** variables can take on $2^8 = 256$ possible values. These values are assigned to characters according to the ASCII character set. The first 128 values correspond to the so-called 7-bit character set and contain the most important symbols, while the last 128 values contain mostly foreign letters and various additional infrequent symbols. The first 32 ASCII characters are exclusively non-printing control characters such as backspace, bell, tab, etc. The most important ASCII values are shown in Table 5.1.

Table 5.1 *Some representative ASCII code values*

Integer value	Character constant	Description
0	'\0'	null character
010	'\n'	line feed
032	' '	space
048	'0'	0
049	'1'	1
...		
057	'9'	9
065	'A'	A
066	'B'	B
...		
090	'Z'	Z
097	'a'	a
...		
122	'z'	z

Thus writing

```
char c;
cout << (c = 65) << (c = 10) << (c = 97)  << endl;
```

yields the output

```
A
a
```

C++ also defines a wide character type, **wchar_t**, with corresponding constants expressed as, for example, L'baa'. This character type is employed by foreign character sets and symbols that are not included in the standard 256-element ASCII character set (these are generally not displayed by **cout)**.

5.7 Casts

As we have already seen, C++ built-in variables are often automatically converted between closely related types. As an example, in

```
int i = 56;
char c = i;
```

the integer **i** will be converted in the variable **c** to the numeral (character) 8, which is the 56th member of the (ASCII) character set.

While the source and destination types are evident in the above example, the conversion procedure for two variables of different types that enter into a binary operator expression can be arbitrarily specified. The general rule implemented by C++ is that if two variables of different types that are automatically converted into each other appear in an operator expression such as

```
char c = 'a';
int i = 10;
cout << (c + i) << endl;             // Output: 107
```

the variable that occupies less memory space (in this case the **char)** is always converted (promoted) to the type that occupies more memory space (an **int**) before the operator is applied.

Conversions can also be forced through an explicit type conversion operator (a cast) as in **cout << char(i) << endl;**. Casts have several additional forms in C++, the oldest of which is inherited from C and is written **(char) i**. In the most recent implementation, the above cast is written as **static_cast<char>(i)**, which explicitly indicates that the conversion of **i** from an **int** to a **char** is mandated at compile-time. Casts with system-dependent behavior and casts that remove the **const** property of variables are segregated into calls to **reinterpret_cast<typename>** and **const_cast<typename>**, respectively, and should be avoided wherever possible. A final form of the cast operator, **dynamic_cast<typename>**, reserves the implementation of the cast until runtime and will be discussed in Section 20.6.

5.8 Operators

To complement our discussion of operator associativity and precedence in the previous chapter, we now survey several important groups of C++ operators. The first of these are the arithmetic operations, which, besides the four standard operators +, –, *, /, include the modulus operator **%**, which is defined such that **m % n** is the remainder when m is divided by n. *It is important to note that the result of the modulus operator is negative for negative numbers, which can give rise to unexpected errors.*

Associated with each arithmetic operator is a corresponding compound assignment operator; for example the statement **i += 2;** is identical to **i = i + 2;**. A further group of four operators are the postfix and prefix operators, **i++**, **i– –** and **++i**, **– –i**, which increment or decrement the value of **i** by unity. The meaning of these operators is best understood from an example. After the statements

```
int j = 4;
i = j++;
```

i = 4, j = 5. On the other hand, the result of

```
int j = 4;
i = ++j;
```

is i = 5, j = 5. That is, if ++ *precedes* a variable name in a statement (the prefix operator), the ++ operator is applied to increment the variable *before* the statement is evaluated, while if ++ *follows* the variable name (the postfix operator), the ++ operator is applied *after* the statement evaluation. The two single-expression statements **++i;** and **i++;** are according to this definition equivalent.

It should be noted that, when statements contain binary operators, different compilers process the right and left subexpressions in different orders. Generally this ordering does not affect the results of computations, but, for the prefix and postfix operators, statements such as

```
int j = i++ + ++i;
```

lead to compiler-dependent results and must be avoided.

Finally, the comma operator is the operator of lowest precedence. It joins two expressions and returns the value of the expression on its right, for example, the value of **k** in the expression **k = (i = 0, j = 1);** is 1.

The C++ logical operations are: **a = = b** (equals) evaluates to 1 if the values of a and b are identical and 0 otherwise, **a != b** (not equals) evaluates to 0 if the values of a and b are identical and 1 otherwise, **a && b** (a and b) evaluates to1 if both a and b are non-zero and 0 otherwise, and **a || b** (a or b) is 0 only if both a and b are 0 and equals 1 otherwise. Finally, the unary operator ∼**a** (not a) is 0 if a is non-zero and is 1 otherwise. *A frequent source of error is to omit the second symbol in* ==, **&&**, *and* ||.

5.9 Control flow

Associated with the logical operations are the control constructs that determine the path of program flow depending on the outcome of logical operations. Recalling that blocks group statements into the equivalent of paragraphs, control statements, which act primarily on blocks, determine whether these paragraphs are read (executed) at runtime and, if so, in what order. Proper use of control statements leads to structured programming in which the program flow is clearly apparent from the code layout. A well-structured program is, therefore, far more easily debugged.

The most fundamental control statements are the **if (logical condition)** {...} and the following **else** {...} statements, which execute code depending on whether the logical condition evaluates to zero or one. To illustrate, the following program employs **if** and **else** statements to branch between two code segments

```
main () {
        int b, a = 1;
        cin >> b;

        if (a == b) {
                a = 1;
        }
// Or:  if (a == b) a= 1;

        else {
                a = 2;
        }
// Or:  else a = 2;
}
```

As indicated, if only a *single* statement follows a control construct, the braces around this statement are optional. This leads, however, to severe errors if the brace is mistakenly omitted around several code lines that follow a control construct so that only the first of the lines is regulated by the control construct. Therefore, a highly recommended practice places the control construct and the subsequent statement *on the same line* when braces are omitted as in the comment lines above.

Iterators are a more complex construct that repeatedly execute one or more statements until a logical condition is fulfilled. In the statement below, the variable **i** is first initialized to zero by the first of the three statements within the parentheses following the **for** statement. Next the logical condition given by the second of the three statements is checked; if this condition is true, the statement

```
cout << i << ' ' ' ';
```

inside the block at the end of the **for** statement is processed. If the condition is instead false, control passes instead to the first program statement following the block. Then **i** is incremented by one according to the last statement, **i++**, within the parentheses, after which the logical condition is rechecked and the iteration repeated until the logical condition evaluates to false

```
for (int i = 0; i < 5; i++){cout << i << " ";}// Output: 0 1 2 3 4
```

5.10 Functions

A function in C++ has the general form

```
type1 functionName(type2 aArgument2, type3 aArgument3 ...) {
      type1 outputValue;               // May be implicit
// ...code of function body
      return outputValue;              // Absent if type1 is void
}
```

and can, therefore, act on many input values, but can have at most one output, termed the return value. The input variables are termed *arguments*; beginning argument names with a lower-case **a** is an optional but helpful programming convention.

A function can be called from any subsequent location in the program (including its own body) with either variables or constants as *parameters*. The parameters are substituted for the arguments and the function body is then executed. As an example, in the code

```
int f(int aI, int aJ){
      aI = aI + 1;
      return aI * aJ;
}

main(){
      int j = f(0, 1);
      cout << f(j, 2) << " " << j;          // Output: 4 1
}
```

aI and **aJ** are the function arguments, while **j** and **2** are the function parameters. When a function is called, new memory locations are reserved for the function arguments, and the values of the function parameters are copied into the memory spaces reserved for the parameters. The variables in the function therefore occupy completely independent memory locations than the variables in the calling program. Hence, if the value of the argument variables in the function body are changed, as in the case of **aI** in the example above, the parameter variables in the calling program are unaffected. This central feature is termed *call-by-value*.

5.11 Arrays and typedefs

An array in C++ comprises an indexed set of variables of the same type. Declaring an array through a statement such as **int v[3];** below both reserves memory for an array of three **int** elements and associates the type of an array of **int** elements with the name **v**. The value of the ith array element can be accessed by following the array name with the index operator **[ ]**, as in the following example

```
main(){
      int v[3];
      for ( int i = 0; i < 3; i++ ) v[i] = i;
      cout << v[1] << endl;                          // Output: 1
}
```

The syntax

```
int v[3] = {1, 2};
```

can be employed to initialize the values of an array *only* when the array is defined. This statement initializes the first element of **v** to 1, the second element to 2, and all remaining elements to 0.

Often it is particularly convenient to specify a second name (alias) for a particular type, especially when the type is complex in nature. This helps ensure that all variables of this type are not declared differently by mistake and also allows the programmer to make a global change in the type of all these variables without extensive reprogramming. The **typedef** keyword accordingly associates a user-defined name with a particular type declaration as illustrated below. Note carefully the location of the new type name **Account**:

```
typedef double Account[10];
typedef int AccountIndex;

main(){
      Account new Account = {3};
      AccountIndex newIndex = 0;
      count << new Account[newIndex]<< endl;
}                                                // Output : 3
```

Clearly such a construct permits the length or type of all related arrays to be readjusted with only one coding change.

5.12 A first look at scientific software development

A scientific program should be developed in several distinct steps. In the first of these, the problem to be solved is precisely formulated. Subsequently, detailed specifications are written for the code – what each piece of code should do, what software will be required, etc. In the third step, the code should be written in small modular segments, each of which is tested separately before inclusion into the body of the program. The program is assembled and tested. At a future time, it will require maintenance and possibly revision.

Below we demonstrate how these steps are implemented in procedural programming. The design and analysis of object-oriented programs are more complex and will be postponed to the next chapter.

Problem definition: In the first stage of program development, a clear statement of the problem is written down. This description should capture the main scenarios (possible outcomes) associated with the problem. For example, if the system does not satisfy a required condition for normal operation (for example, the temperature is too hot, or the angle of flight is too large), the problem statement typically indicates the expected response.

Detailed specification: A program specification embodies a comprehensive strategy for coding the problem definition. The specification can, for example, include the form and content of the input and output, the equations to be programmed, the numerical methods to be employed, the hardware and software to be used, and the manner in which the code will branch among sections for the various scenarios. Often, constructing the graphical user interfaces (GUIs) that the user will be presented with when the program is run is the first step in this analysis. In a modern graphical integrated development environment, these screens can be rapidly generated by dragging and dropping icons representing GUI components into a specialized window. In this manner, the required input and output data can be rapidly identified.

Iterative coding and modular testing: Once the structure and content of the program are established in the specification phase, the code is constructed as a series of modular function units. As each function is programmed, it should be verified independently using a designated suite of test data, which should then be saved as comment lines in the program for future use. A fundamental rule in this process is that *only a single change should be made to the program or a single module added before the program is retested.* In addition, *before any non-trivial change is made, a copy of the code (with an appropriate version number) should be saved so that it will be possible to return to the previous code level in case an inadvertent error is introduced.* A specialized difference checker or an editor, with the capability to display differences between two text files, is also useful in this context. Otherwise, such small mistakes as adding an additional character during editing are in certain cases extraordinarily hard to detect.

Each segment or module of code should preferably contain preconditions that check if the necessary physical requirements for the module are fulfilled by the input data to the module. Postconditions that verify that the output from the module satisfies required physical constraints may be similarly incorporated.

Equations that appear in the program should translate as closely as possible to those contained in the specification document. Thus, either the variable names should be the same in both cases or the names in the program should be more descriptive, compound names.

Each segment of code should be carefully commented as it is written.

Finally, in writing code, clarity should generally be favored over brevity and cleverness. In most cases, far more time is required to develop and later maintain a program than is spent running it. Insuring that a program is easily understandable is therefore generally preferable to implementing efficient but inherently confusing programming constructs. Additionally previously developed and tested code modules should be reused wherever possible.

An illustrative if contrived example that demonstrates the above considerations is given below

```
// This function evaluates the square of the sum of two
// even integers

int evenSquaredSum(int aFirstNumber, int aSecondNumber){

// Precondition: Each number must be even, otherwise -1 is
// returned.
        if ( (aFirstNumber % 2) || (aSecondNumber % 2) ) return -1;
        double sum = aFirstNumber + aSecondNumber;
        double sumSquare = sum * sum;

// Postcondition: The result must be within the limits for int.
        if ( sumSquare > INT_MAX ) return -1;

// Normal case: the result is returned (conversion to an int is
// automatic).
        return sumSquare;
}

int input() {
        int i;
        cout << "Input an even number ";
        cin >> i ;
        return i;
}

main() {
        int i, j;

//      Routine for program input
        i = input();
        j = input();

//      Normal test case
//      i = 6; j = 10;
//      Exceptional cases
//      i = -1; j = 1;
//      i = 100000000, j = 100000000;
//      i = 5; j = 2;

//      Perform the calculation
        int answer = evenSquaredSum(i, j);
        cout << answer << endl;
}
```

The first version of the program should typically solve only the simplest version of the problem and will therefore ignore all non-standard outcomes. Once functioning, it should be validated against a test case, typically a simple analytically tractable problem whose result is known.

Subsequently the code is expanded to include additional scenarios and features. As the program attains higher and higher levels of functionality, new requirements will naturally surface, while alternate coding techniques will be discovered that are more efficient or transparent. Therefore, the program should be developed in an iterative fashion. That is, the experience gained at each stage of the development process should be employed to update the original program specification and then to extend and streamline the actual code.

Final testing: When the program is completed, its accuracy must again be verified against a set of test cases with known results, including those that violate the physical assumptions of the problem or that branch into the code structures that handle exceptional conditions. All test suites should be saved either as comment lines within the program or as separate input files for future program maintenance. If analytic results are unavailable, a second program that solves the same problem with a different numerical or computational approach should be constructed; often the only reliable alternative is to verify every individual code line by inspecting each intermediate result with either a debugger or dedicated **write** statements, which is extremely painstaking and time consuming.

Program maintenance: After a program is completed, periodic updates and revisions will still be required as new requirements surface. Properly commented code together with written documents describing the program structure and operation greatly facilitate this process. Wherever possible, the code should further be carefully structured into high-level modules that can be easily replaced or updated.

5.13 Program errors

Even coding a simple program invariably leads to numerous errors. Understanding the meaning of the error messages generated by the computer greatly decreases the time spent in locating and correcting the error sources. Especially since these messages are often misleading, a working knowledge of the material in this section is of great practical importance. Beginning programmers are also encouraged to record errors and error messages as they arise in order to avoid repeating mistakes.

Compiler errors: The most straightforward problems to isolate and correct are those identified by the compiler. These typically result from incorrect syntax rules or program structure that precludes correct interpretation of the program. Such errors, however, can still be subtle – for example **i = j / * l;** is obviously wrong, but writing **i = j;** without first defining **i** through the statement **int i;** is inherently less obvious.

When a compiler error occurs, the compiler prints an error message containing a line number. Unfortunately, most of these messages bear no relationship to the source of the problem. For example, a single error, such as the above case of not defining the variable **i**, precludes the compiler from resolving the type of **i** and, therefore, from processing any line in which **i** is present. As a result, such an omission usually generates a plethora of error messages that obscure the actual source of the error, especially if examined out of sequence. Similarly, if the semicolon at the end of a line is omitted, the line is joined with the subsequent line, which can produce completely unexpected effects. Variable names are surprisingly often misspelled by beginning programmers or characters such as 1 (one) and l (the letter l) and O (the letter O) and 0 (zero) confused, again evoking cryptic responses from the compiler. Accordingly, the most efficient technique for correcting mistakes is to resolve the first few errors in the compiler error list and then to recompile. Messages resulting from dependencies on the corrected lines will then disappear and the next few errors can be eliminated in the same fashion.

As an example of a compile error (in command line Borland C++), the program

```
#include <iostream.h>

main (){
      j = 2;
      cout << j << endl;
}
```

yields the error message

```
hello.cpp:
      Error E2451 hello.cpp 5: Undefined symbol 'j' in function
      main()
```

upon compilation, where the **.cpp** extension indicates that the error originates from a source code file.

Link errors: A link error generally results when a program attempts to call a function or use a variable that is not defined either in the program file or in files that are linked with the program through include statements, as in the program below

```
#include <iostream.h>

int mySquare(int);        // Declares but does not define the
                          // function

main (){
      int i = 3;
      cout << mySquare( i ) << endl;
}
```

which produces the output

```
Turbo Incremental Link 5.00 Copyright (c) 1997, 2000 Borland
Error: Unresolved external 'mySquare(int)' referenced from
F:\HELLO.OBJ
```

Clearly the missing function definition (function body) requested by the linker must be added to the source code.

Runtime errors: The most difficult errors to locate and correct are inevitably those that occur during program execution. For example, a program may fail because of an error in the numerical algorithm or improper program flow. Alternatively, a construct in the program can be syntactically correct but still yield unintended results; thus if **cin >> i , j;** is coded in place of **cin >> i >> j;**, only the first input value is read into **i**. The program could then yield an incorrect result, an overflow or underflow condition as discussed in the next section, or it could terminate and generate an error message. In the first case, the error is most easily detected if analytic results are known or if a second program is available to recalculate the desired result.

More subtle runtime errors occur if a program incorrectly accesses memory. One possibility is that the program attempts to read from or write to a memory address that is beyond the space allocated to the program by the operating system as in

```
#include <iostream.h>

main() {
      int i[2];
      cout << i[5000000];
}
```

In this event, the operating system often (but not always depending on the compiler and operating system) detects the illegal access and displays a system error message inside a pop-up window indicating the presence of an error (segmentation fault). If, however, the array variable accesses a memory location outside its defined bounds that is normally assigned to another variable with nearly the same value, erratic and difficult to detect errors occur that yield seemingly random fluctuations in the final result when the program is executed with different input values.

5.14 Numerical errors with floating-point types

When calculations are performed with a data type such as **double** or **float**, for which arithmetic is inexact, numerical errors necessarily arise. The behavior of these errors is governed by the calculation being performed and the number of times the computation is repeated. In particular, numerical error comprises perturbation error, which includes truncation and discretization errors, machine error, also known as rounding error, and inferential structure error. The last of these will not be discussed here but expresses the violation of identities, such as $\sin^2(x) + \cos^2(x) = 1$, with a certain characteristic probability when random values are generated for x.

The magnitude of numerical error is often quantified in terms of the *machine epsilon* or machine accuracy, which is the smallest floating-point number, ε_m, such that $1.0 + \varepsilon_m \neq 1.0$. The machine epsilon is equal to 2^{-p}, where p is the number of bits used to store the mantissa. For example, in Dev-C++ the mantissa occupies 23 bits of memory for a **float** and 52 bits for a 64-bit **double** yielding $\varepsilon_m = 1.19 \times 10^{-8}$ and $\varepsilon_m = 2.22 \times 10^{-16}$, respectively (these values are equal to the constants FLT_EPSILON and DBL_EPSILON defined in the **float.h** header file). In contrast, the overflow and underflow thresholds are determined by the largest number that the exponent, which is composed of the remaining bits (minus one sign bit and possibly one so-called "guard bit" used for rounding purposes), can hold. These are given approximately by $10^{\pm 38}$ for a **float** (FLT_MIN and FLT_MAX) and $10^{\pm 308}$ for a **double** (DBL_MIN and DBL_MAX).

A number that exceeds the overflow bound or is obtained by dividing zero by itself is represented by a symbolic value such as **Inf** or **NaN**. These values are named and printed out differently by different compilers and the associated symbolic constants that equal these values also have compiler-dependent names that can be viewed in the **float.h** header file. For Dev-C++, the symbolic constant associated with positive infinity, for example, is **_FPCLASS_PINF**. These symbolic values continue to propagate through the calculation using rules such as **1 / Inf = 0** and **0 / 0 = NaN**. As an example, in Dev-C++ the following lines yield the indicated output

```
#include <float.h>

main(){
      double i = 0.;
      double j = 1./i;
      cout << j << endl;                        // Output : 1.#INF
      cout << 1 / j << endl;                    // Output : 0
      cout << (j == FPCLASSPINF) << endl;       // Output : 1
      cout << j / j << endl;                    // Output : -1.#IND
}
```

A floating-point representation of a number or of the result of a floating point operation with an exact value C yields the approximation $C(1 + \varepsilon_m)$, which contains an *absolute roundoff error* given by $C\varepsilon_m$. If we define the *relative error* as the ratio of the absolute roundoff error to the magnitude of the true value, we obtain a relative roundoff error of ε_m. This value corresponds to the number of significant figures in the result, as it yields an error of at most ½ a unit in the last significant place, also known as an ULP.

Roundoff error exists in any numerical calculation. The net roundoff error of N successive similar calculations typically follows a random walk, and therefore on average is $N^{1/2}\varepsilon_m$. However, if successive errors add as a consequence of regularities in the numerical procedure, the total roundoff error can approach or equal $N\varepsilon_m$.

A more severe form of roundoff error occurs when two nearly equal quantities, which are normally subject to different roundoff errors, are subtracted. If these quantities are written $C(1 + \delta + \varepsilon_m)$ and $C(1 + \varepsilon'_m)$, where $\delta << 1$, their difference is $C(\delta + \varepsilon''_m)$, where $\varepsilon''_m = \varepsilon_m - \varepsilon'_m$ is close to the statistical average

$\sqrt{2}\varepsilon_m$. The corresponding relative precision, which is again the number of significant digits in the result, is, therefore, ε''_m/δ, which is far larger than the machine epsilon. Such catastrophic cancellations should be identified if possible at the specification stage as they occur quite frequently in practice and can lead to serious mistakes. Simple examples are the evaluation of derivatives of nearly constant functions and the evaluation of the quadratic formula for small values of ac.

A second source of error in numeric computations is the *truncation error*, which is the difference between the correct result and the result that would be obtained from a given approximate numerical method for zero rounding error. Assuming that the final result is obtained after an adjustable number of similar steps or iterations, the magnitude of the truncation error will generally approach zero as this number is increased.

An issue closely related to computational error is stability. Consider the computation of the path of a ball starting at $x = y = z = 0$ and rolling in the z-direction along the line given by the maximum of the parabolic surface defined by $y = -x^2$ for all values of z. Any numerical error will grow with time and will lead to rapidly increasing or decreasing values of x. Many numerical methods respond in a similar fashion to small perturbations caused by rounding or perturbation error; in this case, they are termed *unstable* and cannot be safely employed without a careful analysis of the net influence of the error terms.

5.15 Assignments

This assignment is meant to provide a broad overview of some basic features of C++ programming before a more analytic and detailed approach is presented in class. Hand in a paper copy of each program together with its input and output. You can combine all this information into a single document by writing the output to a file instead of to the terminal (refer to the discussion of Section 5.5 and the last problem of this Assignment set) and then reading all the different files in to a single editor window before printing.

(1) Type in and run the following program

```
/* Demonstrates for loop */        //This is termed a c-type comment
#include <iostream.h>              //Enables cout, cin

main()                             //Start of program
   {
   int number,                     //Loop counter
      endNumber;                   //Number of squares
   double squaredNumber;           //Squared Number
   char ch = 'y';                  //Used as a flag to exit the
                                   //while statement

   while (ch == 'y' || ch == 'Y')  //Loops until user does not
                                   //enter y or Y
{
      cout << "Insert the number of squares desired " << endl;
      cin >> endNumber;
```

```
        cout << endl;

     for (number=1; number < endNumber + 1; number++)
                                       //Loop to generate squares
     {
        squaredNumber = number * number;
        cout << squaredNumber << endl;
     }                                 //End of for loop

   cout << endl << "Do you want to continue (y/n)? ";
   cin >> ch;                          //Stays in the while block
   }                                   //until y or Y entered
}
```

(2) The Fibonacci series is defined by $x_{m+2} = x_{m+1} + x_m$. Starting with $x_0 = 1$ and $x_1 = 1$ write down the first few terms of the series by hand. Modify the above program so that it instead computes the first 10 Fibonacci numbers. Submit the modified program and its output.

(3) Type in and run the following program, again submit the program and its output

```
/* Demonstrates functions */
#include <iostream.h>

// If the function is instead defined_after_main(), the
// declaration double mySquare(int x); must appear here!

double mySquare(int aX)            //Function definition follows
{
   return(aX * aX);
}

main()                             //Start of program
{
   int endNumber;
   char ch;                        //Definitions
   do
   {
      cout << "Insert the number of squares desired ";
      cin >> endNumber;
      cout << endl;
         for (int i=1; i < endNumber + 1; i++)
                                   //Squares the numbers
         cout << mySquare(i) << endl;
      cout << endl << "Do you want to continue (y/n)? ";
      cin >> ch;
      while(ch == 'y' || ch == 'Y');
   }                               //The program stays in this
                                   //loop until Y or y is entered
}
```

(4) Copy and run the program below. Clicking on the right mouse button (as indicated in the title bar for the plot) will end the program once you have successfully generated the graph. Now type **disman** at the MS-DOS command prompt. Go to the index of the

manual and read the information about the **metafl( )** function. Modify the program below as indicated so that you produce .tiff or .pdf (if you have Adobe Reader installed) output and submit a printed graph together with the program listing. To print the graph, navigate to the icon for the graphics file you have created using the My Computer icon on the desktop, double click on this icon and select the Print function of your application program

```
/*Demonstrates the DISLIN plotting program */
#include <iostream.h>
#include "dislin.h"             //Includes the DISLIN plotting
                                // routines

double f( double aX) {          //Function to be graphed
   return aX*aX;
}
main(){

   float x[11], y[11];          //11 element arrays to hold the
                                //x-y pairs to be plotted
                                // THESE MUST BE DECLARED FLOAT!!!
   double del = 0.1;            //Spacing between x values
   for (int loop = 0; loop < 11; loop++ ){
                                //from x=0 x Fill the x and y
                                //arrays, starting
      x[loop] = loop * del;
      y[loop] = f(x[loop]);
   }

   metafl("XWIN");              //Initializes the output device —
                                //here the terminal
// metafl("TIFF");              //Other choices
                                //Use to open and print plot in
                                //MS-Paint
// metafl("POST");              //Use to create a postscript plot
                                //file
// metafl("PDF");               //Use to open and print plot in
                                //Adobe Acrobat
   disini();                    //Initializes the plotting package
   name("X-axis","x");          //x and y axis labels
   name("Y-axis","y");
   qplsca(x,y,11);              //Plotting routine
   disfin();                    //Exits DISLIN
}
```

(5) Create a file **input.dat** using a text editor such as notepad – NOT A WORDPROCESSOR – that contains only the following lines

```
0 1
1 2
2 4
3 9
```

Now write the following program and run it. Include the program, the graph, and the **output.dat** file in your assignment solutions

```
/* Demonstrates data files */
#include <fstream.h>    //Includes file handling functionality
#include "dislin.h''    //Includes the DISLIN graphics package

int main ()
{
    float x[4], y[4];    //Array definitions

//This opens fin as an input stream analogous to cin and fout
//as an output stream.
//These streams are then associated with the specified files.
    ifstream fin("input.dat");
    ofstream fout("output.dat");

    for (int loop = 0; loop < 4; loop++) {
//Read the values in from the ASCII file input.dat
         fin >> x[loop] >> y[loop];
//and output new values to output.dat
         fout << x[loop] << " " << y[loop] * y[loop] << endl;
    }
//Apply a simple plotting routine to the result (change XWIN as
//above to print the output)
    metafl("XWIN");
    qplot(x, y, 4);
}
```

(6) Write a program that computes all factorials 1!, 2!, 3!, . . . up to a number input from the keyboard. (Consult the previous assignment to obtain the proper format of the **for** loop that is required in your program.) Part 1: declare all variables to be **int**. Enter the number 17 from the keyboard and, using the fact that an **int** is a 4-byte quantity in the Borland compiler, explain the output values by the program and include your explanation (which should include a manual calculation of the values) with your answer. You may need to refer to the programs of the previous assignment in writing your program. Part 2: Now show that, by changing a single variable declaration from **int** to **double**, you can reproduce the correct result for the first 17 factorials.

(7) Type in and run the following program, which calculates the magnitude of the gravitational field of a point particle with a mass equal to the earth's mass and displays the result as either a contour plot, a color graph, or a three-dimensional line plot

```
// Gravitational field of a point earth.
// Illustrates simple 3-dimensional
// plotting routines.

#include <iostream.h>
#include "dislin.h"

const double km = 1000;    // Note this procedure!
                           // An alternate naming
                           // convention capitalizes
                           // all global constants

const double gravitationalConstant = 6.67e-11;
const double earthMass = 5.97e24;
```

```
const double earthRadius = 6380 * km;

const int matSize = 20;  // This MUST be declared a const int
                         // to be used in the subsequent
                         // type statements!
double gravitationalField(double aX, double aY) {
      return gravitationalConstant * earthMass
            / (aX * aX + aY * aY);
}
main () {
      double position[matSize];  // x and y coordinate positions
      float field[matSize][matSize];
                                 //the gravitational field
      float offset = matSize / 2 - 0.5;
                 //Determines the starting point of the grid
      for (int loop = 0; loop < matSize; loop++) {
         position[loop] = 0.1 * earthRadius * (loop - offset);
      }
      float x, y;
      for (int outerLoop = 0; outerLoop < matSize; outerLoop++){
         x = position[outerLoop];
         for (int innerLoop = 0; innerLoop < matSize;
              innerLoop++) {
            y = position[innerLoop];
            field[outerLoop][innerLoop] =
              gravitationalField(x,y);
         }
      }
      metafl("XWIN");
      disini( );            // Required for 3-dimensional plots
      int iPlot = 2;        // Set to 1 for surface plot, 2 for
                            // color plot
      if (iPlot == 1)       // Surface plot

// The conversion function (explicit cast) float* below (which
// returns the first memory location occupied by the matrix) is
// required by dislin.
         qplsur((float*) field, matSize, matSize);
      else if (iPlot == 2)     //Color plot
         qplclr((float*) field, matSize, matSize);
      else {                   //Contour plot
         int numberOfContours = 30;
         qplcon((float*) field, matSize, matSize,
           numberOfContours);
      }
}
```

By introducing an appropriate **if** statement in the **gravitationalField** function, redesign the above program so that it instead yields the gravitational field magnitude of the actual earth. (One way of doing this is to employ more than one **return** statement in the function.) Recall that the gravitational field for the earth is $GM_E r/r_E^3$ for $r < r_E$, where r_E is the earth's radius. Hand in your program together with surface and contour plots.

(8) First type in and run the random walk program below, which simulates a particle that takes a series of random steps with lengths that are (nearly) evenly distributed on the interval $[-2, 2]$. The resulting histogram of ending positions is displayed as either (1) a

set of markers indicating the number of particles with an ending position in the interval between the grid point and the next largest grid point or (2) a bar graph indicating the number of particles that have final positions within each set of **binSize** of the above intervals starting from the smallest grid point. The width of the distribution increases statistically as the square root of the number of steps, as expected from the properties of a random walk process. The function **srand(time(NULL));** initializes a random number sequence differently every time the program is called. The function **rand( )** then returns an integer value between 0 and the maximum value **RAND_MAX** each time it is called

```
//A simple random walk simulator. Also illustrates scatter
//plots and bar graphs.

#include <iostream.h>
#include "dislin.h"
//#include <stdlib.h>    //Generally required to call rand() —
                         //but not in Borland
//#include <time.h>      //Generally required to call time() —
                         //but not in Borland
//#include <math.h>      //Required (uncomment) when you need
                           to call log() function.

const int matrixSize = 600;

main () {

float histogram[matrixSize] = {0};        //Defines and
                                          //initializes histogram
                                          //values to zero
float gridPositions[matrixSize];          //Grid point positions
int numberOfRealizations = 10000;         //Number of Monte-Carlo
                                          //trials
int numberOfSteps = matrixSize;           //Number of steps per
                                          //run
int offset = matrixSize / 2;              //Initial particle
                                          //position
srand(time(NULL));                        //initializes the
                                          //random number
                                          //generator

for (int loop = 0; loop < matrixSize; loop++)
  gridPositions[loop] = loop - offset;    //Grid point positions
for (int outerLoop = 0; outerLoop < numberOfRealizations;
       outerLoop++) {
  double position = offset;               //Initial particle
                                          //position
  for (int innerLoop = 0; innerLoop < numberOfSteps;
       innerLoop++)
      position = position + 4.0 * (rand()/double(RAND_MAX)
       - 0.5);
  histogram[int(position)] += 1;          //Increments histogram
                                          //at final particle
                                          //position
}
```

```
metafl("XWIN");
int iPoints = 0;                   //Set to 1 for scatter plot, any
                                   //other value for bar plot
if (iPoints == 1)
{
    qplsca(gridPositions, histogram, matrixSize);
                                   //Plots symbols at data values
}
else
{
    const int binSize = 20;        //Number of histogram values that
                                   //are contained in each bin
    const int smallMatrixSize = float(matrixSize) / binSize;
                                   //Number of bars in the
                                   //plot. Note that each
                                   //variable must be const!
    float smallHistogram[smallMatrixSize] = {0};
                                   //The histogram to be used
                                   //in the bar graph
    cout << smallMatrixSize << endl;

    int binNumber = 0;
    for (int outerLoop = 0; outerLoop < matrixSize - binSize;
           outerLoop += binSize) {
      for (int innerLoop = 0; innerLoop < binSize; innerLoop++)
           smallHistogram[binNumber] =            smallHisto-
           gram[bin
               Number] + histogram[binNumber *     binSize +
               innerLoop];
      binNumber++;
    }
    qplbar(smallHistogram, smallMatrixSize);
                                   //Bar graph of histogram
  }
}
```

(a) Modify the above program so that it instead produces a scatter plot of the logarithm of the histogram values for 100,000 trials instead of 10,000 trials. You will need to add a small value (for example, 1 or 0.5) to each histogram element before taking the log in order to avoid log(0). Note that the logarithm approximately describes a parabola, indicating that the distribution of ending positions is Gaussian. Hand in a scatter plot of this logarithmic distribution.

(b) Now modify the program so that instead of each step being evenly distributed on $[-2, 2]$, each step is a random choice between the two numbers -1 and 1. Again generate a scatter plot of the logarithm of the distribution as in part (a) for 100,000 trials. This can be done by changing a single line in the above program. Hand in a scatter plot of the new distribution, together with the resulting program. Note that the difference between the distribution in this part and that in part (a) occurs primarily in the tail region.

Chapter 6

An introduction to object-oriented analysis

Since object-oriented programming incorporates a fundamentally different approach to code structure than procedural programming, its conceptual foundation should be understood in the beginning stages of a scientific computation course in order to prevent subsequent confusion between the two strategies. In this chapter, we consider object-oriented analysis, which designates the manner in which a physical problem is modeled within an object-oriented framework. The translation of the framework into actual C++ code is the domain of object-oriented design and will be examined in the following chapter. Of these two topics, analysis is by far the more complex and confusing initially. Thus we employ two strategies to simplify our presentation. First, we limit our discussion to a straightforward and therefore immediately accessible physical problem that clearly isolates core structural issues. Second, we graphically illustrate each step of the analysis process through the diagrammatic technique of Rational Rose®, which is an industry-leading object-oriented software engineering tool.

6.1 Procedural versus object-oriented programming

To gain a working understanding of the difference between procedural and object-oriented programming, we consider a well-understood example, namely that of calculating and graphing the trajectory of one or more thrown balls. A purely procedural approach to this problem is to acquire a set of input data describing the physical system, and to pass these data into a series of functions, each of which transforms its input data into output data. Control statements can be present that affect the sequencing of the functions, depending on the outcome of certain logical conditions or of user input requests. For the ball example, we might have for a single ball

```
main () {
      double position[50], velocity[50], timeIncrement;
      int numberOfSteps;
      cin >> position[0] >> velocity[0] >> numberOfSteps >>
        timeIncrement;
      throw(position, velocity, timeIncrement, numberOfSteps);
      plot(position, velocity, numberOfSteps);
}
```

In such a program, the initial position and velocity of a ball are stored in the first position of the 50-component arrays **position[50]** and **velocity[50]**. The **throw( )** routine then calculates the subsequent positions and velocities by advancing time over 50 time steps of duration 0.1 seconds. Finally, the resulting arrays are graphed in the **plot( )** routine.

Because such code is typically compact and straightforward for small programming projects, object-oriented code is superfluous. However, extending this paradigm to large programming projects is difficult. First, a detailed program that can be applied to all physically realizable cases generally requires a description of the interaction of the ball with its environment. For example, the ball is in reality thrown by a launching device and its motion is recorded by an appropriate measurement system. Each of these objects in turn has settings that can be changed and that are possibly influenced by interactions with other objects. While the program can be appropriately modified though the introduction of additional functions with multiple arguments, its complexity and interconnectedness rapidly increase.

Further, consider the additional complications that arise if more than one type of ball is present. All of these balls can have different masses and dimensions, and therefore will behave differently during and after launch. These properties together with the trajectories need to be stored for subsequent calculations. Thus many sets of variables, such as **positionOfBall1[50]**, **velocityOfBall1[50]**, **positionOfBall2[50]**, . . . , must be introduced (or alternatively, an additional array dimension must be added to each property of the ball, yielding, for example, **positionOfBall1[40][50]**, **ballMass[40]**, etc.), introducing a potential source of confusion, as all name or index numbers must be identical when referring to a given ball. Similarly, programming errors that result in inadvertent changes to any of the multiple variables must be carefully avoided. Also, if several types of balls exist with differing behaviors, such as large elastic balls and point-like balls, large segments of code may have to be duplicated in a series of independent functions that apply to different types of balls.

Object-oriented programming addresses the issues raised in the preceding paragraphs through encapsulation, polymorphism, and inheritance, each of which is now explained in turn.

Encapsulation and *information hiding* are implemented through the introduction of objects and classes. In a procedural program, the central self-contained program unit is a function, which performs a specific task on a collection of input data through a series of related instructions. However, a physical *object* in the real world possesses a more complex structure. Consider for concreteness a voltage meter. This object has a set of *properties*, corresponding to *internal data* or *member variables*, such as the reading on the display, the setting of the on–off switch, and the position of the dial that selects the voltage scale. Further, it has certain *behaviors*, similarly associated with *methods* or *member functions*, which correspond to its principles of operation. For example, if the meter is switched on, the display will read 0, when the terminals are connected to a 100 V input, the display moves to 1 when the voltage scale is set to 100 V, etc. A description of the meter should therefore ideally correspond to a self-contained program unit that includes both the properties and the behaviors of the physical device.

It is important to emphasize immediately that the complete set of properties and behaviors of the meter could be extremely large. Depending on the nature of the problem, these could include, for example, the internal voltage in each circuit element, or even, in the extreme case, the time evolution of the position and velocity of each atom in the device. However, typically only a few high-level features are of practical interest. Taken together these generate an *abstraction* of the object. In object-oriented programming, the representation of an abstraction of a physical object as a self-contained set of internal variables and internal functions is termed *encapsulation*.

Returning to the voltage meter example, observe next that numerous properties and behaviors of the meter are purposely inaccessible to the user. For example, the user cannot access or change the value of the internal voltage supplied to the display or to the electronic circuits. In object-oriented language, internal variables or functions of an object that can be accessed by any external user are called *public*. Variables, such as the internal display voltage, that are only accessed by components of the meter itself are instead termed *private*. In a programming context, public variables or functions can be accessed from any line in the program, while *private* variables can only be accessed or changed from the internal functions of the object. This *information hiding* segregates internal object properties from the remainder of the program, eliminating interactions among unrelated variables.

Polymorphism is the assignment of multiple context-specific behaviors to a single program entity. In the case of internal functions, this means that the behavior of a function is determined by the object in which it is contained. In our ball example, suppose that there are two types of balls, a **Ball** and a **LargeBall**, the second of which possesses non-negligible angular momentum. Then, launching a ball yields different results, depending on which type of ball is thrown. In a more complex example, the effect of double clicking on a document icon on a computer screen depends on the document type. Enabling the same function name to describe multiple behaviors by binding the function to the type of calling object simplifies programming from both a practical and a conceptual viewpoint.

Finally, we have noted previously that functions promote code reuse through the division of a complex piece of code into independent blocks, each of which can be copied or included into multiple programs. Object-oriented programming, however, incorporates the additional facility of *inheritance*. This enables the variables and functions associated with one class of objects, such as **Ball** objects, to be automatically transferred to a second class of objects, such as **LargeBall**, with largely identical properties. That is with a single statement indicating that **LargeBall** inherits from **Ball**, the **LargeBall** description retains all features of a **Ball** except those that are explicitly redefined or added.

6.2 Problem definition

The first step in developing an object-oriented program is to interpret the problem to be solved as a series of interactions between physical or abstract objects. A related task, object discovery, refers to identifying the objects that interact in the problem. Unfortunately, both of these steps can prove to be surprisingly complex. Numerous external objects can affect the physical system behavior. These should be included in a comprehensive description to insure that the resulting code can be cleanly divided into independent units for reuse in other programs. However, the resources allocated to program development often do not suffice for such a thorough analysis. In the remainder of this chapter, we examine this conflict in the context of our simple physical example.

The generally accepted strategy in the initial problem definition stage is to phrase the problem under consideration as a narrative (short story) that describes the sequence and effects of successive object interactions. For example, in the context of computing the trajectory of a thrown ball, a procedural program based on a set of functions could be based on the following summary:

> One or more balls with certain initial positions and velocities are thrown and subsequent position and velocity values are determined and stored at regular time intervals. The resulting data are then graphed.

However, such a description does not lead to a self-contained set of physical objects. To identify one possible set of relevant objects and object properties, observe first that relevant attributes of the ball might include its mass, moment of inertia (resistance to changes in rotational velocity), elasticity (the relative height after a bounce), and radius. An object, such as a ball launcher that compresses a spring to a variable position and then releases the ball, must be responsible for throwing the ball. Important properties of the ball launcher would include the spring position and the spring constant. The ball's trajectory is subsequently measured by an instrument, such as a position sensor, and the resulting data graphed on a plotter.

This view of the problem yields the more comprehensive problem description:

> Several *balls* with given <u>mass</u> are <u>loaded</u> and <u>launched</u> from a *ball launcher* by <u>compressing</u> a *spring* with a certain <u>spring constant</u> and then <u>releasing</u> the ball. Each *trajectory* is <u>recorded</u> by a *sensor* and sent to a *graphics plotter* for <u>plotting</u>.

Effectively the italicized nouns are physical or abstract objects, while the underlined nouns and verbs are respectively the attributes and actions (for example, variables or functions) associated with these objects.

A subtlety of the problem description is associated with time evolution. In the most realistic implementation, each object independently accesses the internal clock of the computer. For example, once each ball is launched its position should evolve automatically with time, while the sensor records the trajectory after every specified number of machine clock cycles. Unfortunately, code in which two or more processes or "threads" access the CPU independently requires complex, operating system-dependent "multithreading" facilities to handle, for example, resource contention and time-sharing among independent processes. A far simpler procedure is for the main program to control the time evolution of the other objects. In this case, the **main( )** function should be viewed as modeling the sequence of events with time rather than the actions of a physical user. This solution possesses certain structural advantages, since program flow is managed by a single routine. Alternatively, a dedicated "time server" object can be responsible for the time evolution of all the physical objects in the system or an object, such as the sensor, may manage time evolution during certain intervals of program execution.

6.3 Requirements specification

As in procedural programming, once a problem has been suitably defined, the resources and data that the corresponding program will require should be identified. These include the necessary hardware and software, the expected input and output data, and the possible error conditions for the model.

Again, this phase of program development can generally be accelerated by first modeling the graphical user interfaces to be presented to the user. This can be done either on paper or through graphical user interface builders that are contained in nearly all commercial C++ integrated development environments. The problem boundaries, which are the parameter regions that yield incorrect or unphysical behavior, should be isolated. The program response to each distinct error case should be elucidated, which generally requires additional problem descriptions such as:

> When four balls are loaded into the ball launcher, an error message will be displayed and the ball launcher will refuse to accept the fourth ball. Subsequently the launcher will function normally.

The resources required by the program can comprise specialized hardware and software resources, such as dedicated servers, mathematical libraries, internet access, etc. Taken together with the other information acquired during the requirements phase, a reasonably accurate view of the time, expense, and complexity associated with the program development cycle should emerge.

6.4 UML diagrams

While object-oriented programs were once formulated by describing objects on individual index cards or sheets of paper, software packages are now available that automate the development process. These packages enable the user to build visual representations of the participating objects and their interactions after which skeleton code with empty function bodies can be automatically generated. A functioning program is then obtained simply by inserting code into each function body.

While the skeleton code produced by such automated development tools is complex and therefore not appropriate for an introductory presentation, the associated visual interfaces clearly elucidate the object-oriented analysis and design process. We will therefore now analyze below our ball example with the industry-standard Rational Rose® modeling package. Rational Rose® is an implementation of the Universal Modeling Language (UML), which defines a sequence of steps for program development together with a specific set of graphical symbols that represent the entities that comprise each of these steps, namely:

(1) The problem definition and requirements specification are summarized as a set of *use cases* that depict the general manner in which the components of the system interact with the external users (actors) and with each other.
(2) *Sequence and/or collaboration diagrams* then establish the order in which functions are called (messages are sent) among the objects that comprise the system.
(3) A *class diagram* incorporates the objects and functions identified in the sequence/collaboration diagram into distinct classes.
(4) The information contained in the class diagram is converted to C++ skeleton code through a menu item selection.

The Rational Rose® software also provides additional, less fundamental, methods for visualizing and interacting with program structure that are not discussed here.

6.5 Use case diagram

A use case or scenario is a single component of a problem description that summarizes a specific objective for an (animate or inanimate) user of the system. That is, each use case is an answer to the question: "What is one task that some user (an actor) will perform with the system?" In our example, the use cases for normal program flow might be:

Load Balls
Release Balls
Activate Sensor
Observe Path
Generate Graph

Each of these tasks may have associated use cases that represent variations on the normal system flow; these are said to *extend* the normal case. Thus a "Limit Exceeded" use case extends the "Load Balls" use case if the number of balls placed into the ball launcher exceeds the three ball capacity. A use case *includes* a second use case when the functionality of the second case is used by the first case. For example, the sensor may activate electronic subsystems for

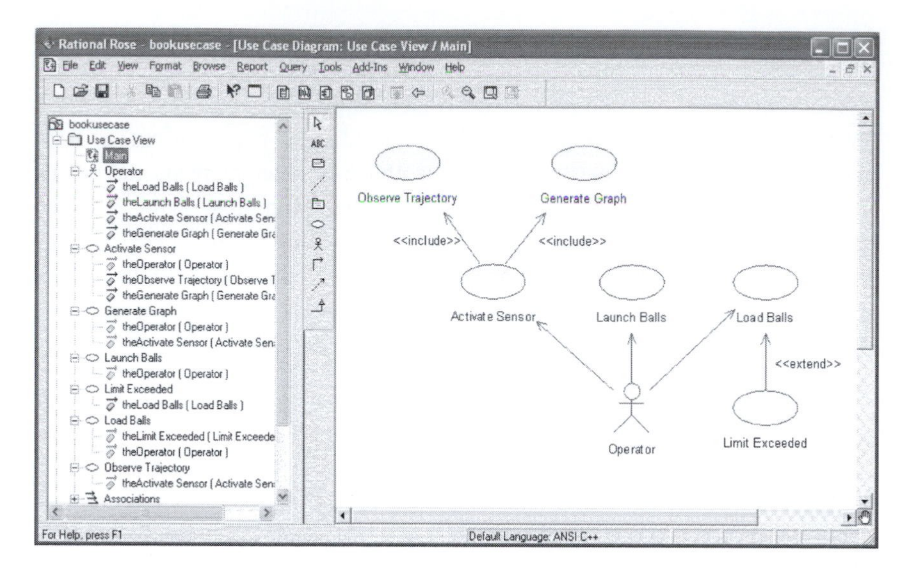

Figure 6.1
Source: Reprinted by permission from Rational Rose Enterprise Edition, © Copyright 2002 by International Business Machines Corporation. All Rights Reserved.

observing the path and plotting the graph that can also be employed by other software components.

While use cases can be collected in a text document, Rational Rose and other object modeling software provide a pictorial representation labeled the use case diagram. Each user is depicted as a stick figure called an actor and the different use cases are represented as labeled ovals. For the model of the preceding paragraph, see Fig. 6.1.

6.6 Classes and objects

Once the relevant high-level system behavior is isolated through use case diagrams, the individual objects that participate in the system and their mutual interactions can be modeled. Objects can be either concrete, such as a ball or a sensor, or abstract such as a time server, matrix, or complex number.

We first present a formal definition of objects and classes. An *object* is an entity that possesses internal variables containing data that define its particular state together with a set of functions that describe its behavior in response to external stimuli (messages). That is, an object is a concrete or abstract entity that has a unique identity (name), certain properties (variables) that characterize its state, and finally a set of specific behaviors (functions). A *class* describes a group of all objects that share common properties and behaviors and provides a form (template) for the creation (instantiation) of these objects.

To concretize the above definition, we consider a particularly simple class that describes the useful properties and behaviors of arbitrary (point-like) balls. Assume that these are the mass, the x- and y-coordinates of the ball's current position and velocity, together with its behavior upon propagation. This yields an abstraction of balls, given pictorially by the class diagram shown in Fig. 6.2.

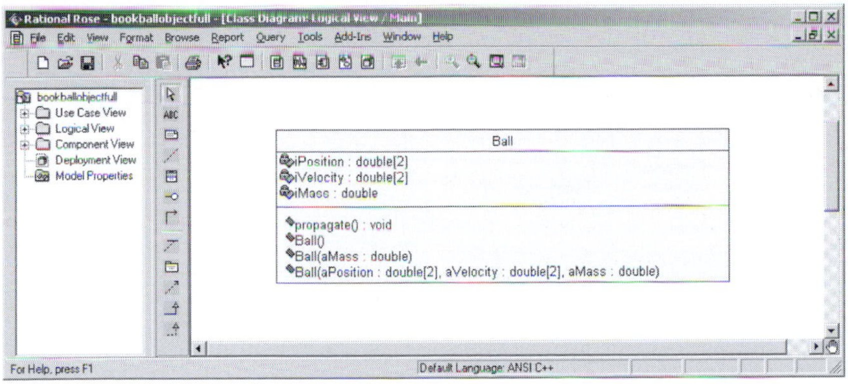

Figure 6.2
Source: Reprinted by permission from Rational Rose Enterprise Edition, © Copyright 2002 by International Business Machines Corporation. All Rights Reserved.

The class diagram has three distinct areas. The class name occupies the top segment of the rectangle, the variables (properties) are placed in the middle segment and the functions (behaviors) are located in the lower segment. Note that the names of internal variables begin with **i** in our naming convention. The icon in the shape of a lock to the left of the variables represents information hiding and indicates that the variables are private (the default specification), and cannot be accessed in the program except by members of the class itself. In contrast, the functions that are marked with a rectangle possess a public access privilege and can therefore be called anywhere in the program.

Note that there are three (overloaded) functions, that have the same name, **Ball**, as the class. These are constructor functions and are employed to generate a **Ball** object while possibly initializing one or more of its internal variables. As discussed later, the first of the three constructors is called through the syntax **Ball B1;**, the second through, for example, **Ball B1(10.0);** and the third through **Ball B1(positionArray, velocityArray, 10.0);**, where **positionArray** and **velocityArray** are arrays.

Normally in a class, internal variables are private by default and are therefore inaccessible outside the class except through the public functions of the class. That is, the internal variables can only be accessed or changed by calling one of the public functions belonging to the class. Therefore, these (private) variables can be thought of as located in an inaccessible core region. The public functions that govern the interaction of the outside world with the variables (the object's state) can then be visualized abstractly as an accessible shell surrounding this region. For the **Ball** class, we accordingly obtain a schematic representation of the form shown in Fig. 6.3, where only a subset of the public functions is included. The public class member functions **getPosition( )** and **setPosition( )**, that respectively allow the private variables **iPosition** and **iVelocity** to be read and changed, do not appear explicitly in the Rational Rose class diagram but are contained in the automatically generated skeleton C++ code if certain menu options are selected.

An analogy that may help conceptualize Fig. 6.3 is that qualities such as a person's degree of thirst or hunger that comprise elements of his state are only accessible by himself; other people or objects can only access these

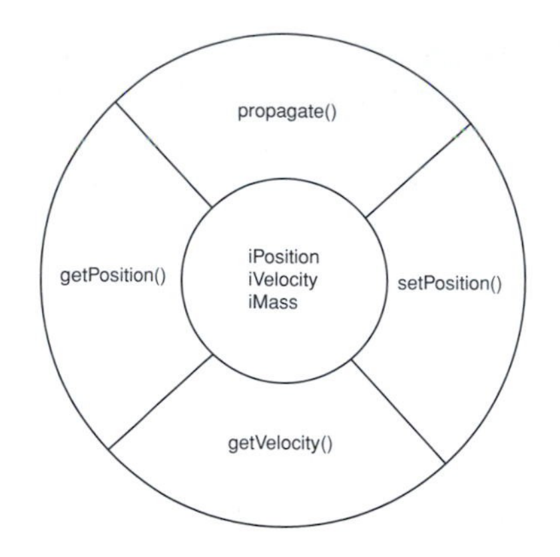

Figure 6.3
Source: Reprinted by permission from Rational Rose Enterprise Edition, © Copyright 2002 by International Business Machines Corporation. All Rights Reserved.

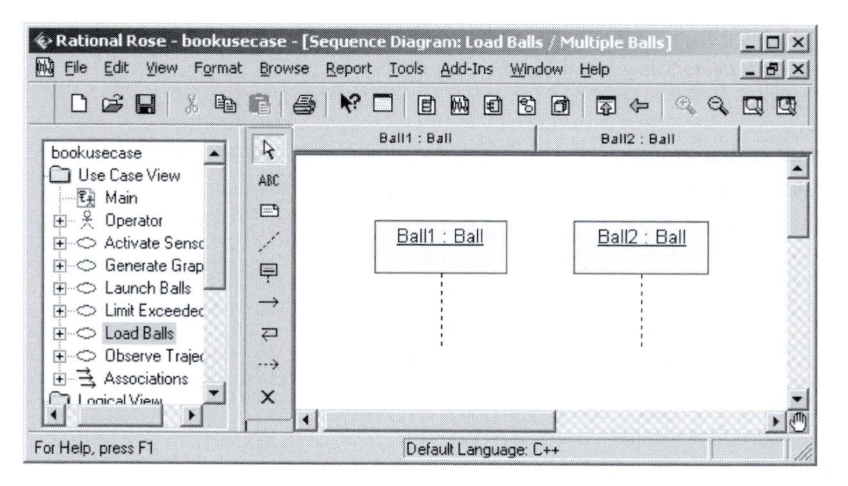

Figure 6.4
Source: Reprinted by permission from Rational Rose Enterprise Edition, © Copyright 2002 by International Business Machines Corporation. All Rights Reserved.

variables by asking appropriate questions, that is directing messages to his public interface.

An object is a realization of a class that is generated by assigning values to the variables present in the class diagram and specifying a unique object name. Thus, the class is like a "form" or "template"; filling in the entries in the form and associating it with a name yields a unique class instance, namely an object. In the UML notation, specific objects are represented by rectangles containing the object name followed by the class name as in the Fig. 6.4.

In this view, two objects, **B1** and **B2**, of class **Ball** have been created. Each object possesses independent values of the internal variables **iMass**, **iPosition**, and **iVelocity** that, however, are not shown. We can view an object as a generalized variable that reserves memory and can be assigned values through its internal data members. In this perspective, a class then corresponds to a generalized *type* (data structure) that groups variables with identical or different types – in the present case one **double** variable and two **double** arrays – together with related functions that act on the variables.

6.7 Object discovery

As stated above, the central problem in object-oriented modeling is to identify the objects that participate in a given problem, together with their properties. Often this information is best extracted from the problem statement supplemented with the use case scenarios. Thus, returning to the problem statement of Section 6.2, namely:

> Several *balls* with given mass are loaded and launched from a *ball launcher* by compressing a *spring* with a certain spring constant and then releasing the ball. Each *trajectory* is recorded by a *sensor* which subsequently graphs the particle positions.

we observe that nouns typically correspond to either classes or their attributes (member variables), while verbs map to behaviors (member functions). This is equally true for the use case diagram of Fig. 6.1, where the nouns imply the presence of classes corresponding to objects, such as **Trajectory**, **Ball**, **Ball-Launcher**, **PositionSensor**, and **Graph**. Here the **Graph** object can be either an abstract software construct that is used to plot trajectories or a dedicated hardware device, such as a plotter that receives and plots information from the sensor. The verbs and nouns in the use case diagram similarly map to functions in the associated class; for example, the use case "Generate Graph" suggests introducing a **plot( )** function within the **Graph** class.

As noted in the introductory discussions of this chapter, the number and type of objects that appear in a program generally depend on the level of physical detail required to solve the underlying physical problem. This is equivalent to determining which nouns in the problem description map to classes and which are associated with class attributes. For example, the spring in our problem description could be an object of a distinct **Spring** class. A spring operating in its linear regime, however, can be considered an integral component of a launcher as it has a simple mathematical description of minimal independent physical interest. Therefore, a more logical procedure is to take into account the properties of the spring through two internal variables, **iSpringDisplacement** and **iSpringConstant**, of a **BallLauncher** object.

The general rule in designing classes is that a class should correspond to a single abstraction; that is, one definite and limited topic. Class members should further be orthogonal, in that the public class provides the smallest number of independent functions that allow the user to access all relevant behaviors of the underlying object. Further, these publicly accessible methods should be those requested by the user as opposed to the programmer. As an example, a car that presents the user with two steering wheels, one for forward and a second for backward operation, would not be desirable, although when the car is built the inaccessible controls (corresponding to **private** variables and functions of the **Car** class) for the two modes of operation are implemented differently. Similarly, although the car designer might find a control that directly sets the RPM of the motor advantageous, this facility would not generally be attractive or advisable to place on the dashboard.

Classes are sometimes separated into different classifications. For example, entity classes correspond to concrete physical entities or to abstract components, such as a **Vector**, that handle or transform these entities. The interaction between external users and the system is handled by boundary classes, such as those associated with graphical user interfaces and plotting routines (**Graph** in our example). Finally, control classes sequence the order of operations in the program.

6.8 Sequence and collaboration diagrams

Once the physically interesting objects have been identified, their interactions (collaborations) can be sequenced; that is, the functions of each object will be called either by the **main( )** program, or by functions of other objects. These calls, which are also known as message passing, are performed in a particular sequence for each scenario (that is for each use case diagram). Each use case diagram thus leads to a sequence diagram in which the objects of the problem are displayed as rectangles at the top of the window. Time advances along timelines that extend below each object toward the bottom of the diagram. A possible sequence diagram for the normal use case in our problem is as shown Fig. 6.5.

The sequence diagram is interpreted as follows: the main program, which here corresponds to the operator of the system, first adds a ball to the ball launcher by calling the **addBall( )** function of an object of class type **BallLauncher**. (This object is constructed with specific values of its internal spring constant and spring displacement that however, like the names of function arguments and object names, are suppressed in the diagram for formatting reasons.) In a final implementation, the user could interface with the program through a dedicated graphical user interface object in place of the **main( )** program. The **main( )** program then calls the **releaseBall( )** function of the **BallLauncher** class that has as input parameters the position of the launcher and the launch angle. In this implementation, the **main( )** program is responsible for time evolution and therefore calls the **propagate( )** function of the ball object to advance the ball's position and velocity in time and then the **observeBall( )** function of **PositionSensor** with the propagated ball as an argument. These last two calls are repeated a prescribed

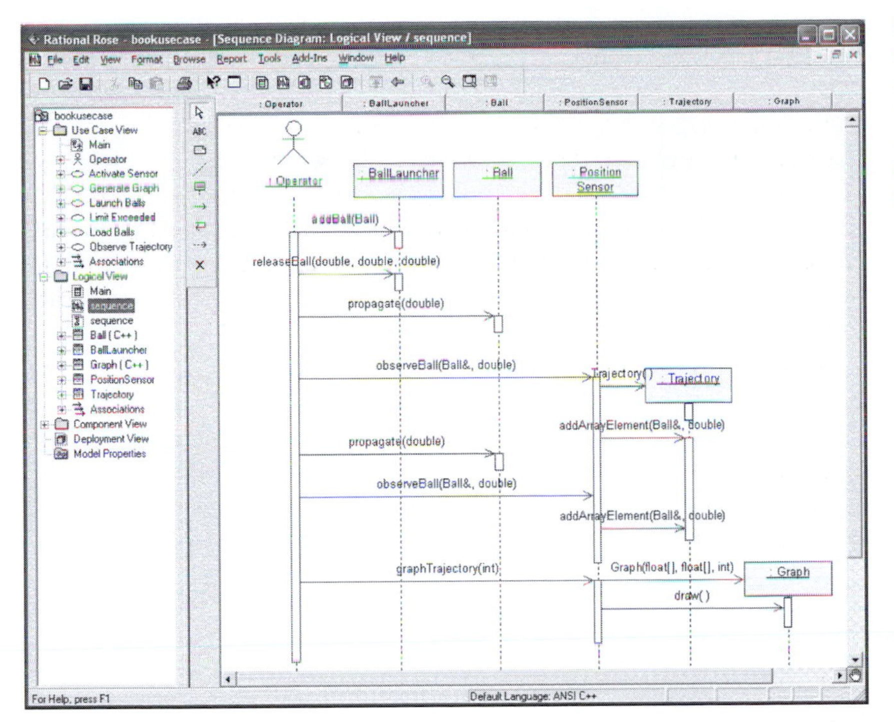

Figure 6.5
Source: Reprinted by permission from Rational Rose Enterprise Edition,

number of times – although this is again not fully depicted in the diagram. The first time the **observeBall( )** function of **PositionSensor** is called, an object of type **Trajectory** is created. This object stores the ball's position and velocity at each time in two separate arrays when its **addArrayElement( )** function is called by the **PositionSensor** object. Finally, when the ball has been propagated the required number of steps, the user calls a function in the **PositionSensor** class that creates a **Graph** object, passes the x- and y- data to be graphed to this object, and then calls the **draw( )** function of **Graph**. It should be noted that descriptive names such as "graph data" or "launch ball" are often provided in the sequence diagram in place of actual function names. The names and signatures are then entered into the class diagram at a later point in the development process.

The data in the sequence diagram can also be presented as a collaboration diagram, which depicts the interactions between objects as connecting lines. Numbers placed along the lines designate the order in which functions are called through a numeric prefix, while arrows extend from the object that calls a function (the client) to the object containing the called function (the supplier), as seen in Fig. 6.6.

A collaboration diagram clearly shows the links between objects and therefore assists in understanding the global structure of a program as opposed to its time

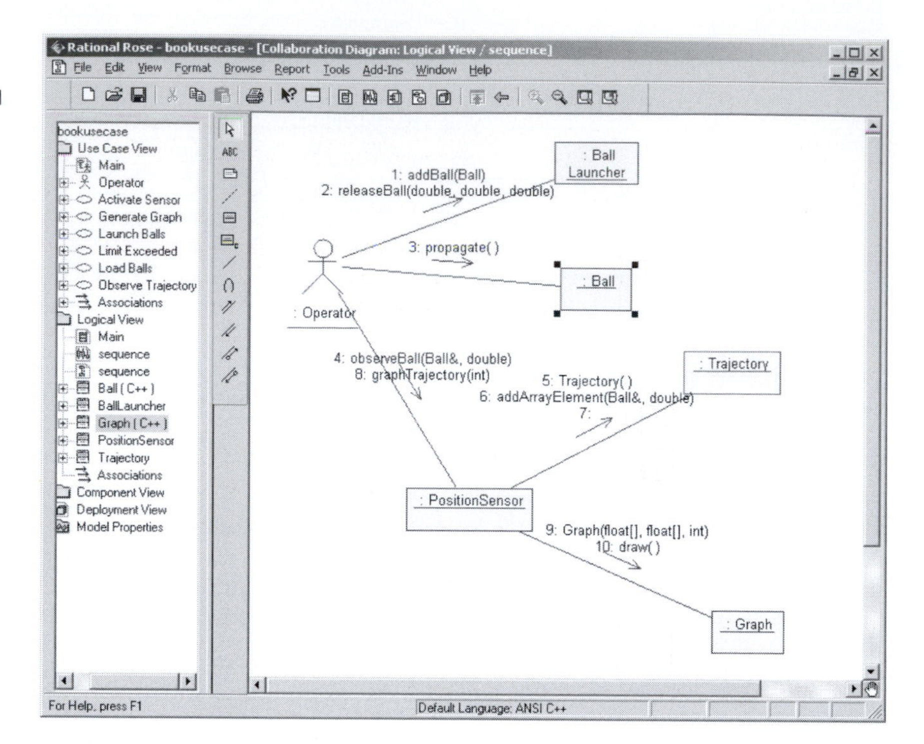

Figure 6.6
Source: Reprinted by permission from Rational Rose Enterprise Edition, © Copyright 2002 by International Business Machines Corporation. All Rights Reserved.

evolution. In Rational Rose the user can create the collaboration from a sequence diagram or the sequence from a collaboration diagram through a menu selection.

6.9 Aggregation and association

Object-oriented analysis distinguishes between *aggregation* or *containment*, in which one class includes a second class as a member variable (also called a "has-a" relationship), and *association*, in which two classes are separate entities, although one class calls functions in a second class. For example, a **PositionSensor** could store the trajectory data of each observed particle in dedicated internal circuitry. In this case, the **Trajectory** object is an inseparable component of the **PositionSensor** and is thus contained or aggregated in the sensor. Aggregation is represented in a Rational Rose class diagram by a line joining two classes marked by a closed (containment by value corresponding to exclusive ownership) or open (containment by reference indicating non-exclusive ownership) diamond at the containing object. A number termed the *multiplicity indicator* on each side of this line indicates how many objects of each type can participate in the relationship. That is, if a **PositionSensor** can contain only one trajectory while a trajectory can be stored in only one **PositionSensor**, we would obtain Fig. 6.7.

On the other hand, the **PositionSensor** and **Graph** objects could have an association relationship if the **Graph** object corresponds to a separate networked

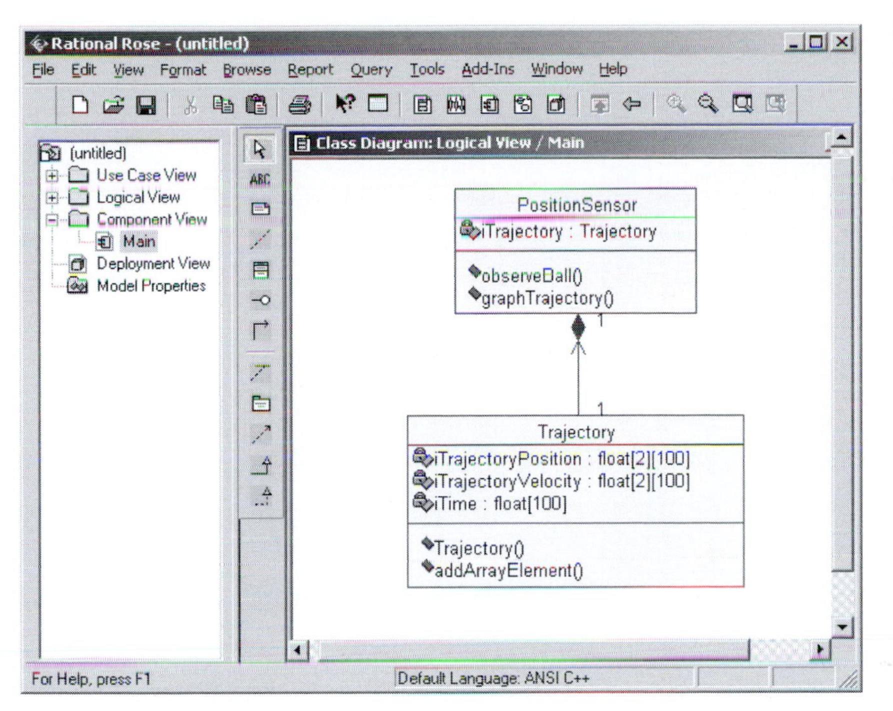

Figure 6.7
Source: Reprinted by permission from Rational Rose Enterprise Edition,

plotter. Here the **PositionSensor** initializes and then calls functions of the plotter, but the plotter is not a "part of" or owned by the sensor apparatus. Such a relationship is depicted in the class diagram as a single line between classes, cf. Fig. 6.8.

The arrow in the diagram indicates that the **PositionSensor** calls functions in (navigates to) the **Graph** class and is therefore always the client class, while the **Graph** class has no functions that access the **PositionSensor**.

Finally, collecting the above diagrams and displaying the variable names, function names and function signatures for all the classes, we obtain the full class diagram, Fig. 6.9.

Note that the diamond appearing between **Ball** and **BallLauncher** in Fig. 6.9 is open. This indicates that the balls are contained as (reference) member variables in a **BallLauncher**; they can be detached from the **BallLauncher** as the **BallLauncher** does not exclusively own the balls.

As noted previously, once the class diagram is complete, the program may be written simply by selecting a menu item and inserting appropriate implementation code into the resulting empty function bodies.

6.10 Inheritance

Inheritance enables new, specialized classes, termed derived classes, to be assembled from a preexisting, more generic, base class. A derived class shares the

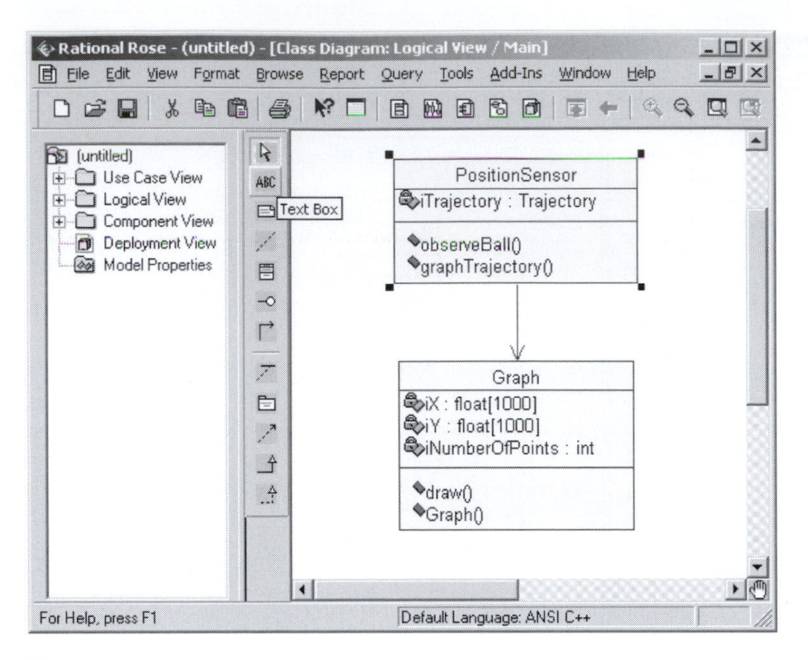

Figure 6.8
Source: Reprinted by permission from Rational Rose Enterprise Edition,

structure and/or behavior of one or more base classes by acquiring the attributes and behaviors (internal variables and functions) of these classes except for those that are explicitly redefined. Semantically, a derived class exhibits an "is-a" or a "kind-of" relationship to its base classes, as opposed to the "has-a" relationship that exists when a class contains a second class as a member variable. If classes inherit through several levels of derived classes, variables and functions that are shared by several derived classes are moved to the highest applicable level of the inheritance tree.

To illustrate, consider a large ball whose moment of inertia and angular velocity are physically significant quantities. If an associated **LargeBall** class inherits from the **Ball** class, a different **propagate( )** function should be supplied as well as constructors that initialize the additional variables. However, the **position** and **velocity** variables of the **Ball** class are unchanged and therefore should be automatically present in the **LargeBall** class. As inheritance is represented in Rational Rose by a triangular arrow pointing from the derived class to the base class, a possible class diagram, also called the inheritance diagram, is therefore as shown in Fig. 6.10.

To understand this diagram fully, note the symbols to the left of the three variables in the **Ball** base class (these have been chosen for example purposes

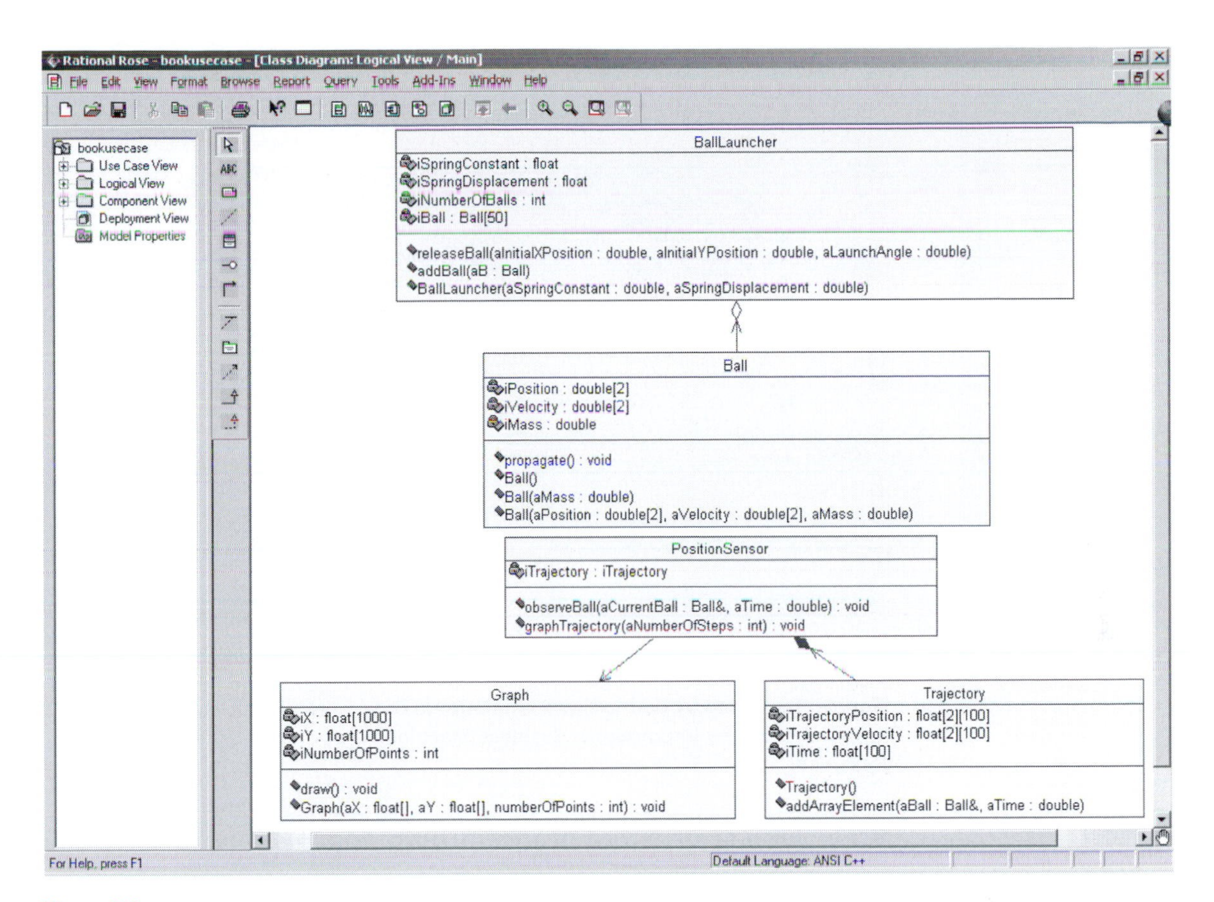

Figure 6.9
Source: Reprinted by permission from Rational Rose Enterprise Edition, © Copyright 2002 by International Business Machines Corporation. All Rights Reserved.

only). The first of these, **iPosition**, is marked with a blue rectangle, indicating that the variable possesses a *public* access privilege. Such a variable can be read or set anywhere in the program that a **Ball** object can be accessed and is also automatically present in the **LargeBall** derived class. The second variable, **iVelocity**, is marked with a key icon indicating *protected* access. Such a variable is present in, and is therefore accessible from, any derived class but is not visible elsewhere in the program. The third variable, **iMass**, is marked (for illustrative purposes only) with a lock icon, indicating *private* access. This variable is visible within the class in which it is contained but is not accessible from any other program unit. Therefore, in the list of the member variables in the **LargeBall** class in Fig. 6.11 (the "Show inherited" check box is selected by default), the public variable **iPosition** and the protected variable **iVelocity** are inherited but not the private variable **iMass**.

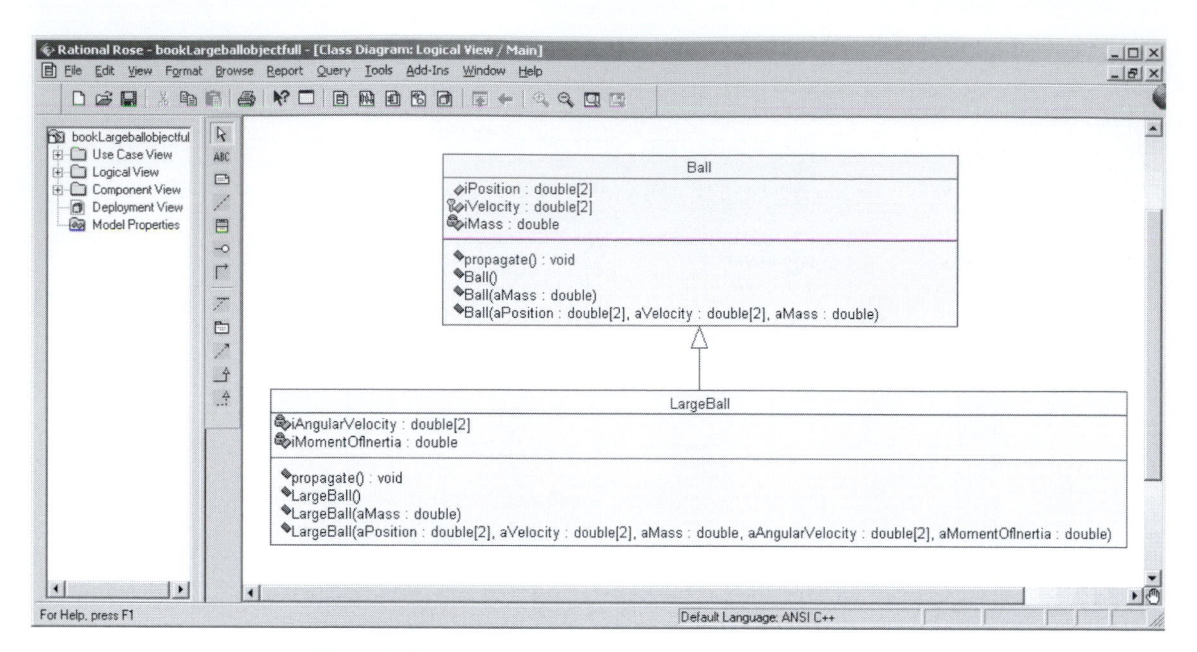

Figure 6.10

Source: Reprinted by permission from Rational Rose Enterprise Edition, © Copyright 2002 by International Business Machines Corporation. All Rights Reserved.

6.11 Object-oriented programming approaches

Object-oriented programming styles inevitably reflect the personal viewpoint of each programmer, although stylistic differences are reduced in a full description in which each significant participating physical system is identified and mapped to an object. Often, however, the system model is abridged in order to decrease the overall programming effort. This effectively yields a programmer-specific hybrid of object-oriented and procedural techniques.

To illustrate, we simplify our launched ball example by grouping logically related functions into a few classes. While these classes no longer represent the properties of actual physical objects, they still structure the independent variables and functions appearing in the procedural program into separate logical units. The purpose of the **Ball** class is then to provide a single generalized array of data and methods that are logically related to the **Ball** as opposed to describing the actual physical object. Thus we include in the **Ball** class the functions and variables that influence or describe the ball motion, including **propagate( )** and **release( )** functions and the trajectory data. Eliminating the objects that actually provide these behaviors from consideration leads to conceptual difficulties, as a ball certainly does not launch itself or record information about its trajectory. This loss of clarity is, however, often offset by the decreased code size.

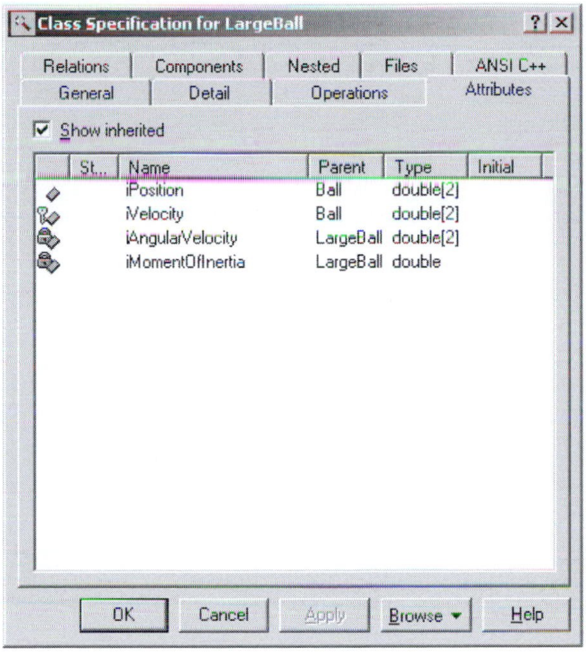

Figure 6.11
Source: Reprinted by permission from Rational Rose Enterprise Edition, © Copyright 2002 by International Business Machines Corporation. All Rights Reserved.

Associated with the abridged description is a use case diagram that captures the essential functionality of the physical system in fewer use cases at the cost of ascribing unphysical properties and behaviors to objects (see Fig. 6.12). A simplified sequence diagram corresponding to the above use case is shown in Fig. 6.13. In such a description, a **Graph** object reads and plots the trajectory information stored in a **Ball** object. A class diagram might be as shown in Fig. 6.14. (Note that we have added for illustrative purposes additional internal variables to the **Graph** class, namely axis labels and a **bool** variable that takes the value **true** or **false** depending on whether these labels are to be displayed.) Clearly, if a program will only be run a limited number of times and the possibility of code reuse is small, so that the details of all the objects that are present in the system are of limited interest, an abridged implementation is considerably more practical.

6.12 Assignments

In the following problems, do not attempt to write down code for your functions, simply introduce a plausible set of functions and internal variables. UML diagrams should be sketched by hand and need not be excessively detailed, for example for the class diagrams, you do not need to include the public, private, or protected icons.

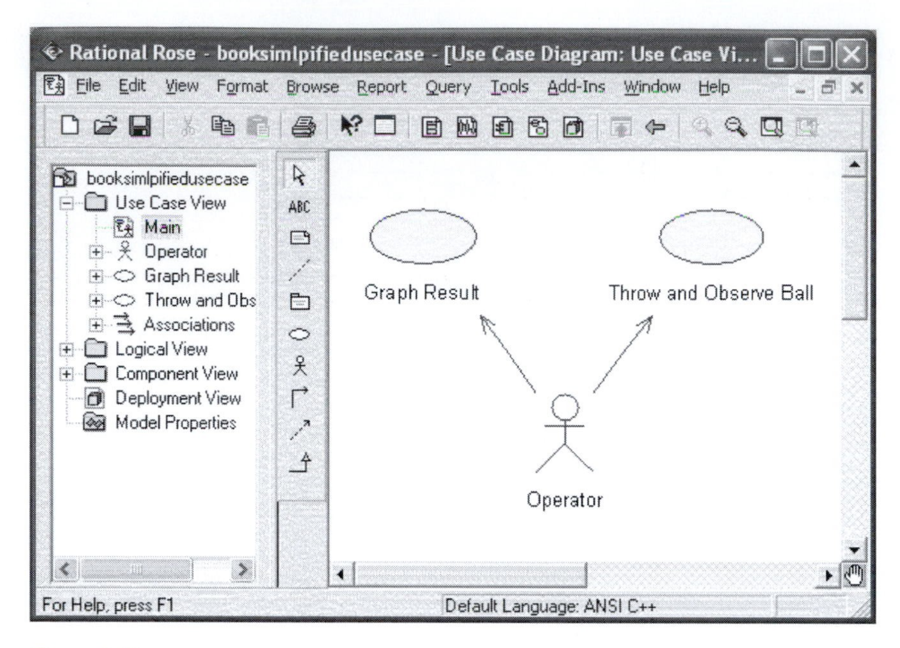

Figure 6.12
Source: Reprinted by permission from Rational Rose Enterprise Edition, ©

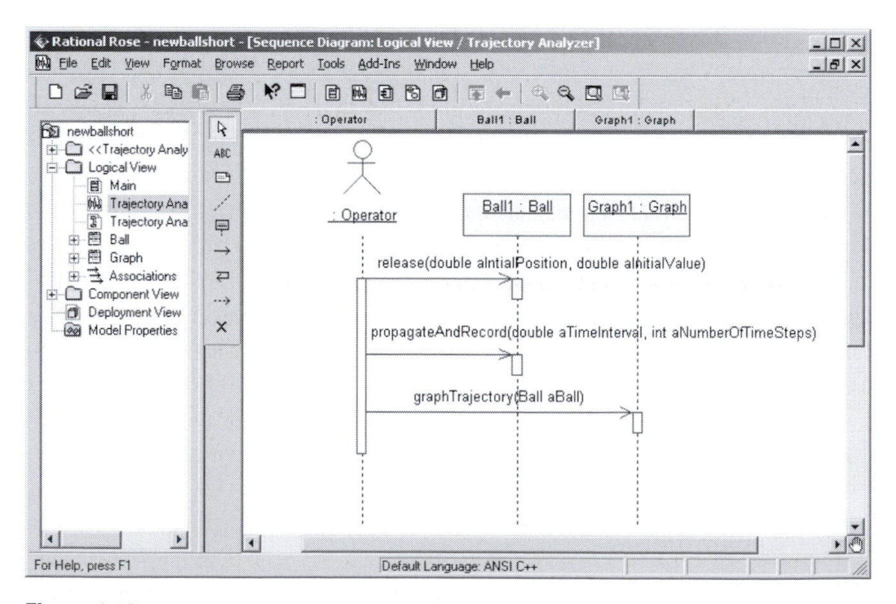

Figure 6.13
Source: Reprinted by permission from Rational Rose Enterprise Edition, ©

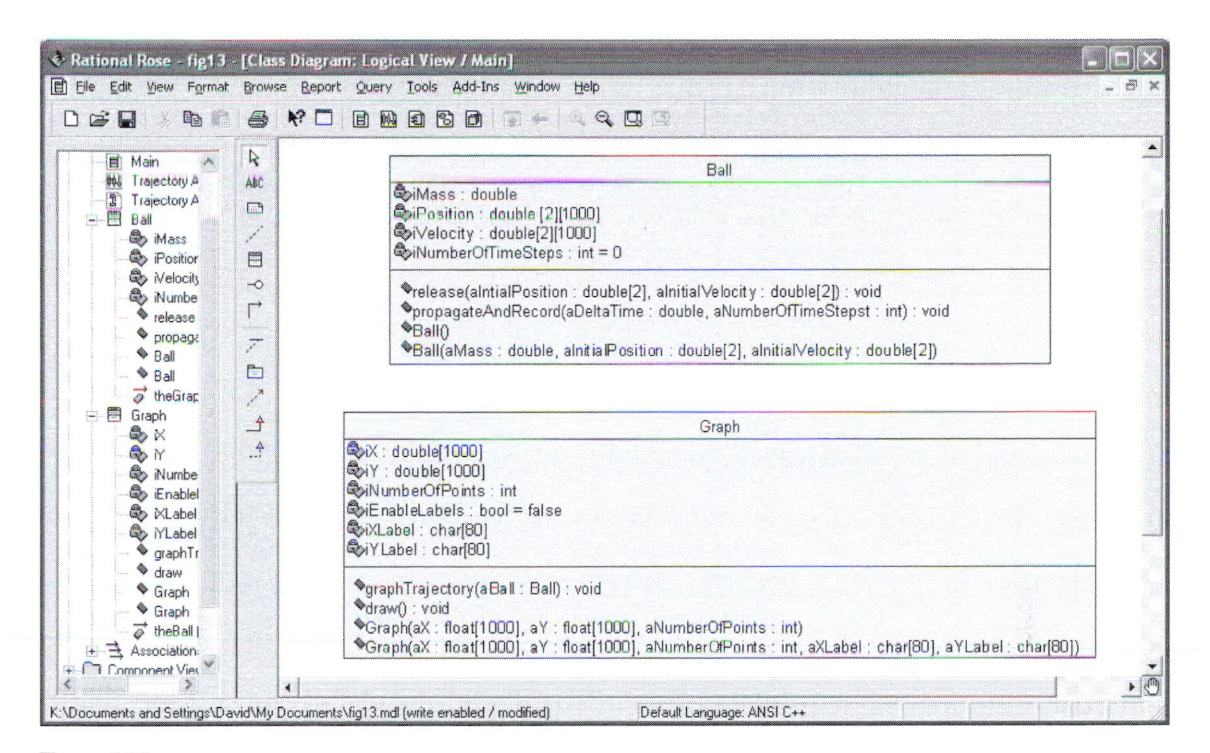

Figure 6.14
Source: Reprinted by permission from Rational Rose Enterprise Edition, © Copyright 2002 by International Business Machines Corporation. All Rights Reserved.

(1) Consider a light source. When the source is switched on, it produces light with intensity proportional to its internal voltage. Draw a class diagram that shows the name of an appropriate class describing the light source, its internal variables, and internal functions. Give as well a representation of the class as a circle as in Fig. 6.3. For one of your functions you could, for example, specify **void attachToPowerSupply (double powerSupplyVoltage);**.

(2) An abstraction of a radio has the following properties. A radio has an on–off switch, a station dial, and a volume dial. The result of turning the volume dial is to increase the audible volume, while turning the station dial changes the internal frequency that the radio receives. Generate a class diagram for a **Radio** class based on this abstraction.

(3) An adjustable light source has all the attributes and properties of a normal light source but also contains a rotary dial that allows the internal voltage of the light to be continuously adjusted. Draw an inheritance diagram that shows how the adjustable light source can inherit and extend the properties of the light source in the first problem. Be sure to specify the access modifiers (for example public, private, or protected) of the base class variables and functions.

(4) This is an example of an abstract class: a set is an unordered collection of elements, each of which may not appear more than once. Construct a class diagram for a set of

double elements that stores its elements in a 1000-component array. The set should have procedures for returning the number of elements, for storing an additional element, for determining if an element is present in the set, and for removing an element that matches a given value.

(5) Construct a class diagram for a simple calculator with one memory register that stores a **double** and one main memory location that also stores a **double**. The calculator will work as follows: the user first enters a value into the calculator's main memory location. He presses a button placing the value into the memory register and then enters a second value into the main memory location. Next, he presses a calculate button that adds this value to the value stored in the register and returns the result into the main memory. Finally, he presses a display button to display the result on the output screen.

(6) Consider the following physical system:

> An underwater remote controlled device is designed to be operated by remote control. When a button is pressed on the remote controller, the device is fueled. A second button then starts the motor and a third button puts the device in motion. Subsequently, radio contact is maintained between the controller and the device such that the controller registers the device position every 0.5 seconds. The controller radios back the changes in the motor velocity and steering setting required to maintain the desired path. Finally, when a fourth button is pushed on the controller, the device returns to the surface.

Develop a full use case diagram, a sequence diagram, and a class diagram for this model. Then, following the chapter discussion, devise abridged versions of these diagrams.

Chapter 7
C++ object-oriented programming syntax

Object-oriented analysis is followed by object-oriented design, which refers to the conversion of the system model into code. In this chapter, we summarize only central language features, relegating a discussion of more complex object-oriented syntax to the final chapters of the text.

7.1 Class declaration

A class declaration, which does not reserve memory space and may accordingly be repeated an arbitrary number of times, simply instructs the compiler that a certain name is associated with an user-defined class

```
class Trajectory;
```

Once a class is declared, its name becomes functionally equivalent to any built-in type identifier, such as **int** or **float**. Therefore, the compiler can subsequently resolve, for example, the function declaration

```
void myFunction(Trajectory aT);
```

However, the compiler obviously does not yet possess any information regarding the components of the class, hence it cannot resolve the second of the two statements below

```
Trajectory T;
T.plot();          // Error: T.plot() not defined
```

7.2 Class definition and member functions

A class definition specifies the elements of a class as well as its name and may additionally include the code (bodies) of its member functions. A class definition must be terminated with a semicolon. An example of a class definition in which the function body is not specified is (note the required semicolon at the end of **void plot( );**)

```
class Trajectory {
        public:
        float iPosition[100], iVelocity[100];
        int iNumberOfPoints;
        void plot();          // This statement cannot be repeated!
};
```

(The reason that a class can be defined without specifying the function bodies is that the total memory space associated with a class is that required to store the type information associated with its internal variables, together with the values of the beginning memory addresses of the locations at which data containing the function commands are stored. Since once the amount of memory associated with an entity is known to the compiler, the entity is automatically defined, the size of the function bodies is irrelevant in establishing the definition)

The statement **void plot( );** cannot be repeated within the class definition, since each statement inside the definition leads to the compiler reserving memory for an address of a function record (stored data containing the function commands). Repeating this statement would therefore incorrectly assign two memory locations to the same function name.

The body (definition) of the **plot( )** function must then be supplied a single time at global scope after the class declaration. The name of the function must then be preceded by the class name **Trajectory** followed by the *scope resolution operator* **::** which indicates that the **plot( )** function is a member of the **Trajectory** class, that is

```
void Trajectory::plot() {
        metafl("XWIN");
        qplot(iPosition, iVelocity, iNPoints);
}
```

On the other hand, if the body of the function is supplied within the class definition we would have instead

```
class Trajectory {
public:
int iNumberOfPoints;
float iPosition[100], iVelocity[100];
void plot() {
        metafl("XWIN");
        qplot(iPosition, iVelocity, iNPoints);
        }
};
```

The compiler handles functions with bodies that are placed inside a class definition differently (the functions are inlined, cf. Section 9.11) than externally defined functions, yielding faster execution times at the cost of larger object files and slower compilation speeds.

Generally, function bodies should not be included in class definitions, to avoid subtle problems. For example, in the program below, the class **First** must know that the class

Second has a member variable **i** before the body of its **print( )** function can be processed, while the class **Second** must similarly be aware that **First** possesses a member variable **i** before its **print( )** function can be processed. Accordingly, for at least one of these classes, the body of the **print( )** function must be specified after the other class has been defined.

```
class Second;
class First;                //The compiler now knows Second and First
                            //are classes

class First {
        public:
        int i;
        void print( Second aSecond );
                            //so it can resolve Second here
};

class Second {
        public:
        int i;
        void print( First aFirst );
                            //and it can resolve First here
};                          //The compiler is now aware of First and
                            //Second's members

void First::print( Second aSecond ) {
        cout << aSecond.i << endl;
                            //so it can resolve aSecond.i here
}

void Second::print( First aFirst ) {
        cout << aFirst.i << endl;
                            //and it can resolve aFirst.i here
}

main() {
        First F1;
        F1.i = 3;
        Second F2;
        F2.i = 5;
        F1.print( F2 );                 //Output: 5
        F2.print( F1 );                 //Output: 3
}
```

A *very common error* in writing code for classes is *to redefine one or more internal member variables within a member function* as below

```
class C {
        int i;
        public:
        void setI (int aI ) {
                int i = aI;             // Error!
        }
};
```

The new variable **i** hides the internal class member variable of the same name, so that changes to the new variables do not affect the internal member variable, which remains uninitialized.

7.3 Object creation and polymorphism

A class, like a built-in variable type name, can be thought of as a blank form with entry fields that accept values. Each realization (instance) of the class is a separate copy of this form with a unique identifying name into which a set of (possibly random) values have been placed. A specific realization of the class can be termed an object variable, just as a specific realization of a built-in type, such as **int**, is termed a built-in variable. *Declaring an object with the type of a user-defined class is functionally identical to declaring an object of a built-in type, such as int; thus a class implements a user-defined data type.* Accordingly, the syntax for the creation (instantiation) of objects is the same as for that of the creation of variables, for example the statements

```
Trajectory T1;
Trajectory T2;
```

instantiate two objects of type Trajectory. *Note that since these statements are type declarations and do not call functions, parentheses do not appear after* ***T1*** *and* ***T2*** *in the above statements; this is a major source of confusion!* Just as writing **int i;** creates an **int** variable containing a random value, the values of all internal variables in **T1** and **T2** above are random. To assign meaningful values to *public* variables, we can employ the *member-of* operator, **.**, which selects a member from the class with a specified name. In this manner, we can assign values to the first two position variables, the first two velocity variables, and the **iNumberOfPoints** variable in **T1** through the code

```
main () {
        Trajectory T1;
        Trajectory T2;
        T1.iNumberOfPoints = 2;
        T1.iPosition[0] = T1.iVelocity[0] = 0;
        T1.iPosition[1] = T1.iVelocity[1] = 1;
        T1.plot( );
}
```

While we have placed these lines inside the **main( )** function, they could equally well be located in any other function or in the global space outside all function bodies after the **Trajectory** class definition.

An object that only contains *public* member variables can also be conceptualized as a generalized array in which, unlike a standard array, the elements can possess different types and are accessed through their names (possibly followed by an array index if the internal variable is an array type) instead of through an array index. Consistent with this view, these variables can be assigned values at the point of definition, following the same syntax as arrays. Thus the **numberOf-Points** variable and variables **position[0]** and **position[1]** of **T2** are set to 2, 0, and 1, respectively, while all other variables are set to zero through the definition

```
Trajectory T2 = {2, 0, 1};
```

[Note: this syntax fails if a user-defined default constructor exists or if any class members are not public!]

Viewing an object as a generalized array clarifies to a certain extent the effect of setting one object equal to a second object of the same type. In this case (although such a manipulation is illegal for a built-in array), each member variable of the second object, irrespective of access privilege (**public**, **protected** or **private**) is copied to the corresponding element of the first object as in

```
class C {
public:
        int i[2];
};

main() {
        C C1, C2 = {1, 2};
        C1 = C2;
        cout << C1[1] << endl;              // Output: 2
}
```

A similar element-by-element copy occurs when an object is passed to or returned from a function, or when a class is initialized with a second class at the point of definition

```
C C3 = C1;
```

which can also be written as

```
C C3(C1);
```

Classes implement polymorphism in a particularly convenient manner as the behavior of a function is manifestly bound to the type of the calling object. For example, suppose a **Circle** class exists that contains the center position and radius of a circle together with a **plot( )** function that plots the circle. Then, if **C1** is a **Circle** object, the statement **C1.plot( );** draws a circle, while **T1.plot( );**, where **T1** is a **Trajectory** object, draws a trajectory. This facility parallels real-world behavior and therefore yields concise and understandable code.

7.4 Information hiding

C++ realizes information hiding through three keywords, **public**, **private**, and **protected**, that are followed by a colon in class definitions. All subsequent class members acquire the specified access privilege until a new access keyword is encountered. *Internal variables and functions default to private until a* ***public:*** *or* ***protected:*** *keyword is specified.* A private class member is only accessible to functions in the same class, while a public class member is accessible anywhere in the program. In the latter case, however, the member-of operator is required (except in certain specialized cases) when accessing the member from code outside the class. This is illustrated by a simple but critically important example

```
class InformationHidingExample {
      int iPrivateReadWrite;                    // Private by default
      public:
      void setPrivateReadWrite(int aPrivateReadWrite) {
                                                // Set member function
              iPrivateReadWrite = aPrivateReadWrite;
              iPrivateReadOnly = aPrivateReadWrite *
                              aPrivateReadWrite;
      }
      int privateReadWrite() {return iPrivateReadWrite;}
                                                // Get member function
      private:
      int iPrivateReadOnly;
      public:
      int privateReadOnly() {return iPrivateReadOnly;}
};

main() {
      InformationHidingExample I1;
      I1.setPrivateReadWrite( 4 );
      cout << I1.privateReadWrite() << endl;        // Output: 4
      cout << I1.privateReadOnly() << endl;         // Output: 16
}
```

Consider first the **iPrivateReadWrite** variable. Since no keyword is specified before the declaration of this variable, it is private by default. However, it can still be assigned a value as in the third line in the **main( )** function through the public **setPrivateReadWrite( )** member function, which is therefore called a set member or writer function. Its value can be further accessed by the public get member or reader function **privateReadWrite( )**, as in the fourth line of **main( )**. A private internal class member with only a get member function, such as **iPrivateReadOnly** above, is read-only, in other words write-protected, from *outside* the class, while a member with only a set member function is write-only (read-protected). *Note the naming conventions for these functions*, namely that the internal variable name is composed of a prefix **i** followed by the actual variable name. Then set member functions prefix the variable name with **set** while get member functions possess the name of the variable (many programmers prefer, however, to prefix the names of get member function with the word **get**). These conventions should be consistently followed. Note that the internal variable identifier cannot coincide with that of the get member function;

that is if the internal variable is named **i**, a name collision occurs if the get member function is also named **i** (of course in this case the internal member variable should be named **iI** according to the above procedure).

Note that the variable **iPrivateReadOnly** appears in the **setPrivateReadWrite( )** function but is only defined several lines later. This illustrates a singular feature of class definitions. As a rule, when the compiler has processed a program up to a certain token, it cannot access information contained later in the program than this token. However, the compiler analyzes a class definition before resolving any member function bodies that are enclosed within the definition, which enables backward referencing in this particular case.

A common convention is to place the part of the class that the user can access, namely the **public** interface, first in the class declaration followed by the **protected** and **private** sections. Therefore, the internal variables, which are normally **protected** or **private**, appear at the end of the class body. (Further, the constructor is often placed first in the declaration, followed by the destructor.)

7.5 Constructors

Now consider the read-only internal variable **iPrivateReadOnly** in the program of the previous section. Since it cannot be changed from outside the class, when an object of type **InformationHidingExample** is created, this variable will contain a random value that can only be changed by member functions of the class, as illustrated by the **setPrivateReadWrite( )** function. To set this variable to a meaningful value, such as zero, at the point that the object is created, we might attempt to replace its declaration within the class body by

```
int iPrivateReadOnly = 0;       // WRONG!
```

This illegal construct is a common error of beginning programmers. Instead, C++ provides a facility termed the *constructor* that can be employed to initialize internal member variables when an object is first defined. Constructor functions can have any number of arguments, including zero, and are only called when an object is defined as in the example below

```
InformationHidingExample I1( aArg1, aArg2, ...);
```

A zero-argument constructor (or a multiple-argument constructor all of whose arguments have default values) is labeled a *default constructor* and *is called through a normal definition statement*, namely

```
InformationHidingExample I1;
```

As before, parentheses are absent after the object name when calling the default constructor. C++ supplies a default constructor if no user-defined constructor is supplied. However, this constructor generally does *not* initialize the internal class variables, which therefore contain random values.

A class can have any number of constructors as long as each constructor has a unique sequence of arguments, which can differ by number, type, or both. These functions are only employed to define an object and consequently do not have a return type. Further, they have the same name as the class since a definition statement contains the class name.

Two non-equivalent options exist for initializing an internal variable in the body of a constructor function. That is, to initialize both internal variables in the **InformationHidingExample** class to zero through a zero-argument default constructor, we can initialize either the variables in the constructor body

```
class InformationHidingExample {
...
        InformationHidingExample() {
              iPrivateReadOnly = iPrivateReadWrite = 0;
        }                                          // Default constructor
...
};
```

or prior to the constructor body through an *initialization list*

```
class InformationHidingExample {
...
        InformationHidingExample() : iPrivateReadOnly(0) ,
              iPrivateReadWrite(0) {}          // Default constructor
...
};
```

(Of course, one variable could be initialized in the body and a second in the initialization list.)

An initialization list is generally preferred for reasons of clarity and efficiency. Further, the initialization list is processed *before the code in the constructor body is executed* and, more importantly, *before the class object is actually constructed*. Consequently, the initialization list *must be used to initialize* ***const*** *internal variables*, since if the object were to be constructed prior to initialization, these variables would contain random bit patterns that could not be subsequently changed.

A two-argument constructor might take the form

```
InformationHidingExample(aPrivateReadOnly, aPrivateReadWrite) :
      iPrivateReadOnly(aPrivateReadOnly) {
      iPrivateReadWrite = aPrivateReadWrite;
}
```

However, when introducing constructors with argument values, one must be extremely attentive to the fact that *if the user supplies a constructor with any number of arguments, a default constructor is no longer automatically generated by the system*. That is, the compiler assumes that if only non-default constructors are supplied by the programmer, the intention is to represent physical quantities that do not have characteristic values. Unfortunately, *features in the code that at first sight do not seem to require a default constructor fail if one is not present.* For example, as will be illustrated in Section 7.7, if a class, **C**, contains an object of a second class, **D**, as an internal member variable, the default constructor of **D** will be called if no **D** constructor is present in the initialization list. Therefore an initialization list is *required* if **D** lacks a default constructor. As another example, if a default constructor is absent in a base class, user-defined constructors must be supplied in all derived classes. Accordingly, good programming practice always ensures the presence of a default constructor.

Finally, we note that while **print** statements can be included in a constructor to facilitate debugging, in the final version of the program a constructor should be only used for its intended purpose, namely to allocate and initialize class objects (and possibly for type conversion as will be explained in Section 20.9).

7.6 Wrappering legacy code

Since scientific program packages are often highly complex or application specific, many have not been ported from FORTRAN or C code into C++. Legacy C code can, however, be rewritten as object-oriented code for use in large programming tasks by wrappering related routines into C++ classes, although slight differences in syntax between C and C++ must sometimes be addressed. (Compiled FORTRAN code can be accessed and wrappered as well from C++, but with far greater difficulty, as discussed in Appendix D.) Both source code and precompiled object code can be wrappered, however we will only discuss a simple source code example here.

Consider the following procedural DISLIN graphics code

```
float x[1000], y[1000];
int numberOfPoints;
//... Assign values to x, y, numberOfPoints
metafl("XWIN");
qplot(x, y, numberOfPoints);
```

To wrapper these lines into a class, the plotting routine must be packaged together with the data that it processes. That is, the arrays **x** and **y** together with **numberOfPoints**, the number of (x,y) points to be graphed, must be converted into internal class members. At the same time, a constructor must be supplied so that data can be easily transferred into the internal variables and a **draw( )** function should be defined that plots the data. Accordingly, an appropriate class construct is

```
class Graph { public:
        int iNPoints;
        float iX[1000], iY[1000];
        Graph(float aX[ ], float aY[ ], int aN) : iNPoints(aN) {
              for (int loop = 0; loop < iNPoints; loop++) {
```

```
                        iX[loop] = aX[loop];
                        iY[loop] = aY[loop];
                }
        }
        void draw() {
                metafl("XWIN");
                qplot(iX, iY, iNPoints);
        }
};
```

Now sets of data points can be stored and later graphed as follows:

```
#include <iostream.h>
#include <dislin.h>

// insert Graph class here

main () {
        float x[2], y[2];
        x[0] = y[0] = 0;
        x[1] = y[1] = 1;
        Graph G1(x, y, 2);
        y[1] = 3;
        Graph G2(x, y, 2);
        G1.draw();
        G2.draw();
}
```

Observe that in contrast to the procedural case, two **Graph** objects rather than four arrays are required to store the position and velocity data. The previous **Trajectory** class can now be rewritten as

```
class Trajectory {
public:
int iNumberOfPoints;
float iPosition[100], iVelocity[100];
void plot() {
        Graph G1(iPosition, iVelocity, iNumberOfPoints);
        G1.draw();
        }
};
```

7.7 Inheritance

As we have noted earlier, a particularly convenient feature of object-oriented programming is that functionality in one class that is common to a second, derived, class can be directly incorporated into the derived class through the mechanism of inheritance. In particular, a derived class inherits all the non-private variables and functions of a base class that are not explicitly redefined in the derived class. However, additional complexity arises from the manner in which the access privileges (**public**, **protected**, or **private**) of the base class members propagate to the derived class. The simplest form of inheritance is *public inheritance*. For a class **D** to inherit from a class **C** in this manner, its class definition commences with **class D : public C** {, as illustrated in the example below. This syntax can be easily remembered since the elements of the base class **C** are always constructed before those of the derived class **D** so that **C** forms a type of initialization list for **D**. In public inheritance, the access privileges of all elements in the base class are preserved in the derived class unless an element is explicitly redefined in the derived class with a different access privilege. A public variable in the base class is public in the derived class, while a private variable in the base class *cannot be accessed* in the derived class. Thus, the following code generates a compiler error

```
class C {
        public:
        double iVelocity;
        private:
        double iPosition;
};
class D : public C { public:
        setPosition(double aPosition, double iVelocity) {
                iVelocity = aVelocity;
                iPosition = aPosition; // Error - iPosition is
                                       // private to class C!
        }
};
```

since the variable **iPosition** is a private member variable of class **C** and therefore cannot be accessed from within any other class. A less employed form of inheritance is *private inheritance*, in which all public members of class **C** become private members of class **D**.

A critically important feature of inheritance follows from the statement that a derived class is a specialized form of the base class so that a derived class object is by definition also a base class object. That is, suppose a derived class **AirFilter** is a derived class of the base class **Filter.** Clearly an **AirFilter** is a particular type of **Filter**, therefore it can be employed anywhere in the program where a **Filter** is expected. However, except in certain specialized cases discussed in Chapter 6, the object then behaves as a **Filter**, not as an **AirFilter**, as illustrated by the following generic example

```
class C {
        public:
        void print () {cout << "C";}
};

class D : public C {
        public:
        void print () {cout << "D";}
};

main () {
        C C1;
```

```
        D D1;
        C CArray[2];
        CArray[0] = C1;
        CArray[1] = D1;          // Valid: D1 is also a C object!
        D1.print();              // Output: D
        CArray[1].print();       // Output: C
}
```

The compiler reserves an amount of memory appropriate to a **C** object for each element in the array **CArray**. Therefore, when a derived **D** object is placed into the array, its additional features are discarded by the target object, leaving only its **C** properties.

A derived class constructor constructs its base class components by calling the default base class constructor *before* the derived class properties are constructed. A user-defined base class constructor can be employed in place of the default constructor but then must accordingly be placed in the initialization list of the derived class constructor. As noted earlier, this step is *required* if the base class lacks a default constructor. These considerations are illustrated in the program below

```
class D {
        public:
        int i;
        D(int aI) : i(aI) {} // Note: default constructor absent
};

class C {
        public:
        D iD;
        C(D aD) {iD = aD;}   // Error: initialization list required
//      C(D aD) : iD(aD) {}  // OK
};

main() {
        D D1(1);
        C C1(D1);
        cout << C1.iD.i << endl; // Output: 1
}
```

Replacing the single-argument **C** constructor with the commented-out line containing an initialization list enables the above program to compile properly.

7.8 The 'protected' keyword

There are two standard procedures for accessing base class variables, such as **iPosition** in the program below, from within a derived class. The first is to supply public get and set member functions for the **iPosition** variable within class **C**; these will then be accessible from within class **D** as well as from elsewhere in the program. Alternatively, access to an internal variable can be restricted to classes that are derived from the base class through the **protected** keyword. A variable or function that is protected in a base class is accessible and remains protected in all public derived classes, but is inaccessible (private) to all other program units. In the code

```
class C {
        protected:
        int iPosition;
};

class D : public C {
        public:
        setPosition(int aPosition) {
                iPosition = aPosition;          // OK: iPosition is
                                                // protected in C!
        }
};

main () {
        D D1;
        D1.setPosition(0);
        cout << D1.iPosition << endl;           // Error!
}
```

the variable **iPosition** is therefore accessible from within both the base and the derived class, but not from elsewhere in the program. Substituting **protected:** for **public:** in the line **class D : public C** { yields **protected** inheritance in which both public and protected members of the base class become protected members of the derived class, while private members of the base class remain private and therefore inaccessible from within the derived class.

7.9 Assignments

Part I

In problems 1–11 give the output of the code if the program will function properly. If there is an error or if a potential error condition, such as overflow, underflow, incorrect memory access, etc., arises, so that the program result or runtime state is unpredictable, indicate instead where the error is and what type of error has been made. Assume **#include <iostream.h>** is entered as the first line of each program. Some programs may contain more than one error.

```
(1) class A {
            int iI, iJ;
            print() {cout << iI << '\t' << iJ << endl;}
    };

    main () {
            A A1;
```

```
        A1.iI = A1.iJ = 2;
        A1.print();
}
```

(2)
```
class B;
class A{
        int iI;
        public:
                int i() {return iI;}
                void setI(int aI) {iI = aI;}
                void print(B aB ) {cout << aB.i() << endl;}
};

class B{
        public:
                int iI;
                int i() { return iI; }
};

main() {
        B B1;
        B1.iI = 2;
        A A1;
        A1.setI( 3. );
        A1.print(B1);
}
```

(3)
```
class C {
        int i[2];
        public:
        print() {cout << i[0] << '\t' << i[1] << endl;}
        C() {
                i[0] = 1;
                i[1] = 2;
        }
        C(int aI) {
                i[0] = aI;
                i[1] = 2 * aI;
        }
};

main () {
        C C1(2);
        C C2();
        C1.print();
        C2.print();
}
```

(4)
```
class C{
        public:
        int iI;
        void C() { iI = 3; }
};

main() {
        C C1;
```

```
        cout << C1.iI << endl;
}
```

(5)
```
class C {
        int iI;
        public:
        int i() {return iI;}
};

main () {
        C C1;
        C1.iI = 2;
        cout << C1.i() << endl;
}
```

(6)
```
class C {
        int iI;
        public:
        int i() { return iI; }
        C() { iI = 3; }
};

class D : public C {
        public:
        D() { iI = 2; }
};

main(){
        D D1;
        cout << D1.i() << endl;
}
```

(7)
```
class A;

void A::print () { cout << "In function" << endl; }

class A {
        void print ();
};

main () {
        A A1;
        A1.print();
}
```

(8)
```
class C {
        protected:
            int iI;
            C() { iI = 3; }
        public:
            int i() { return iI; }
};

class D : public C {
        public:
```

```
            D() : C() { }
};

main(){
        D D1;
        cout << D1.i() << endl;
}
```

(9)
```
class B {
        private:
        int iB;
        public:
        void setIB( int aB ) : iB(aB) {};
};

main () {
        B B1;
        B1.setIB(3);
}
```

(10)
```
class C {
        protected:
        int iC;
};

class D : public C {
        int iD;
        public:
        int d(){ return iD; }
        void setD(int aD) {iD = aD; iC = 0;}
};

main () {
        D D1;
        D1.setD(5);
        cout << D1.d() << endl;
}
```

(11)
```
class C {
        public:
        int iI;
        void print() {cout << iI * 2 << endl;}
};
class D : public C {
        public:
        void print() {cout << iI << endl;}
};

main () {
        D D1;
        D1.iI = 2;
        C C1[2];
        C1[0] = C1[1] = D1;
```

```
        C1[1].print();
}
```

Part II

(1) Hand in the program code for parts (a), (b), and (c) together with the graphs for parts (a) and (c).

(a) Write a program that contains the two classes **Trajectory** and **Graph** presented in Section 1.6 together with a **main()** function (a test driver) that generates a **Trajectory** object. Place the position–velocity pairs (0, 0), (1, 1), and (2, 4) into the first three array positions in this object. Call the **plot()** member function of your **Trajectory** object from within **main()** to plot these three points.

(b) Next, change the three-element **Graph** constructor so that it instead becomes a one-element constructor with a **Trajectory** argument, for example **Graph(Trajectory aT)**, and rewrite the body of the constructor so that you reproduce the same result as that of part (a) above. Remove the **plot()** function in the **Trajectory** class so that your program takes the form

```
#include <iostream.h>
#include "dislin.h"

class Trajectory {
    public:
    float position[100], velocity[100];
    int numberOfPoints;
};

class Graph {
    public:
    int nPoints;
    float x[100],y[100];
    Graph(Trajectory);
    void draw();
};

main () {
    Trajectory T1;
    ... place values into T1 ...
    Graph G1(T1);
    G1.draw();
}
```

(c) Now generate a derived class **TimeTrajectory** that inherits (using public inheritance) the member variables and functions of **Trajectory** but adds a new float array variable, **time[100]**. Modify your **main( )** function to store three time–position–velocity triples: (0 1 3), (1 3 5), and (2 8 10). However, *do not modify* the **Graph** object that you created in part (b). Show that you can graph the resulting **TimeTrajectory** object exactly as you graphed the **Trajectory** object in part (b) by calling the unmodified **Graph** constructor of part (b) from **main( )** with your new

TimeTrajectory object as the argument in place of the previous **Trajectory** object argument. This works because a **TimeTrajectory** is a **Trajectory** (of a specialized type). Note that you have written this program by making one change at a time (hopefully).

(2) Construct a system consisting of two objects. The first is a **Resistor** object with a private member variable **double iResistance**, public get and set member functions **double resistance( )** and **void setResistance(double aResistance)**, a one-argument constructor **Resistor(double aResistance)** that sets **iResistance** to **aResistance** when a new **Resistor** object is constructed, a zero-argument constructor **Resistor( )** that constructs a 100 ohm **Resistor**, a set member function **void setResistance(double aResistance)** and two public member functions **void displayVoltage(double aCurrent)**, and **void displayCurrent (double aVoltage)**. The first of these calculates the current if the voltage over the resistor is specified, while the second calculates the voltage when the current is specified. Both of these functions then display the line

The voltage in the resistor is *** **[the numerical value of the voltage] and the current is** *** **[the value of the current].**

Now write the code for a **PowerSupply** object that has a private internal **double** variable **iVoltage**, a private internal variable **iResistor** of class type **Resistor**, a two-argument constructor **PowerSupply(double aVoltage, Resistor aResistor),** a private internal **bool** variable **isOn** (a variable defined as **bool b** can be assigned the two values **b = true;** or **b = false;** (these are displayed by **cout** as 1 and 0 respectively), and public get and set member functions for **iVoltage**. The set member function **setVoltage** should write to the console

The voltage in the power supply is *** **[the numerical value of the voltage] volts.**

When a **PowerSupply** object is constructed, **isOn** should be set to **true**. Write a **void enablePowerSupply(bool aIsOn)** function that can be used to turn the power supply on or off by setting the internal variable **isOn** to **true** or **false**. Couple the **PowerSupply** object to its associated **Resistor** object (the internal **Resistor** variable) by insuring that whenever the power supply voltage is changed (note here that when the power supply is switched off the voltage falls to zero) the value of the current through the resistor is printed out (do this through appropriate additional modifications to the **setVoltage( )** member function in **PowerSupply**). Check your program by writing a **main( )** program that instantiates (creates) a **Resistor** with a resistance of 5 ohms and then instantiates a 10 V **PowerSupply** connected to this resistor. Then turn off the power supply from the **main( )** program. Note: in more sophisticated C++ code, the resistor would not be a simple member variable of the **PowerSupply** class.

That is, the class declarations and main program should be (you need only supply the appropriate function bodies)

```
#include <iostream.h>
class Resistor {
     double iResistance;
```

```
    public:
    Resistor(double aResistance) // Note: a default (zero-
        //argument) constructor must be supplied!
    Resistor()                    // Default constructor
    double resistance()
    void setResistance(double aResistance)
    void displayCurrent(double aVoltage)
    void displayVoltage(double aCurrent)
};

class PowerSupply {
    double iVoltage;
    bool isOn;
    Resistor iResistor;
    public:
    void enablePowerSupply(bool aIsOn)
    void setVoltage(double aVoltage)
    double voltage()
    PowerSupply(double aVoltage, Resistor aResistor)
        // Construct iResistor in the initialization list
};

main () {
const bool on=true, off=false;
Resistor R1(5);
PowerSupply P1(10,R1);
P1.enablePowerSupply(off);
}
```

Submit the program together with your output.

(3) Draw the collaboration and sequence diagrams for problem 2 above, in analogy with the diagrams given in the previous chapter.

(4) In a semiconductor, the concentration of free electrons, n, and the concentration of holes (atomic sites in the lattice that are missing an electron), p, are related according to the bimolecular recombination law $np = n_i^2$, where for undoped Si the intrinsic carrier density, n_i, at room temperature equals $1.5 \times 10^{10}/\text{cm}^3$. Write a **Semiconductor** class with three **double** variables **iElectronDensity**, **iHoleDensity**, and **iIntrinsicCarrierDensity**. Provide get and set member functions for the first two of these variables, but insure that **iIntrinsicCarrierDensity** is read-only. You will want to set both the electron and hole density through the set member functions when only one of these two quantities is specified. Provide a one-element constructor that only sets the intrinsic carrier density. Test this program by writing a **main()** program that sets the free electron density to $1.0\times10^{10}\text{cm}^{-3}$for Si and then prints out the values of the internal variables; that is the structure of your program will be

```
class Semiconductor {
    double iElectronDensity, iHoleDensity,
      iIntrinsicCarrierDensity;
    public:
        double electronDensity()
        double holeDensiy()
```

```
            double intrinsicCarrierDensity() //Readonly
            void setElectronDensity(double aElectronDensity)
            void setHoleDensity(double aHoleDensity)
            void print()
            Semiconductor(double aIntrinsicCarrierDensity)
};

main() {
            Semiconductor Silicon(1.5E10);
            Silicon.setElectronDensity(1.e10);
            Silicon.print();
}
```

Again hand in your program and the output.

(5) Write a program that contains a class **Rectangle** with two private double precision members **iLength** and **iWidth**, public set and get member functions for these two members, a two-argument constructor that sets the length and width to any two user-specified values, and a void function **area( )** that computes the area and then prints this value by inserting it into the **cout** output stream. Write a **main** function that creates a **Rectangle** with a length of 10 and width of 20 and compute its area.

(6) Form a class **Vector** that wrappers **double** arrays to implement bounds checking in the following manner. The private member variables stored in **Vector** (data members) are the array **double iArray[1000]** and the integer **iArraySize** that correspond to the number of actual stored array elements. The public class functions that you should implement are:

double getArrayElement(int aPosition) – returns the array element at **aPosition** or terminates the program through a call **exit(0)** that is found in **stdlib.h** (that is, you must write **#include <stdlib.h>** at the beginning of your program) if **aPosition** is either less than 0 or greater than **iArraySize**−1.

int getArraySize( ) – a get member function for the internal variable **iArraySize.**

void addLastElement(double aArrayElement) – calls **exit(0)** if **iArraySize** is 1000, otherwise adds the element **aArrayElement** to the end of the array and increments **iArraySize** by one.

void changeArrayElement(int aPosition, double aElementValue) – calls **exit(0)** if **aPosition** is larger than **iArraySize** or less than zero, otherwise changes the value of **iArray[aPosition]** to **aElementValue**.

Vector(double aArray[1000], int aArraySize) – a constructor that sets the first **aArraySize** elements of **iArray** to the values given in **aArray** and calls **exit(0)** if **aArraySize** is less than 1 or larger than 1000**.**

void printArray( ) – a function that sends the values of all **iArraySize** elements of the array to **cout**.

Now check your program by writing a **main( )** function that first applies the **Vector** constructor to instantiate (construct) a 4 element **Vector** object with the four successive internal array values {**1, 2, 4, 9**} (that is, **aArray[0]=1, aArray[1]=2** . . .). Next, apply the **changeArrayElement( )** function to change the second element to 5, then add a new element 10 to the end of the array using **addLastElement( )**. Invoke **getArraySize( )** and **getArrayElement(3)** from **main( )** and send the return values, namely the size of the array and the fourth element of the array to **cout**. Finally call **printArray( )** from within the main program to print out the full contents (the first five elements) of the array. Hand in your program listing and the output.

Chapter 8
Control logic and iteration

At this point, we have surveyed the foundations of the C++ language and object-oriented programming. The remainder of the text is devoted to a careful study of the C++ language and to scientific applications. The topic of this chapter is control constructs, which are logical operations that enable branching among different paths through a program. In the next chapter we analyze functions, which provide an alternative method for redirecting program flow.

8.1 The *bool* and *enum* types

To implement logical program control a computer language must define a representation of true and false. In C++, if the memory space associated with any variable or object contains only zero bits, the variable evaluates to a logical false, otherwise its logical value is true. Thus the logical values of all the following variables are false: **int i = 0;**, **double d = 0.0;**, **char c = '\0';** (the last of which has the numeric (ASCII) value 0 and is called the null character). These representations of false are automatically converted into each other when a variable can be legally assigned to a variable of another type so that writing, for example, **double d = 0.0; char c = d;** yields a zero character.

The 1998 C++ standard introduced a built-in data type, **bool**, which accepts the values **false** and **true**. Assigning any non-zero value to a **bool** variable yields a value of 1. Sending (piping) the *manipulator* **boolalpha** to **cout** prints out the values of succeeding **bool** variables as either **false** or **true**; that is, the program

```
bool b = false;
bool c = 3;
cout << b << '\t' << c << endl;
cout << boolalpha << b << '\t' << c << endl;
```

yields the output

```
0       1
false   true
```

The **bool** type is closely related to the **enum** type, which is also frequently employed in control structures. An **enum** type is a particular form of an integer type that additionally specifies a set of literal (that is, named) integer constants called *enumerators* that can be assigned to variables of the type. An enumerator variable can be used anywhere an integer type is expected; if this is done, it is automatically (implicitly) converted to the integer type. The concept is best understood though an example. A general form of the **enum** statement (a common convention is to capitalize **enum** variables) reads

```
enum suits { HEARTS, SPADES, CLUBS = 4, DIAMONDS };
```

In this case, a variable of type **suits** can be assigned the four literal values **HEARTS**, **SPADES**, **CLUBS**, or **DIAMONDS**, as well as any other integer value. The numerical equivalent of **HEARTS** is 0, of **SPADES** 1, **CLUBS** 4, and **DIAMONDS** 5, which are the values obtained when a **suits** variable set to one of these constants is converted to an **int**. The above features are illustrated by the program

```
enum suits { HEARTS, SPADES, CLUBS = 4, DIAMONDS};
suits b = 3;
suits c = DIAMONDS;
cout << b << '\t' << c << endl;
if (c == DIAMONDS) cout << "C is a Diamond" << endl;
```

which produces the output

```
3       5
C is a Diamond
```

An **enum** can also be anonymous, which means that it lacks a dedicated typename. In this case, implicit integer values are associated with an identifier; that is, the definition

```
enum { notEnabled, isEnabled } switchSetting;
```

creates a variable **switchSetting** of type **enum** that can be set equal to the values **notEnabled** and **Enabled**.

If class has a public member variable of **enum** type, to refer to the variable from outside the class its name must be prefixed with the class scope as in

```
class C{
public:
     enum iSwitchSetting {notEnabled, isEnabled};
     iSwitchSetting iControl;
};

main(){
     C C1;
     cout << (C1.iControl = C::isEnabled) << endl;      // Output: 1
}
```

The two lines in the class definition can as well be combined into a single statement

```
enum iSwitchSetting{ notEnabled, isEnabled } iControl;
```

Enumerators can be useful as return values or arguments of functions, since they restrict the meaningful values of the inputs and outputs. The following somewhat contrived code, for example, determines if the voltage applied to a circuit is sufficiently large enough to turn the circuit on

```
enum isEnabled{off, on};

class Circuit{
public:
      int iVolts;
};

isEnabled test (Circuit aC){
      if (aC.iVolts > 0.2) return on;
      else return off;
}

main(){
      Circuit C1;
      C1.iVolts = 0.1;
      cout << test(C1) << endl;
}
```

8.2 Logical operators

Complex logical expressions can be formed from single logical variables through logical operators. These are **&&** (and), || , (inclusive or), where the vertical line key is located above the backslash (\) key on a standard keyboard, **!** (not), == (logical equality) and **!=** (logical inequality) together with the ordering operators < (less than) <= (less than or equal to), > (greater than) and >= (greater than or equal to). *A space is not permitted between the two symbols in any compound operator such as <=, == or ++*. The behavior of logical operators can be understood by examining the following program, which generates the logical truth table for the statement **(A && B)**, which evaluates to **true** whenever both variables **A** and **B** are **true**

```
for (bool a = 0; a < 2; a++) {
      for (bool b = 0; b < 2; b++) {
            cout << boolalpha << "A = " << a << '\t' <<
               "B = " << b << '\t' <<
               "A and B = " << (a && b) << endl;
       }
}
```

This yields

```
A = false    B = false    A and B = false
A = false    B = true     A and B = false
A = true     B = false    A and B = false
A = true     B = true     A and B = true
```

In a similar fashion, **A || B** returns **true** if **A** is **true** or **B** is **true** or both are **true**. The logical operation **A = = B** returns true if and only if **A** and **B** have the same values.

The notation for logical operators invariably leads to severe errors that *must be memorized* to insure accurate coding. *The first and most troublesome error occurs when the assignment operator = is employed in place of the logical equality operator ==*. When this occurs within any control construct, such as **if(B = C)**, three problems arise. First, the value of **B** is set to the value of **C**, after which the logical statement is then evaluated with the value of **C** as an argument. Since **C** is usually non-zero, the logical statement typically evaluates to **true** whether or not **B** and **C** are identical and control passes to the statement following the control (**if**) condition. A related error is *the use of a single ampersand in place of two ampersands for or a single vertical bar in place of two vertical bars*. In either case, bitwise operators are invoked in place of the logical operators, as discussed further in Section 20.15, which again typically results in a logical value of **true**.

8.3 *if* statements and implicit blocks

Logical operations are generally employed by C++ *control constructs* that direct program flow among blocks according to the value of their logical argument. The simplest of these statements is the **if** statement

```
if (A){
...statements ...
}
```

which executes the block of code labeled **statements** when its argument, the logical expression **A**, evaluates to **true**. If a block in any control construct contains a single statement, the braces can be omitted, but the block is still implicitly present

```
if (A) statement;
```

Although the above compound statement is often written on two lines, this is very error prone and should be avoided when space permits. The three major sources of error are first, introducing an additional semicolon after the control condition

```
if (A); statement;             // WRONG!
```

which executes the control condition followed by a null statement. Consequently, **statement** is always executed, regardless of the value of **A**. Next, while the code designated by **statement** is not enclosed by curly braces, it still implicitly belongs to a separate block. Therefore in

```
if (A) int i = 1;
```

the variable **i** is destroyed after the line executes and is unavailable in the remainder of the program. Finally, *the curly braces are often omitted by mistake in* *multi-line* **if** *statements*. In this case, only the first of the statements following the **if** statement is influenced by the control condition.

8.4 *else, else if,* conditional and switch statements

The logical control of an **if** statement can be extended by appending an **else** statement that is executed if the assertion in the **if** statement is false. The **else** statement can be followed by a further **if** statement in which case the **else** and **if** keywords are generally placed on a single line. That is, we can write:

```
int grade;
cout << "0 - 9 prints C, 10 - 19 prints B\
  and 20 - 29 prints A "<< endl;
cin >> grade;
grade = grade/10;
if (!grade) cout << "C" << endl;   //Executed if grade/10 = 0
                                   //so that !grade is 1.
else if (grade == 1) cout << "B" << endl;
else if (grade > 2) cout << "Input Error" << endl;
else cout << "A" << endl;
```

An abbreviated form for **if ... else** is the ternary conditional operator **? :** which is defined such that, for example, the compound statement

```
if ( a <= b ) x = a;
else x = b;
```

which sets the variable **x** to the smaller of the values **a, b** is identical to

```
x = (a <= b ) ? a : b;
```

Repeated **else if** statements can be simplified through the **switch** construct. The equivalent code for the **grade** example above with a **switch** statement is as follows

```
int grade;
cout << "0 - 9 prints C, 10 - 19 prints B and 20 - 29 prints A"
  << endl;
cin >> grade;
switch(grade / 10) {
      case (0): cout << "C" << endl; break;
      case (1): cout << "B" << endl; break;
      case (2): cout << "A" << endl; break;
      default: cout << "Input Error" << endl;
}
```

The **break** statements in the above program transfer control to the statement immediately following the **switch** block. Removing these statements leaves control within the block, in which case the **default:** statement is always executed for any value of **grade**. A **switch** construct can also be applied to character variables as in **char c; cin >> c; switch(c)** { **case ('a'): ...** }. Since the syntax of the **switch** statement is complex, an example should normally be consulted during coding.

While not recommended, program control can be passed to a line prefixed by a label such as **start:** in the example below through a **goto** statement located within the same function block. However, control cannot be passed into a statement in a block that follows any initialization statement. A related restriction is that a definition in one branch of a multiple branch conditional statement cannot be accessed in a second branch. The following code is invalid for both these reasons

```
main() {
     int i;
     cin >> i;
     if (i == 1) goto start;
     if (i == 0) {
        int j = 4;
        start: j = 3;               // Error: initialization of j
                                       // bypassed
        cout << j << endl;
     }
     else cout << j++ << endl;         // Error: j defined in
                                       // separate branch
}
```

8.5 The *exit( )* function

The **exit( )** function terminates a running program. (In older compilers this function is enabled through inclusion of the **stdlib.h** header file.) Typically the function is used in conjunction with a logical condition as in

```
cin >> grade;
if (grade > 30) exit(0);
```

It should be noted that while an argument, normally set to zero to indicate abnormal program termination, must be supplied to the **exit( )** function, the program terminates regardless of the value of its argument, which is generally captured by the operating system and is invisible to the user.

8.6 Conditional compilation

Often entire sections of code must be included in or excluded from the compilation process as, for example, to incorporate header files or to insure that definitions are not read twice when several files are included. To assemble and transform code before the compiler is activated, *preprocessor directives* are employed. These directives begin with a pound sign (#) and are not terminated with a semicolon. Therefore, two such statements cannot be placed on the same line. The most important directives are:

(1) The **#include** statement which reads in code from a second file into the current file at the position at which the statement appears. It may adopt either of two forms, the first of which is

```
#include "includeFile"
```

in which case the compiler searches for a file named **includeFile** first in the directory from which the program is being run and then, for most compilers, in the include file subdirectory(s) of the compiler's main directory. Alternatively, writing

```
#include <includeFile>
```

leads to the compiler attempting to locate **includeFile** first in the compiler's include directories and then in most cases, in other directories such as the user's directory. In either case, additional directories can be added to the directory in which the corresponding **#include** statement searches through an appropriate compiler switch, typically –I.

(2) The statement **#define A B** sets a variable or an expression, **A**, to a numeric or text expression, **B**. All instances of **A** are then replaced by **B** *before* the code is compiled. In the case that **B** is a constant value, **A** is therefore replaced by a constant. In contrast, **A** in the statement **const typename A = B** is a non-modifiable lvalue that has an accessible storage address. As a further example, writing

```
#define square(x) x*x
int i = 3;
cout << square(i);
```

replaces all instances of **square(i)** by **i * i** prior to compilation. In C++, such applications of **#define**, which circumvent type checking and scope resolution, have been superseded by **const** variables and **inline** functions (cf. Chapter 9.11).

(3) The **#ifdef** or **#ifndef**, **#else**, and **#endif** statements are principally employed to ensure that the same code lines are not compiled twice. In particular, if a block of code is surrounded by preprocessor directives as follows

```
#ifndef_MYFLAG
#define_MYFLAG
int i;
...additional definitions and code statements
#endif
```

it will be compiled only once even if included into a given program multiple times. Note that if the literal expression _**MYFLAG** appears anywhere in the program body, it will be eliminated by the precompiler.

8.7 The *for* statement

The **for** statement repeats a single or compound statement followed by an incrementation step until a certain logical condition is fulfilled. The **for** statement has the form

```
for ( initialization (I); termination (T); expression (E) )
  {statements (S);}
```

where **I**, **T**, **E** represent any number (including zero) of statements separated by commas and S is a code segment. The order in which the statements are executed is: **I, T, S, E, T, S, E, . . . , T**.

The simplest **for** statement is

```
for (int loop = 0; loop < 5; loop++ ) cout << loop << ' ';
                                            // Output: 0 1 2 3 4
```

In this context **++loop** can be used interchangeably in place of **loop++.** While single letters such as **i** or **j** are often employed as loop variables, this yields less-transparent code together with potential collisions with similarly named variables elsewhere in the program. Therefore, dedicated names such as **loop**, **loopInner**, **loopOuter** are highly recommended, although for formatting reasons many **for** statements in this book employ single-letter loop variables. Except in older compilers a loop variable that is defined in the initialization statement is considered to be defined *inside* the body of the loop and is therefore destroyed when the **for** block is exited, even if the loop is terminated prematurely. *To access the value of the variable* ***loop*** *after the* **for** *block terminates, define* ***loop*** *outside the loop body as follows*

```
int loop;
for ( loop = 5; loop > 0; loop-- ) cout << loop << '';
                                            // Output: 4 3 2 1
cout << endl << loop << endl;               // Output: 1
```

To illustrate a more complicated **for** statement

```
for (int i = 1, j = 10; i + j <20; i++, j += 2)
   { ++i ; cout << i ; }                 // Output: 246
```

Modifying the loop variable **i** within the body of the **for** loop (as in the statement **++i;** above) is however inherently dangerous and should be avoided.

Infinite (non-terminating) **for** loops omit the condition statement, as in **for (loop = 0; ; loop++){. . . }** and **for (; ; ;){ . . . }**. Such constructs are meaningful if, for example, the statements enclosed within the block contain code that waits for a certain user or system response before exiting the loop.

8.8 *while* and *do. . . while* statements

Alternatives to **for** loops are provided by the **while** and **do . . . while** statements that execute code repeatedly as long as a given logical condition is fulfilled. A program that is equivalent to the simple **for** loop above is

```
int i = 0;
while ( i < 5 ) {
       cout << i++ << " ";
}
```

or alternatively

```
int i = 0;
do {
        cout << i++ << " ";
} while (i < 5);
```

A terminating semicolon is required for a **do...while** statement but not for the **while** statement. These procedures differ if the logical condition following the **while** is initially false. That is, if **int i = 6;** is substituted for **int i = 0;** above, the value of **i** will be displayed once in the second program but not in the first program.

8.9 The *break* and *continue* statements

Finally, C++ provides statements for prematurely terminating control structures and iterations. The **break** statement, which can only be placed within a control structure, transfers program control to the first statement following the end of the structure. A **continue** statement in an iteration ends the current iteration. The termination condition is then evaluated and, if false, the subsequent iteration commences as illustrated below

```
for ( int loop = 0; loop < 5; loop++) {
        if ( loop == 1 ) continue;
        if ( loop == 3 ) break;
        cout << loop << ' ';                    // Output: 0 2
}
```

The **break** statement can be used to exit a control construct manually. For example, placing

```
cout << "Enter 0 to terminate ";
cin >> r;
if (r == 0) break;
```

inside a control loop enables the user to terminate the loop from the keyboard; the variable **r** is then termed a sentinel. An alternative procedure is

```
int r = 1;
while (r != 0) {
        statements;
        cout << "Enter 0 to terminate ";
        cin >> r;
}
```

8.10 Assignments

Part I

In problems 1–11 give the output of the code if the program will function properly. If there is an error or if a potential error condition, such as overflow, underflow, incorrect memory access, infinite loop, etc., arises so that the program result or runtime state is unpredictable, indicate instead where the error is and what type of error has been made. Assume **#include <iostream.h>** is entered as the first line of each program. Some programs may have more than one error.

(1)
```
main () {
  int j = 1;
  if (j == 1) int i = 4;
  cout << i;
}
```

(2)
```
int myFunction () {
  int k = 1;
  for ( int loop = 0; loop <= 10; loop += 2)
    k = k * loop;
  return k;
}
main () {
  int test = 10, test2;
  if ( test ) test2 = 2 * myFunction();
  cout << test2 << endl;
}
```

(3)
```
main () {
  int test1 = 2;
  if (test1 = 0) cout << test1;
  else cout << ++test1 << endl;
}
```

(4)
```
main () {
  int i = 1;
  do {
    if (i < 5) {
    if (i >= 3) continue;
    i += 2;
    cout << i << endl;
    }
    else cout << ++i << endl;
   } while ( i < 6 );
}
```

(5)
```
main() {
   enum E{a, b = 3, c = 6, d};
   E = d;
   cout << E;
}
```

(6)
```
enum myEnum {red=2, green, yellow};
int myFunction(myEnum aMyEnum) {return aMyEnum;}
main () {
  myEnum myEnum1 = yellow;
  cout << myFunction(myEnum1) << endl;
}
```

(7)
```
main() {
    double d = 3;
    switch ( d ) {
      case 4: cout << "A" << endl;
      case 3: cout << "B" << endl;
      case 2: cout << "C: << endl;
      case 1: cout << "D" << endl;
      default: cout << "error" << endl;
    }
}
```

(8)
```
main() {
   int i = 10;
   {
      if (i != 0) break;
      i = 3;
   }
   cout << i << endl;
}
```

(9)
```
main () {
   int i = 3;
   #define j;
   #ifndef j;
   i = 4;
   #endif
   cout << i;
}
```

(10)
```
main () {
   bool i = true;
   bool j = false;
   bool k = i;
   cout << (!i && j || !k) << endl;
}
```

(11)
```
main () {
   int test1 = 3;
   int test2 = 0;
   cout << (test1 || test2) << endl;
}
```

Part II

(12) Generate a logical truth table for "**A** or (**B** and **C**)", where **A**, **B** and **C** are logical variables.

(13) Input an integer n from the keyboard and then compute the sum of all the squares from 1 to n using a **for** loop. Assume that the user will insert a value of n between 1 and 100 (you must check for this). Then rewrite the program above using a **while** construct and again using a **do . . . while** construct. Insure in each case that the result is the same as in the case of the **for** loop (this may require additional control constructs).

(14) Write a program using **if** and **else** (and **else if**) and then a program using **switch** statements that prompts the user for an integer from the keyboard and then prints out

the letter A if the number entered is in the range 90–99, the letter B if the number entered is 80–89, and the words incorrect input otherwise.

(15) In this problem you will develop a simple simulation of radioactive decay that will be plotted out using both an x–y plot and a histogram. First create a set of plotting classes (these will be used in future assignments). The class definition is here (you must fill in the bodies based on the material in previous exercises):

```
#ifndef PLOTROUTINES                  // Insures that the code is
                                      // not included twice.
#define PLOTROUTINES
class Graph {
  public:
    int nPoints;
    float x[10000],y[10000];
    Graph(float[ ], float[ ], int);
                                      // Used to initialize x and y
    void draw();                      // Calls qplot
};
// You can fill in the routines below if desired, otherwise
// eliminate the next two lines. Note that a numerical value
// for the dimension is not needed here. If these lines are
// present, you must terminate the draw and Graph declarations
// with a semicolon in the class definition as done above
Graph :: Graph(float aPosition[ ], float aVelocity[ ], int
        aNumberOfPoints)
void Graph :: draw()
class Histogram {
   float iValues[100];  // 100 is the largest histogram array
   float iGridPositions[100];
   int iMatrixSize;
   public:
     Histogram(int aMatrixSize)
                           // Should initialize all iValues to 0
                           // and set iGridPositions[0],
                           // iGridPositions[1] to 0, 1, 2...
     void addValue(int aI) {iValues[aI]++;}
                           // Adds an additional sample to the
                           // histogram.
     void plotFullHistogram()// Calls qplsca to plot the
                           // histogram (see previous
                           // assignment)
};
#endif
```

Save the above file in the directory in which you are compiling programs as **plotroutines.h**. Now write a new program file, for example **decay.cpp**, that represents the equation

$$\frac{\Delta N(t)}{\Delta t} \approx \frac{\partial N(t)}{\partial t} = -\frac{N(t)}{\tau}$$

where τ is the decay constant. Here we are taking a number **numberOfSteps** of steps of time Δt over a time interval of length **timeInterval** so that Δt = **timeInterval / numberOfSteps**. We repeat the calculation for a number of runs given by

numberOfAttempts and display a histogram of the number of runs that yield each possible outcome for the total number of remaining particles.

Hand in the graphical result of this program as well as your code for 10,000 instead of 1000 realizations.

```
#include <iostream.h>
#include <stdlib.h>
#include "dislin.h"
#include "plotroutines.h"

class RadioactiveSubstance {
   private:
     int iNumberOfParticles;
     double iLifetime;
   public:
     RadioactiveSubstance(int aNumberOfParticles, double
       aLifetime)
     void print()     // print out the number of particles
     int numberOfParticles(){ return iNumberOfParticles;}
                      // get member function
     int setNumberOfParticles(int aNumberOfParticles)
                      // set member function
     void advanceTime(int numberOfSteps, double timeInterval)
       // Implement the following algorithm:
       // 1) If deltaTime / iLifetime is larger than 0.1, exit
       // using exit(0) in stdlib.h (the time step is too
       // large)
       // 2) Otherwise, loop over all numberOfSteps time
       // steps. In each step check for each remaining
       // particle if rand() / double(RAND_MAX) is
       // less than deltaTime / iLifetime. If it is, decrease
       // the number of particles by 1. If the number of
       // particles is 0, write ''No particles are left'' to
       // the terminal and return control to the main program
       // (use return).
};

main() {
   srand(time(NULL));
   int numberOfAttempts = 1000;
   int numberOfSteps = 100;
   double timeInterval = 4;
   int numberOfParticles = 1000;
   double lifetime = 4;
   RadioactiveSubstance R1(numberOfParticles, lifetime);
   Histogram H1(100);
   for (int loop = 0; loop < numberOfAttempts; loop++) {
      R1.setNumberOfParticles(100);
      R1.advanceTime(numberOfSteps, timeInterval);
      H1.addValue(R1.numberOfParticles());
   }
   H1.plotFullHistogram();
}
```

(16) A particular realization of a so-called logistic map is given by the recursion $x_{n+1} = \mu x_n(1 - x_n)$ for $0 < x_0 < 1$. This construct exhibits a stable state for $\mu < 3$ and period doubling into a chaotic regime for $3 < \mu < 4$. That is, in the $\mu < 3$ regime, any initial value will reach a fixed point $x_* = (\mu - 1)/\mu$ as the number of iterations becomes large, while for somewhat larger values of μ the system will oscillate between two fixed values. The number of values that the system visits becomes larger as μ increases still further and eventually becomes infinite. In this chaotic regime, infinitesimal changes in the starting value generate large and unpredictable variations in the output value after a large number of iterations. To study the logistic map, implement the following program. Hand in your program together with the resulting (eight) graphs and histograms for a starting x-value of 0.5 and values of μ given by 2.6, 3.3, 3.52, and 3.9.

```
#include <iostream.h>
#include <stdlib.h>
#include "dislin.h"
#include "plotroutines.h" // Note that this line can be
                          // repeated because of the
                          // preprocessor directives in
                          // plotroutines.h!

class LogisticMap {
   float iGrowthConstant;
   float iMapValues[10000];
   float iStep[10000];
   float iStartValue;
   int iTotalSteps;
   int iNumberOfBins;
   Histogram iHistogram; // Holds a histogram over all visited
                         // x-values during the iteration.
   public:
     LogisticMap(int, float, float, int) :
       iHistogram(aNumberOfBins)
     // You must construct the Histogram object before the
     // body of the constructor since Histogram does not have
     // a default constructor!
     void evaluate()
     // Here you must implement the logistic map.
     // Successive values of x are placed in the array
     // iMapValues[loop] At the same time, use the addValue
     // member function of the Histogram class with an
     // argument iMapValues[loop] * iNumberOfBins to build
     // up the histogram values stored in iHistogram.
     void printGraph(int iNumberOfPoints)
                 // Use the draw() function of the Graph
                 // class
     void printHistogram() // Use the plotFullHistogram()
                           // function of the Histogram class
                           // to display the values stored in
                           // iHistogram.
};
```

```
main () {
   float growthConstant, startValue;
   int numberOfGraphPoints; // Number of iteration results
                            // appearing in the graph.
   cout << "Input growth constant (space or CR) start value
           (space or CR) and the graph size ";
   cin >> growthConstant >> startValue >> numberOfGraphPoints;
   cout << endl;
   int totalSteps = 5000;            // Total number of
                                     // iterations
   int numberOfHistogramBins = 30;   // Number of histogram
                                     // bins - must be less
                                     // than 100
   numberOfGraphPoints = min(numberOfGraphPoints, totalSteps);
   LogisticMap L1(totalSteps, startValue, growthConstant,
         numberOfHistogramBins);
   L1.evaluate();
   L1.printGraph( numberOfGraphPoints );
   L1.printHistogram();
}
```

Chapter 9

Basic function properties

A function implements repetitive tasks that are performed on a specified set of input variables. Since the variables defined within a function are largely isolated from the remainder of the program, a function can be tested individually and reused in other programs. A modular program consists almost exclusively of functions and control structures and thus represents a physical problem as a logical flow among individual subtasks. Within the context of object-oriented programming, internal member functions principally implement the behaviors of objects. That is, calling such a function transforms the internal state of the object and possibly the external environment according to the values passed to the function through its input arguments. Code for a function should be short, for example less than two pages of code, and perform a single, well-defined task.

9.1 Principles of function operation

Depending on the details of the program flow, a function in a C++ program may be invoked any number of times (including zero) during program execution. Since a function can, however, require substantial computer resources, memory is not automatically allocated for the variables defined in the function body when a program is run, unlike global variables or the variables defined in the **main( )** program. Rather, at the start of the program the machine language instructions associated with the function body are stored in memory. The function name in the calling program is then associated with the 4-byte (for a 32-bit machine) memory location of the start of this instruction set, as can be verified by running the following simple program

```
void print() { cout << "test"; }
main () {
        cout << print << endl;
}
```

The output of this program is a hexadecimal number such as 0040115E, which is the starting location of the function's instruction set (record). Note that **print** in

main() *does not contain parentheses*; the effect of the parentheses is to activate the instructions starting at this location.

A general function is of the form **returnType myFunction(type1 aP1, type2 aP2, . . . , typeN aPN)**, where the variables or objects **aP1, . . . , aPN** are the function arguments. The function is then called from elsewhere in the program through a statement such as **x = myFunction(p1, p2, . . . , pN);** in which **p1, . . . , pN** are termed formal parameters. A function may have any number (including zero) of arguments but can only possess zero or one return values. A function that lacks a return value has a return type of **void**, except for constructor (and destructor) functions as noted in Section 7.5. (One must be aware however that in many situations C++ assumes a default type of **int** when no type is specified; for example, the definition **const i;** is interpreted as **const int i;**. Therefore a function that actually returns an **int** can be defined without a return type so that **f( ) {return 5;}** is identical to **int f( ) {return 5;}**. Such a function declaration or definition should not be confused with a **void** function.) A function without arguments *must* be called with parentheses, as discussed above.

When a function is called, the operating system reads the machine language instructions stored sequentially in memory from the starting location onward. These commands reserve space in memory for the argument variables and then copy the values of the formal parameters into this newly allocated memory, which effectively means that the first statements executed when a function is called define and initialize the argument variables. The statements in the function block are executed until the end of the block (for a **void** function) or a **return** statement is encountered. In the latter case, the function passes the return value and control back to the calling program. Finally, the operating system deallocates the memory occupied by the function variables.

The most significant consequence of this procedure is that the statements in the function act on *copies* of the function parameters in the calling program that are held in physically separate memory locations (the function arguments are defined within the function block). Therefore, changes to the values of the function arguments inside the function cannot affect the values of the formal parameters in the calling routine. Hence in the program

```
void change( int aI ) { aI = 2 * aI; }
main () {
      int i = 3;
      change ( i );
      cout << i << endl;                  // Output: 3
}
```

the value of the argument **i** does not change inside **main( )**. This behavior is termed "call by value."

9.2 Function declarations and prototypes

Just as in the case of a variable, for the C++ compiler to interpret (parse) a call to a function in source code it must be able to establish the number and type of the function arguments as well as the return type. Through this "type checking," the compiler determines if the function obeys the syntax rules that guarantee its successful operation at runtime. A *declaration* statement that enables type checking but does not reserve physical memory (and, therefore, can be repeated arbitrarily many times within a program block) is called the function *prototype* or *signature*. For the above program, a prototype statement can be written either as **void change(int aI);**, where any name can in fact be employed in place of **aI**, or simply as **void change(int);**. In either case, the final semicolon is *required*. Obviously, such a prototype cannot be a definition that leads to memory allocation, since an undetermined amount of memory is required to store the contents of the function body.

Once a prototype is supplied, the compiler can type check any call to the function. Therefore, the actual function *definition* that supplies the function body and therefore establishes the memory requirements of the function can be located anywhere in the program. The definition can in fact even be placed in a separate program that is later compiled or linked together with code containing the function prototype. Thus, we can rewrite our previous program as follows (this format is actually required in the C language)

```
void change( int );

main () {
        int i = 3;
        change ( i );
        cout << i << endl;                   // Output: 3
}

void change( int aI ) { aI = 2 * aI; }
```

9.3 Overloading and argument conversion

Since C++ type checking resolves both the number and the types of function arguments, a function is uniquely specified by this information together with its name. In fact, in the compilation process the argument types are *mangled* together with the function name to produce a new name for the function in the object file that is a composite of the function name and the argument types. Therefore, functions with the following three prototypes are treated by the compiler as completely different entities as they acquire different mangled names

```
int change(int);
int change(double);
double change(int, double);
```

Writing **change(1)** invokes the first of the three functions, while **change(2.0)** calls the second. This is termed function *overloading* and is a form of polymorphism in which the types and number of variables on which a function acts determine its behavior. We cannot however add to the above list a fourth function, such as **int change(int, double);**, with the same number and types of arguments as an existing prototype but with a different return value, since the C++ compiler lacks a facility for resolving a function call based on the type of variable that will ultimately store the return value.

Automatic type conversions are performed between the function parameters in the calling program and the function arguments. Here it is again helpful to visualize the first (implicit) operation executed at runtime in, for example, the function **void f(int aI);** as the variable definition and initialization **int aI = pI;**, where **pI** is the function parameter in the call **f(pI)** to the function. From this viewpoint, calling the function with a **double** parameter such as 2.5 clearly sets the argument **aI** in the function body to 2 after automatic conversion.

Type conversion through the arguments of overloaded functions is non-trivial. If multiple versions of a function have the same numbers of arguments, as in the two single-argument versions of the **change** function, rules must be supplied to determine which of these will be employed if the function is called with an argument of a third type such as, for example in **char c = 'a'; change(c);**. While a full discussion of these rules is beyond the scope of this text, simple logical considerations often can be used to guess correctly the outcome of a given call. In our example, since the size of a **char** is closer to that of an **int** than to a **double**, the **void change(int);** version of the function is employed as verified below

```
void change( int aI ) { cout << "one"; }
void change( double aI ) {cout << "two"; }

main () {
        char c = 'a';
        change ( c );
}                                        // Output: one
```

9.4 Built-in functions and header files

The C++ compiler and its include files provide extensive libraries of functions and predefined constants. However, before using these functions, their argument and return types must be known for the reasons outlined in the previous section. For example, the C++ compiler includes a built-in absolute value function with the prototype

```
int abs(int);
```

A very common mistake is to use this function with a **double** argument with the expectation that a **double** rather than an **int** will be returned as in

```
double a, b, eps = 1.e-5;
cin >> a >> b;
if (abs(a - b) < eps) { ...statements ...}
```

Unless the absolute value of the difference between **a** and **b** exceeds 1, it is converted through the **abs** function argument to 0, and the logical condition in the **if** statement argument will evaluate to true.

The correct function for the above program is

```
double fabs(double);
```

which is enabled through the preprocessor directive

```
#include <math.h>
```

The **math** library contains a number of important mathematical functions that operate on **double** arguments. These include **pow(x, n)**, which raises **x** to the (integer or non-integer) power **n**, **sqrt(x)**, **fabs(x)**, **ceil(x)**, which rounds **x** upward, **floor(x)**, which rounds **x** downward, **exp(x)**, **log(x)**, **log10(x)**, **sin(x)**, **cos(x)**, **tan(x)**, **tanh(x)**, **asin(x)**, the arcsin function, and other associated inverse functions. As well, the **math.h** library defines several important global constants such as π (generally but not always labeled **M_PI**).

To determine the exact set of functions and constants present in an include file, which can vary among compilers or compiler versions, the header file should be examined directly. To locate the **math.h** header file in your Borland C++ installation, navigate to the X:\Dev-Cpp\include directory through, for example, the My Computer icon on your desktop (for Borland C++ this directory is X:\borland\bcc55\include) where X: is the drive letter on which the product has been installed. Double clicking on the icon for math.h opens the notepad editor. Browsing through the file reveals the lines

```
double sin (double);
double cos (double);
double tan (double);
double sinh (double);
double cosh (double);
double tanh (double);
double asin (double);
double acos (double);
double atan (double);
double atan2 (double, double);
double exp (double);
double log (double);
double log10 (double);
double pow (double, double);
double sqrt (double);
double ceil (double);
double floor (double);
double fabs (double);
```

Similarly, the file contains the following partial list of predefined constants

```
/* Traditional/XOPEN math constants (double precison) */
#ifndef __STRICT_ANSI__
#define M_E            2.7182818284590452354
#define M_LOG2E        1.4426950408889634074
#define M_LOG10E       0.43429448190325182765
#define M_LN2          0.69314718055994530942
#define M_LN10         2.30258509299404568402
#define M_PI           3.14159265358979323846
#define M_PI_2         1.57079632679489661923
#define M_PI_4         0.78539816339744830962
#define M_1_PI         0.31830988618379067154
#define M_2_PI         0.63661977236758134308
#define M_2_SQRTPI     1.12837916709551257390
#define M_SQRT2        1.41421356237309504880
...
```

Professional programmers frequently employ this technique to extract comprehensive and up-to-date information on the language implementation from include files.

9.5 Program libraries

Having examined a typical system header file, we now discuss personal header files and program libraries. Recall first that header files must guard against repeated definitions when a program is assembled from several separate files. Further, the program should compile for any order of the include statements. These requirements can be satisfied by inserting appropriate preprocessor directives.

If only one or a few files of a multifile program are typically changed at each program development step, the compilation time can be minimized by separately compiling each file into an object file. The object files are combined by the linker. Only files that are altered are subsequently recompiled. Additionally, the object files can be inserted into a program library. Functions in a program library are stored in an alphabetical lookup table, which can be rapidly searched by the linker, decreasing link times.

The procedure for generating header files and program libraries is best understood through an illustrative if contrived example. Suppose we wish to generate a program **mySquare.cpp** of the form

```
// mySquare.cpp - Preliminary outline
main () {
      float i;
      printSquare( i );
      printFourthPower( i );
}
```

The functions **printSquare( )** and **printFourthPower( )** both call a third function **square( )** that is a member of a **SquareCalculator** class and write results to **cout**. Accordingly, we start by generating the three header files below with a .h extension that only include function declarations and possibly class definitions. All header files that contain definitions guard against multiple inclusion through **#ifndef ...#endif** and **#define** statements as in Section 8.6

```
//square.h
#ifndef_square
#define_square
class SquareCalculator {
        int iValue;
public:
        SquareCalculator(int);
        int calculate();
};
#endif

//printsquare.h
void printSquare( int );

//printFourthPower.h
void printFourthPower( int );
```

Next, we place the function definitions corresponding to each header file and the **main( )** program into separate code files with .cpp extensions. These can each now include any number of header files. In our example

```
//square.cpp
#include "square.h"

int SquareCalculator::calculate(){return iValue * iValue;}
SquareCalculator::SquareCalculator(int aValue) : iValue(aValue) {}

//printsquare.cpp
#include <iostream.h>
#include "square.h"

void printSquare ( int aI ) {
        SquareCalculator SC(aI);
        cout << SC.calculate() << endl;
}

//printfourthpower.cpp
#include <iostream.h>
#include "square.h"

void printFourthPower ( int aI ) {
        SquareCalculator SC(aI);
        cout << SC.calculate() * SC.calculate() << endl;
}
```

The main program is

```
//mysquare.cpp
#include <iostream.h>
#include "printsquare.h"
#include "printfourthpower.h"

//      The following section is uncommented when only mysquare is
//      compiled.
//      #include "printsquare.cpp"
//      #include "printfourthpower.cpp"
```

```
//      #include "square.cpp"

main () {
        int i = 4;
        printSquare( i );
        printFourthPower( i );
}
```

Recall that the use of double apostrophes in the name of the include file directs the compiler to search for the header file first in the user's current directory and then in the compiler's include file subdirectories.

At this point, if the multifile program is generated inside a development environment such as Dev-C++, a *project* is typically created, while command-line based C++ environments simply employ additional flags during compilation and linking to combine the different object files. A better command-line alternative in which advanced compilation and linker commands are grouped into a *make* file that automatically determines which source code files require recompilation is somewhat complex and will accordingly not be discussed here.

Creating a Dev-C++ project: A project is a collection of related files that are compiled separately and then linked together. If one file in a project is changed only that file and files that are perceived to be dependent on it are normally recompiled (in Dev-C++ this is implemented through automatic generation of a make file entitled **Makefile.win** that you can locate and examine in your program directory. This file can then be run from the command line if desired by first renaming it **makefile** and then issuing the commands **set PATH=X:\dev-cpp\bin;%PATH%** followed by **make**.) To create a project, select the first "Project . . . " icon on the upper toolbar and then select "Empty Project" together with the "C++ Project" radio button when prompted. Subsequently enter a location for the project file. Next, select the third to the last "Add to Project" icon on the upper toolbar and add **square.cpp**, **square.h**, **printsquare.cpp**, **printsquare.h**, **printfourthpower.h**, **printfourthpower.cpp**, and **mysquare.cpp** to the project. You can verify that they have been incorporated correctly by depressing the Project tab in the left-hand windowpane. Now again select Tools → Compiler Options from the menu bar and uncheck the upper check box entitled "Add the following commands when calling compiler" (do not remove the text in the associated windowpane since *you will need to reselect the check box* when you are again compiling a single non-project file). Finally, selecting the fourth "Rebuild all" icon on the lower button bar will correctly compile and link together your files. Subsequently if only a subset of the files are changed, depressing the "Compile and Run" button will recompile only these files and any other files that depend on them.

To employ DISLIN within a project, depress the last, "Project Options," button on the upper toolbar, then select the "Parameters" tab and enter the text

```
"X:\dislin\dismgc.a" -luser32 -lgdi32
```

into the rightmost windowpane entitled "Linker." Similarly, to include any library file, such as the **square.a** file created through the mingw **ar** command described below, depress the "Add Library or Object" pushbutton on this windowpane and navigate to the location of the desired file.

Command-line linking and library generation: To assemble an executable file from a number of source code files using a command line C++ compiler, each file can be compiled into a separate object file by specifying the **–c** (compile only) switch and the resulting object files are subsequently linked together. In mingw or other standard g++ based environments the following sequence of commands is entered from within a command (MS-DOS) window

```
g++ -c square
g++ -c printsquare
g++ -c printfourthpower
g++ -c mysquare

g++ -o mysquare.exe square.o printsquare.o printfourthpower.o
  mysquare.o
```

The corresponding Borland C++ commands are

```
bcc32 -c square
bcc32 -c printsquare
bcc32 -c printfourthpower
bcc32 -c mysquare

bcc32 mysquare.obj square.obj printsquare.obj printfourthpower.obj
```

In both cases the executable file **mysquare.exe** is generated.

To create an optimized library of object modules called **square.a** in mingw or any other g++ based compiler, enter the command

```
ar rvs square.a mysquare.o square.o printsquare.o
  printfourthpower.o
```

where the **rvs** flags place the files into the library, replacing any previous versions if present, print the names of the included files and finally add an object file index to the library for faster linking.

A program is linked with the object modules in the library by entering

```
g++ -o myProgram.exe myProgram.c square.a
```

In Borland C++ a library file is obtained by invoking the **tlib** program

```
tlib square +square.obj , report
tlib square +printsquare.obj,printfourthpower.obj , report
```

which generates the library file **square.lib**. The optional argument "**, report**" in the lines above produces a file **report.lst** in the current directory that contains a listing of all the object modules currently resident in the library module. A listing of all switches (options) recognized by the **tlib** program can be found in standard fashion by typing **tlib** at the command prompt without any additional parameters.

Linking with the functions stored in the library is finally performed by entering

```
bcc32 myProgram -L square.lib
```

Finally, note that in our example, **mysquare.cpp** can be compiled and linked without creating separate **.obj** modules for the function files by uncommenting the three **#include** lines in the program that read the **.cpp** files for the three required mathematical functions. While the header file **square.h** is then included three times, the single *definition* in this file, namely that of the **SquareCalculator** class is surrounded by preprocessor commands and is accordingly only processed once by the compiler.

In multifile programs, **typedef**, **enum**, **const**, and **inline** definitions are all local to the file in which they are placed; for these to be used in more than one file they must be placed in a header file and included in each individual component file. A non-**const** variable that is defined in one file and employed in a second file must be declared in the second file with matching type using the **extern** keyword (without an initializer, which would convert the declaration into a definition). For example, if we place a global variable **int j = 10;** in the file **mysquare.cpp** in the program above, it can only be accessed in the file **square.cpp** if the declaration **extern int j;** is present in this file. If the **extern** declaration statement appears inside a block in a given file, the variable **j** acquires the scope of the block within the file.

9.6 Function preconditions and postconditions – the *assert* statement

Design by contract is a software engineering technique in which the programmer specifies for each function the physically valid ranges for the input and output values. These preconditions and postconditions (the "contract") are checked upon function entry and exit by appropriate logical statements. A convenient method for implementing such requirements is provided by the **assert( )** statement, which terminates program execution if its argument evaluates to a logical false as in the following example

```
#include <assert.h>

main () {
      int i = 1;
      assert (i == 2);          // An error test
}
```

This generates the standard output

```
Assertion failed: i == 2, file assert.cpp, line 5
```

For production runs, including the preprocessor directive **#define NDEBUG** prevents the compiler from processing all subsequent assert statements.

An example of preconditions and postconditions using assert functions is

```
#include <assert.h>
#include <math.h>

// This returns -1 if the temperature is below freezing,
// otherwise it returns the temperature in Celsius.

double degreesAboveFreezing(double aT) {
        // precondition: The input temperature in Kelvin is above
        // absolute zero.
        double result;
        assert (aT > 0);

        if (aT > 273) result = aT - 273;
        else result = -1;

        // postcondition: The result is -1 or greater.
        assert (result >= -1);
        return result;
}
```

A more flexible procedure for implementing preconditions and postconditions is provided by **try** and **catch** blocks. When a **throw** statement is invoked within the **try** block, program control passes to the **catch** block with the closest valid match to the arguments of the **try** block. The program continues after the **catch** block terminates unless an **exit** statement is placed inside the **catch** block. Note that a variable defined within a **try** block will be destroyed when leaving the block. Thus, we could rewrite the **degreesAboveFreezing( )** function in the following form

```
#include <math.h>
#include <stdio.h>          // includes the exit() function.

// This returns -1 if the temperature is below freezing,
// otherwise it returns the temperature in Celsius.

double degreesAboveFreezing(double aT) {
      // precondition: The input temperature in Kelvins is above
      // absolute zero.
      double result;

      try {
            if (aT < 0) throw "Temperature cannot be negative";
            if (aT > 273) result = aT - 273;
            else result = -1;

            // postcondition: The result is -1 or greater.

            if (result < -1) throw result;
      }
```

```
        catch (double aResult) {
                cout << "result = " << result << endl;
                exit ( 0 );
        }

        catch (char message[80]) {
                cout << message << endl;
                exit( 0 );
        }

        return result;

}
```

9.7 Multiple return statements

While a function in C++ can only return a single variable, it can possess arbitrarily many **return** statements. For example, a function that returns the character 'y' when the user inputs a 1 from the keyboard and the character 'n' otherwise could be written

```
char get() {
        int i;
        cout << "Insert a value" << endl;
        cin >> i;
        if (i == 1) return 'y';
        return 'n';
}
```

After any **return** statement is executed, the function immediately terminates.

9.8 Functions and global variables

While a function C++ only returns zero or one variable, it can modify any number of *global variables*, that is variables defined in the global scope outside all function blocks. Such variables are visible and accessible to the bodies of all subsequently defined functions (unless hidden by a second variable with the same name). Thus the **change( )** function below modifies two global variables that are subsequently accessed by the **main( )** program

```
int i, j;
void change() { i = 2; j = 3; }

main() {
        i = 0;
        j = 1;
        change();
        cout << i << '\t' << j <<endl;          // Output: 2 3
}
```

Such a technique is, however, inherently error prone, since an inadvertent change to a global variable in any section of the program (which could be thousands of lines long) propagates to all other program sections. This generates unforeseen side effects that are separated by a large distance in code from the offending line and are therefore very difficult to connect to the actual error source.

9.9 Use of *const* in functions

An argument in a function that is declared **const** cannot be changed within its body. Further, an internal member function of a class can be prevented from modifying the data members of the class it belongs to by placing the keyword **const** between the end of the parameter list and its code body as in

```
class C {
        public:
        int iI;
        void print( const aJ ) const { cout << iI * aJ << endl;}
};
```

The compiler will generate an error message if the **print( )** function attempts to change either the argument **aJ** or the internal variable **iI** of the **C** class. A function that cannot alter internal class members is termed a **const** member function. As a rule, get member functions should be **const** member functions.

If a **const** member function is overloaded by a non-**const** member function with the same name, the **const** member function is called if the object that invokes the function is declared **const** (indicating that its internal data members cannot be changed), otherwise the non-**const** member function is called

```
class C {
        public:
        int iI;
        void print( const int aJ ) const { cout << iI * aJ <<
          endl;}
        void print( const int aJ ) {cout << ++iI * aJ << endl;}
};

main() {
        const C C1 = {1};
        C1.print(1);            // Output: 1
        C C2 = {1};
        C2.print(1);            // Output: 2
}
```

If the first **print( )** function is not present, a warning message is generated by the compiler and the output **2 2** is obtained.

Since each program in this book focuses as much as possible on a single aspect of C++, the **const** keyword is generally omitted. However, declaring input function arguments **const** is highly recommended since an argument marked **const** indicates an input value that should not be changed by the function, preventing unintended coding errors. Note that, since variables that are passed by value to a function are isolated from the remainder of the program, functions with both **const** and non-**const** (value type) arguments can accept either type of variable as parameters.

9.10 Default parameters

Often an internal variable in a function has a standard value that the calling program only overrides in exceptional cases. In this case, a function can be defined or declared with default parameters that have certain assigned values unless a different value is explicitly specified. Otherwise, the function is called with a reduced number of arguments as in

```
void printRoomTemperature( double aTemperature = 25 ) {
        cout << aTemperature << '\t';
}

main () {
       printRoomTemperature();                          // Output: 25
       double newRoomTemperature = 24.5;
       printRoomTemperature(newRoomTemperature);        // Output: 24.5
}
```

Default parameters must be placed last in the function. Further, *the default parameter values can only be specified once*; that is, either in one and only one declaration statement or in the function definition. That is, we can write

```
void myFunction( int a, int b );
void myFunction( int a, int b = 1 );
main () {
       myFunction(3);                                   // Output: 3   1
}
void myFunction( int a, int b ) { cout << a << '\t' << b << endl;}
```

but we cannot additionally introduce the default parameter either into the first function declaration or into the function definition.

9.11 Inline functions

The body of a function that is declared inline is automatically substituted into each function call before compilation. That is, suppose we write

```
inline int square( int x ) { return x * x; }
```

Then a call to this function such as **square(4 * y)** in the source code is replaced by **(4 * y) * (4 * y)** before compilation. This results in faster executable code since memory does not have to be allocated and subsequently deallocated upon each call to **square**, but at the cost of increased compile times and larger object files.

A function whose body is supplied *within* a class definition such as **calculate()** in

```
class Square {
        int iX;
        public:
        Square( int aX ) : iX(aX) { }
        int calculate() { return iX * iX; }
};
```

is compiled as an inline function with certain exceptions, such as if the body contains **for** loops. Generally, small functions that are not defined within a class body should be declared **inline**.

An inline function does not have a prototype since the compiler must have the body of the function already accessible the first time the function is called so that it can perform the necessary substitutions.

9.12 Modular programming

A modular program consists solely of declarations and definitions, control logic, and function calls as illustrated by the following simple example

```
class NumberPair {
        int iNumber1, iNumber2;
        public:
        NumberPair ( int aNumber1, int aNumber2) :
             iNumber1(aNumber1), iNumber2(aNumber2) { }
        int number1() { return iNumber1; }
        int number2() { return iNumber2; }
};

int getInput() {
        int r;
        cin >> r;
        return r;
}

int multiply(NumberPair iNP) {
        return iNP.number1() * iNP.number2();
}

void print(int aI) {
        cout << aI << endl;
}

main (){
        int i = getInput();
        int j = getInput();
        NumberPair NP1(i , j);
        int result = multiply(NP1);
        print(result);
}
```

Modular programs are easily analyzed and corrected, since each individual function can be tested separately.

9.13 Recursive functions

Recursive functions are functions that call themselves. This construct is permitted since the function prototype precedes the body in a function definition. Therefore, within a function body the compiler can already type check a call to the function itself.

As an example, consider the factorial function, which is defined recursively by n! = n(n-1)! together with 1! = 1

```
double factorial( const int aN ) {
       double temp;                 // a floating point type is
                                    // required for large aN
       if (aN == 1) temp = 1;
       else temp = aN * factorial(aN - 1);
       return temp;
}
```

Consider the statement **int k = factorial(3);**. In this first call, the variable **temp** is identified with **3 * factorial(2)**, but this expression cannot be evaluated until a return value is obtained from **factorial(2).** However when **factorial(2)** executes, it produces a new variable named **temp** in a separate memory space, which is set to **2 * factorial(1).** Finally when **factorial(1)** is called, 1 is returned, allowing **factorial(2)** and then **factorial(3)** to complete. This procedure therefore incurs the overhead of three function calls and at one point reserves memory space for all three functions simultaneously. Often, however, this inefficiency is outweighed by increased programming simplicity.

9.14 Assignments

In the problems below give the output of the code if the program will function properly. If there is an error or if a potential error condition, such as overflow, underflow, incorrect memory access, etc., arises so that either the program result or the runtime state is unpredictable, indicate instead where the error is and what type of error has been made. Assume **#include <iostream.h>** is entered as the first line of each program. Some programs may have more than one error.

```
(1) int myFunction( int aI) {
         aI = 3;
         return aI;
    }

    main () {
         int aI = 1;
         cout << myFunction (aI) << '\t' << aI << endl;
    }
```

(2)
```
int myFunction (int a = 4, int);

int myFunction (int a = 4, int b) {return b*a;}
main() {
    cout << myFunction(2, 5) << endl;
}
```

(3)
```
int[3] myFunction (int aI) {
    int a[3];
    a[1] = aI;
    return a;
}

main () {
    int aI = 2;
    int b[3] = myFunction(aI);
    cout << b[1] << endl;
}
```

(4)
```
void print ( int aI ) { cout << aI << endl; }

main () {
    int i = 4;
    print (int i);
}
```

(5)
```
#include <assert.h>

class C {
    int iC;
    public:
    c() {assert (iC > 0); return iC; }
    void setIC(int aC){ iC = aC; }
};

main() {
    C C1;
    C1.setIC(-10);
    C1.c();
}
```

(6)
```
main() {
    int i = 10, j;
    j = i / -40.0;
    float k = -30;
    int p = sqrt(abs(k));
    cout << p << '\t' << k << '\t' << j << '\t' << i << endl;
}
```

(7)
```
void print(int aI) { cout << "This is " << aI << endl; }

void print(double aI[ ]) {cout << "This is (2) " << aI <<
  endl;}

main () {
   double i = 3;
```

```
   print(i);
}
```

(8)
```
void print(int aI , int aJ = 5);

main() {
   int i = 10;
   print( i );
}

void print(int aI, int aJ = 5) {
   cout << aI << '\t' << aJ;
}
```

(9)
```
int myFunction ( int aJ ) {
   if (aJ == 5) return aJ;
   else return aJ * myFunction(++aJ);
}

main() {
   cout << myFunction(2);
}
```

(10)
```
const int i = 20;

void f( int i ) { cout << i << endl; }

main () {
   int i = 10;
   f(i);
}
```

(11)
```
void f( int i ) { cout << i << endl; }

main () {
   const int j = 10;
   f(j);
}
```

(12)
```
class C {
   public:
   inline void f(int i);
};

main () {
   C C1;
   C1.f(3);
}

inline void C::f(int i) { cout << 2 * i << endl; }
```

(13)
```
int aI = 3;

void print();

main () {
```

```
    aI++;
    print();
  }
  void print() {
    cout << "In function " << aI << endl;
  }
```

(14) Using the math.h and DISLIN libraries, write a program that graphs the values of the function sin (2*M_PI*0.9*x) + sin(2*M_PI*1.1*x) from $x = 0$ to $x = 10$. Note that M_PI (for example π) is automatically defined for you by the math.h library. Use 1000 points in your graph. Hand in the graph and the program.

(15) Write an object-oriented program that prints out all combinations $C(q,p)$, where $C(q,p) = P(q,p)/p! = q!/(p!(q-p)!)$ where P is the number of permutations of k objects out of a group of n distinguishable objects. Hand in the program and a table of combinations from $q = 0$ to $q = 10$ in the form

```
1
1 1
1 2 1
1 3 3 1
...
```

To compute the combination function, use the following class

```
class CombinatorialCalculator {
  int iP, iQ;
public:
  void setIP(int aP)
  void setIQ(int aP)
  int p()
  int q()
  float computeCombination()
};
```

(16) Write a modular (non-object oriented) program that incorporates a boolean function (a function that returns a **bool** argument) **isFibonacci** that returns **true** if its input parameter is a Fibonacci number (see Assignment Part I) and **false** otherwise. The program should print out the word true if a given input number is a Fibonacci number and the word false otherwise. That is, the structure of the program should be

```
main() {
int n = getInput();
bool c = isFibonacci( n );
print( c );
}
```

Run this on a few test cases and hand in the program and sample input/output.

(17) Use the function in the preceding problem to generate a second function with three integer arguments that determines if the product of these three arguments is a Fibonacci number. Write a modular program in the form of the program of the previous problem and hand in this program together with sample input/output.

(18) Program a Fibonacci class with two private internal variables **currentValue** and **nextValue** and a function **float calculate( )** that returns successive values of each number in the Fibonacci sequence defined by $F_{i+1} = F_i + F_{i-1}$ with $F_0 = 0,\ F_1 = 1$. By supplying an appropriate constructor, write code that will generate the first five Fibonacci numbers (the first number you should print out should be F_1). The **main()** function in your code should contain the lines

```
Fibonacci F1;
int numberOfValues = getInput();
for (int loop = 0; loop < numberOfValues; loop++) {
        F1.calculate();
        F1.print();
}
```

(19) A continued fraction expansion for the square root of 2 is given by

$$\sqrt{2} - 1 = \frac{2-1}{\sqrt{2}+1} = \frac{1}{1+\sqrt{2}} = \frac{1}{2+(\sqrt{2}-1)} \approx \frac{1}{2+\left(\frac{1}{2+\frac{1}{2+\ldots}}\right)}$$

Terminate this series by placing a zero in the place of the ellipsis (. . .) after n fractional terms and develop a program employing recursive functions that writes out the value of the square root of 2 as a function of n for values of n between 1 and 10.

(20) A particularly simple abstraction of a gas views the gas as a material with a pressure-dependent volume. In this abstraction, the actual chemical composition of the gas is unimportant as only its physical properties are of practical interest.

Develop a class that implements a physical abstraction of a gas in which the only relevant properties are its Van der Waals constants a and b and the problem is to determine both the pressure and the volume of the gas given a fixed temperature and number of moles, n. The relationship between these quantities is given by the Van der Waals equation of state

$$\left(P + \frac{an^2}{V^2}\right)(V - bn) = nRT$$

For oxygen gas, the Van der Waals constants are $a = 0.027\ \mathrm{l}^2 \cdot \mathrm{atm/mol}^2$ and $b = 0.0024$ L/mol, while the gas constant $R = 0.082\ \mathrm{l} \cdot \mathrm{atm} / (\mathrm{mol} \cdot \mathrm{K})$. For these values, use your program to graph the dependence of pressure on volume for $n = 1$ mole of gas for volumes from 1 to 10 liters. Use 100 points in your graph. Hand in your program and the graph.

Your program should take the following form

```
#include <iostream.h>
#include <dislin.h>

const double R = 0.082;

class Gas {
public:
    double iA;
```

```
    double iB;
    Gas(double aA, double aB)
};
class VanDerWaalsCalculator {
public:
    Gas iGas;
    float iVolume[100];
    float iPressure[100];
    float iTemperature;
    float iNumberOfMoles;
    int iNumberOfPoints;

    VanDerWaalsCalculator ( Gas aGas, double aMinimumVolume,
         double aMaximumVolume, double aTemperature, double
         aNumberOfMoles, int aNumberOfPoints)

// Use the Van der Waals formula to generate the pressure
// vector

    void generatePressure()

// Graph the pressure as a function of volume.
    void draw()
};

main() {
    Gas Oxygen(0.027, 0.0014);
    int numberOfPoints = 100;
    float minimumVolume = 1;
    float maximumVolume = 10;
    float temperature = 300;
    float numberOfMoles = 1;
    VanDerWaalsCalculator VDW(Oxygen, minimumVolume,
       maximumVolume,
    temperature, numberOfMoles, numberOfPoints);
    VDW.generatePressure();
    VDW.draw();
}
```

Chapter 10
Arrays and matrices

10.1 Data structures and arrays

In this chapter, we examine arrays, which correspond to vectors in scientific programming. Although arrays therefore occur very frequently in practice, one should be aware from the beginning that they comprise a single built-in (native to the compiler) member of a general group of data structures. A data structure is a collection of related quantities with common ordering properties. While we will defer more detailed study of data structures to later chapters, a list of common occurring structures serves to illustrate the concept:

A *bag* is an unordered collection of objects, for example a bag of balls could be specified schematically as <**Ball1, Ball1, LargeBall1**>.

A *set* is a bag in which no object can appear more than once.

A *list* is a collection of objects such that from each object it is possible to navigate to a succeeding object.

An *ordered list* stores the objects of a list in a particular order (sequence) such that it is possible to access an object through its position instead of only through its neighboring object.

A *sorted list* stores the elements of a list in sorted order according to a given ordering operation and preserves this order as new objects are added or removed.

A *key set* is an unordered collection of objects that have a key such as a word or number that is employed to locate rapidly a particular element of the set.

A *stack* is a container with the property that elements can only be added to or removed from the top of the stack. Elements in the stack are ordered so that the last item added to the stack is the new top item.

A *queue* resembles a stack except that the only object that can be removed from the container at any given time is the object that was first added to the container.

An *array* is a collection of similar objects that is accessible through an integer number termed the index. Every element of an array must be of the same type.

10.2 Array definition and initialization

Recall again that a construct in C++ is defined once the compiler is able to determine the amount of memory that it requires. Following each definition statement, the compiler issues instructions that can potentially be used to reserve the amount of memory space required by the construct's type. Accordingly, when a (compiler allocated) array is defined the *array type and the number of array elements* must be provided to the compiler and *cannot be changed at runtime*. This implies that the array size in the dimension statement must be an expression formed from positive integers and **const int** variables such as

```
const int i = 3;
const int j = 4;
double a[2 * i * j];          // 24-element double array
```

Here the elements of **a** are uninitialized and therefore contain random values. If the array size were not a constant integer, it could be adjusted during runtime as in

```
int i = 3;
cin >> i;
double a[i];                  // Compile-time error
```

A dynamic memory allocation procedure that essentially enables the above procedure will be introduced in a later chapter. However, if the same array dimension is required for many successive runs of the program, standard compiler allocation is more efficient and less error prone.

As noted in Section 5.11, array elements can be initialized at the point of definition through the syntax

```
int a[3] = {1, 2};
```

If the first elements of an array are initialized in this manner, all further elements are set to zero. Therefore, to initialize all the elements in an array to zero, we can write

```
int a[3] = {0};
```

If an array is declared constant, the array elements cannot be changed after the array is defined. All elements of the array must therefore be initialized at the point of definition through a statement such as

```
const int a[3] = {1, 2, 3};
```

Otherwise the array elements would contain random values that could not subsequently be set to meaningful data.

10.3 Array manipulation and memory access

An array name in C++ has a fundamentally different interpretation than a variable name as can be seen by running the simple program

```
main () {
      int a[2] = {1, 2};
      cout << a << '\t' << a[0] << '\t' << a[1] << endl;
}
```

which yields an output such as

```
0012FF84   1   2
```

The origin of the first value in the output is that the name of the array variable, **a,** is an alias (that is, an alternate name) for the memory location at which the array starts in much the same way as a function name is an alias for the starting memory location of its instruction record. (Recall that memory locations are by default printed out in hexadecimal notation when piped to cout.)

To access a value stored in the array, the index operator **[ ]** is employed. Writing **a[0]** yields the **int** value stored at the starting location in memory, 0012FF84 in the above example, while **a[m]**, where **m** is an integer, returns the **int** value stored at a position shifted from the starting location by **m** times the size of the memory occupied by a single element of the **int** array type. From this definition of the index operator it becomes apparent that the indexing of the elements of an n-element array, **a**, namely, **a[0], a[1], . . . , a[n−1]** starts from zero and that all array elements must be of the same type. However the index operator is defined without reference to the array size as specified in the array definition. That is, the array operator can be used to access memory locations located at *any* positive or negative integer offset from the starting address of **a**. Largely because of this property, an array can be initialized without reference to its size as follows

```
int a[ ] = {1, 2, 3};
```

Such a construct initializes **a** to a three-element array, which is the amount of memory the compiler requires to store the specified values.

Except in a definition statement, where the purpose of the array index is not to address an array element but to determine the extent of memory allocation, the index of an array can be any const or non-const integer expression such as

```
a[i * (j + 1) * 3] = b[j + 2];
```

The programmer must clearly be very careful that such an expression does not evaluate to a value that falls outside the defined array limits.

A frequently occurring error is is to write

```
a = b;          // ERROR: a is a fixed address, not an lvalue.
```

when the intention is to set the elements of **a** equal to the elements of **b**. Since **a** represents the starting address of the array's memory, which is fixed during program execution and is therefore an rvalue, the above statement yields a compiler error. To equate each array element instead requires a statement such as

```
for (int i = 0; i < 10; i++) a[i] = b[i];
```

Because the index operator will accept any integer value as its argument, any memory location in the computer can, if permitted by the operating system, be accessed through an array variable. Three errors, however, commonly arise when the array index equals or exceeds the argument of the index operator (the array size) appearing in its definition statement. The least troublesome of these occurs if the array index is abnormally large. The associated memory location is then nearly always outside the memory space that the operating system has reserved for the running program. The operating system consequently intercepts the attempted illegal memory access, terminates the program, and typically issues a generic error message in a pop-up window. To locate the offending source line then requires introducing either numerous statements that write out values of intermediate variables during program execution or a debugger, which executes up to the line containing the error. Since the array index is far from a physically reasonable value, such errors can also generally be found without excessive effort by direct inspection of the source code.

A second possible effect of employing an improper array index occurs if the memory location accessed by the array variable falls into a region that is reserved for program operation but either is not initialized or is associated with a variable of an incompatible type (for example, a **double** instead of an **int**). In this case, the value returned by the index operator will be large and random. This type of error is again often simply corrected, as the program will generate either unphysical results or an error message that can, however, appear to originate from a location in the program several lines past the actual error.

The third, and by far the most dangerous possibility arises if the array index is only slightly larger than the array size, and a second initialized variable of a compatible type is resident at the corresponding memory position, as in

```
main() {
      double b[2] = {0.02, 6.0};
      double a[2] = {0.01, 3.0};
      cout << b[0] << endl;
      cout << b[-1] << endl;
      cout << a[2] << endl;
      double wrongNorm = sqrt(a[0] * a[0] + a[1] * a[1] +
            a[2] * a[2]);
      cout << wrongNorm << endl;
}
```

which yields

```
0.02
3
0.02
3.00008
```

The error manifest in **wrongNorm** is indicative of many scientific calculations in which localized distributions, such as pulses or particle wavefunctions, that are small near their right and left endpoints are stored in adjacent arrays. While such mistakes will eventually be noticed because the result will change unpredictably when the program input and thus the content of the adjacent array are varied, they can be misinterpreted as the intrinsic error of the numerical algorithm. Many compilers (but not the Dev-C++ or the free Borland C++ compiler) offer a switch that enables array bounds checking. Specialized programs are also available that detect common memory errors.

10.4 Arrays as function parameters

Since an array is an rvalue, a function cannot return an array (it could not be assigned to a second array variable in the calling program). However, an array can be an argument of a function. In contrast to most other types of variable arguments, if one of the elements of the array parameter is altered within the function, the corresponding array element in the calling program changes. This behavior, which is termed "pass by reference," is demonstrated by

```
#include <math.h>

void zero(int aA[ ]) {
      cout << aA << endl;
      aA[0] = 0;
}

main() {
      int p[ ] = {1, 2, 3};
      cout << p << endl;
      zero(p);
      cout << p[0] << endl;
}
```

which yields the output

```
0012FF80
0012FF80
0
```

In the second line of **main()**, the starting memory address of the array **p** is written out. That this *address* is passed as a parameter to the function **zero( )** follows from the second output line. In other words, the starting memory location associated with the array argument **aA** in the function is the same as that of **p** in the calling program. Since the arrays **aA** and **p** are therefore identical, changing the value of an element of **aA** within the function body causes a corresponding change in **p** in the calling program. Duplication of the array memory followed by element-by-element copying is therefore avoided, which is especially appropriate for large arrays.

A common compile-time error is to pass an array element to a function of an array argument; that is to write, for example, **zero(p[0])** in place of **zero(p)**. Of course, **p[0]** is the integer value stored at the first memory location in the array, not the array itself. Note finally that the array dimension need not be specified in the function parameter list, as the behavior of the index operator is unaffected by the size of the array that it acts upon.

10.5 Returning arrays and object arrays

While a function cannot return an array, an object is handled by the compiler in effectively the same manner as a built-in variable and can be assigned to a second object. Therefore, object return values are permitted and an array can therefore be returned from a function if it is wrappered in an object

```
class Matrix {
        public:
        int iA[10];
};

Matrix f() {
        Matrix M1;
        M1.iA[0] = 10;
        return M1;
}

main (){

        Matrix B = f();
        cout << B.iA[0] << " " << f().iA[0] << endl;
                                                  // Output: 10 10
}
```

The near equivalence of user-defined and built-in types also enables arrays of objects to be defined in exactly the same manner as arrays of built-in variables. For example

```
class Position {
        public:
        double iPosition;
};

main () {
        Position P1[2];
        P1[0].iPosition = 1;
        P1[1].iPosition = 2;
        cout << P1[1].iPosition << endl;        // Output: 2
}
```

This construct appears frequently in object-oriented programming.

10.6 *const arrays*

As stated earlier, a function argument that is declared **const** cannot be altered within the function body. For variables that are passed by value, the effect of a constant argument definition is limited to the function body, since a change to the function argument in any case does not alter the value of the function parameter in the calling program. For array arguments and other argument types that are passed by reference, however, the **const** keyword prevents unwanted side effects from occurring in case the argument is inadvertently changed inside the function. As an example, if we write

```
void print(const int aI[ ], const int n) {
      cout << aI[n] << endl;
}
```

attempting to assign a new value to any element of **aI** or to the variable **n** within the function body yields a compiler error.

A non-const array can be passed as a formal parameter to a function, such as the print function above, with a **const** array argument. However, unlike pass by value arguments, a **const** array can only be used as a parameter in a function call if the corresponding array argument in the function is declared **const**. That is

```
void print(int aI[ ]) {
      cout << aI[0] << endl;
}

main () {
      const int a[2] = {1, 2};
      print(a);                       // Error: aI not a const array
}
```

yields a compiler error since the function argument **aI** is not declared **const** in **print( ).** Otherwise, the **const** property of the array in the calling program would be circumvented through changes to elements in the function body.

10.7 Matrices

An array can be defined with two or more dimensions. As an example, a two-dimensional array with two rows and three columns that stores **int** values is defined as

```
int M[2][3];
```

Such an array corresponds to a two-dimensional matrix. Similarly, a $2 \times 3 \times 4$ three-dimensional array is defined as

```
int N[2][3][4];
```

A multidimensional array can be initialized at the point of definition in either of the two identical ways

```
int M[2][3] = {{1, 2}, {4, 5, 6}};
```

or

```
int M[2][3] = {1, 2, 0, 4, 5, 6};
```

both of which correspond to the matrix

$$M = \begin{pmatrix} 1 & 2 & 0 \\ 4 & 5 & 6 \end{pmatrix}$$

Note that each term in braces in the first definition statement represents a single matrix row; any row elements that are left unspecified are therefore automatically initialized to zero.

A helpful technique for understanding the effect of matrix definitions is to reverse the order of the symbols in the definition statement; that is

```
int M[2][3];
```

should be thought of as

```
(int[3])[2] M
```

Reading this from right to left, we see that **M** is a two-element array of three-element arrays of integers. Therefore, in the initialization statement involving two sets of inner braces, the elements in the first set of braces initialize the three-component array **M[0]**, while the second elements initialize the components of **M[1]**.

10.8 Matrix storage and loop order

That the definition **int M[2][3];** generates a *two-element array of three-element integer arrays* each of which occupies 12 bytes of memory space has fundamental implications in scientific applications that are now analyzed in depth.

First observe that the function **sizeof(M)** returns $6 \times 4 = 24$ bytes while **sizeof(M[0])** is $3 \times 4 = 12$ bytes. The elements of the three-element int array **M[0]** are **(M[0])[0]** = **M[0][0]**, **(M[0])[1]** = **M[0][1]**, and **(M[0])[2]** = **M[0][2]**. These are followed in memory by the three-element array **(M[1])[0], . . . , (M[1])[2]**. Generalizing this to an n-dimensional array, we conclude that the *rightmost indices vary most rapidly in memory* (it should be remarked that in some other programming languages such as FORTRAN, the leftmost indices instead vary most rapidly).

One consequence of the storage order of the matrix elements is that when a matrix is simultaneously defined and initialized, as in a function argument or in

```
int M[ ][3] = {{1, 2, 3}, {4, 5, 6}};
```

the *first* dimension does not have to be specified. All other matrix dimensions, however, must be supplied, since, for example, in the illustration above, the compiler must establish that the fourth memory location is associated with the first element of the second row of the matrix, **M[1][0]**. This can already be achieved if **M** is defined as an array of three-element integer arrays.

A further implication of the so-called "row-major storage order" of the matrix in memory is that the following code

```
for (int i = 0; i < n; i++)
     for (int j = 0, j < m; j++) M[i][j] = N[i][j];
```

is generally far more efficient than

```
for (int j = 0; j < m; j++)
     for (int i = 0, i < n; i++) M[i][j] = N[i][j];
```

since, in the latter case, *memory is not addressed sequentially by the innermost, most rapidly varying, for loop*. If the matrix dimensions are large, each successive element might accordingly require accessing high-level cache memory and not fast memory, such as low-level cache or the CPU memory registers. In the most unfavorable case, accessing each or every few new memory elements induces a page fault in which portions of the matrix are exchanged (swapped) between hard disk and RAM memory. The execution time then increases by orders of magnitude (however modern optimizing compilers will in many cases reorder the sequence of loops automatically).

Traversing matrix elements in an optimal fashion can require careful planning. As an example, the standard code for multiplying two matrices is

```
for (int leftLoop = 0; leftLoop < leftDimension; leftLoop++) {
      for (int rightLoop = 0; rightLoop < rightDimension;
           rightLoop++) {
      C[leftLoop][rightLoop] = 0.0;
           for (int innerLoop = 0; innerLoop < innerDimension;
                    innerLoop++)
                 C[leftLoop][rightLoop] += A[leftLoop][innerLoop]*
                    B[innerLoop][rightLoop];
      }
}
```

Clearly the problem with this code is that, while the elements of **A** are traversed linearly (stride 1), the elements of **B** are accessed suboptimally (stride **secondDimension**). This can generally be avoided if the values of the matrix elements are initially read into the transpose, **BT**, of the matrix **B** instead of into **B** itself. In this case, the formula for **A** * **B** in the program above is replaced by

```
C[leftLoop][rightLoop] += A[leftLoop][innerLoop] *
      BT[rightLoop][innerLoop];
```

which results in stride 1 access for both **A** and **B**.

An additional improvement can be achieved by *blocking*, which refers to the subdivision of a problem into smaller units, each of which fits within a fast memory structure. A simple implementation of blocking for the above problem is provided by the code, which assumes for simplicity that both the matrices **A** and **B** are square matrices with size **nM** = 4

```
main () {
    int A[4][4] = {{1,2},{3,4}};
    int BT[4][4] = {{4,3},{2,1}};
    int C[4][4] = {0};
    int cMatrixElement;
    int nM = 4;
    int blockSize = 2;
    for (int leftBlock = 0; leftBlock < nM; leftBlock +=
          blockSize) {
       for (int innerBlock = 0; innerBlock < nM; innerBlock +=
             blockSize) {
          for (int rightLoop = 0; rightLoop < nM; rightLoop++) {
             for (int leftLoop = leftBlock; leftLoop < min(leftBlock
                   + blockSize, nM); leftLoop++ ){
                cMatrixElement = 0;
                for (int innerLoop = innerBlock; innerLoop <
                      min(innerBlock + blockSize, nM); innerLoop++)
                   cMatrixElement += A[leftLoop][innerLoop] *
                     BT[rightLoop][innerLoop];
                C[leftLoop][rightLoop] += cMatrixElement;
             }
          }
       }
    }
}
```

The time-critical part of the calculation is now performed in the innermost loop, which multiplies reduced-size square matrices with a row size given by **blockSize** = **2**. The size of these reduced matrices should be adjusted to insure that the matrix multiplication can be performed within the fastest available memory. As the parameter **blockSize** is increased so that the multiplication requires successively higher levels of cache memory, the time required to perform the calculation undergoes a series of nearly discontinuous jumps. Since very few optimizers determine the hardware memory configuration, implementation of blocking strategies must generally be performed by the application programmer.

Finally, it should be noted that a surprisingly common error in writing nested loops is to write

```
for (int j = 0; j < m; j++)
       for (int i = 0, i < n; j++) M[i][j] = N[i][j];
```

where **i** in the incrementation step of the second **for** loop has been mistakenly replaced by **j**, resulting in an infinite loop.

10.9 Matrices as function arguments

A matrix can be employed as a function argument in the same manner as an array. Further, the one-dimensional array components of the matrix can be passed to functions that require array arguments. As an example

```
const int n = 2;

void f1(double aV[ ]) {
   aV[0] = 1;
}

void f2(double aM[ ][n]) {
   aM[0][0] = 2;
}

main () {
   double M[n][n] = {0};
   f1(M[1]);                                          // M[1][0] is now 1
   f2(M);                                             // M[0][0] is now 2
   cout << M[1][0] << '\t' << M[0][0] << endl; // Output: 1 2
}
```

As in the case of an array parameter, in the call **f2(M)** in **main( )**, only the matrix name, which evaluates to the starting memory location, is passed. Therefore, the matrix argument **aM** in **f2( )** occupies the same memory space as the matrix **M**, yielding pass by reference semantics.

10.10 Assignments

Part I

Identify the errors in the following programs and indicate if they are compile-time, link-time or runtime errors. If the program does not contain any errors, write down its output.

(1)
```
main() {
    int i = 5;
    int j[10] = {10, 5};
    int k[20];
    for (int j = 0; j < 4 * i; j++) k[j] = j;
    cout << k[j[i]] << endl;
}
```

(2)
```
void myFunction(int aI[10]) {cout << sizeof(aI) << endl;}

main(){
   int aI[10] = {0};
   cout << sizeof(aI) << endl;
   myFunction(aI);
}
```

(3)
```
void myFunction(int aA[2]) {
   cout << aA[2] << endl;
}
```

```
main() {
  int A[2][2], step = 0;
  for (int loopOuter = 0; loopOuter < 3; loopOuter++) {
         for (int loopInner = 0; loopInner < 3; loopInner++)
                A[loopInner][loopOuter] = step++;
  }
  myFunction(A[0]);
}
```

(4)
```
main() {
  int a[3] = {1,2,3};
  int b[3];
  b = a;
  if (b == a) cout << "The two matrices are equal" << endl;
}
```

(5)
```
void myFunction (int a[ ]) {cout << a[2] << endl; }

main () {
  int a[] = {0,1,2,3,4};
  myFunction(a+2);
}
```

(6)
```
void myFunction (char aM[][2] ) {
  cout << sizeof(aM) << endl;
}

main () {
  char M[2][2] = {0};
  myFunction(M);
}
```

(7)
```
main () {
  int a[2][2] = {1, 2, 3, 4};
  cout << a[0][1] << endl;
}
```

(8)
```
main() {
  int j = 3;
  float a[j];
  a[0] = 1;
  cout << a[0];
}
```

(9)
```
main() {
  int A[ ][2] = {1, 2, 3, 4 };
  cout << A[1][0] << endl;
}
```

Part II

(10) Write a function that takes as an argument an integer of 8 or fewer digits and returns a second integer composed of the digits in reverse order (for example if the integer argument is 1234, the return value is 4321). *Hint*: A rapid way to solve this problem employs the statements

```
rightDigit = oldInteger % 10;
oldInteger /= 10;
```

(11) To calculate the molecular weight of a compound in grams/mole, the atomic weights of each element in the compound are multiplied by the number of atoms that appear in the compound's chemical formula. The resulting values are then summed.

Construct a simple object-oriented molecular weight calculator based on the following problem description:

First, construct an **Element** class. This class is composed of a **char elementName[3]** string that stores the element symbol (for example Na) and a **double** variable that stores the atomic weight of the element. In the first part of the program, the user enters a series (up to 20) of element names and atomic weights and these are stored in the **main( )** program in an array, **elementArray[20]**, of **Element** objects. This element list is passed to the **MolecularWeightCalculator** object through its constructor. The **computeMolecularWeight( )** method of this object is then called, which prompts the user for the elements in the formula and their multiplicities. The elements are then looked up in the **elementList** vector and the molecular weight is computed, stored in an internal variable of the **MolecularWeightCalculator** class, and printed out through a **print( )** function.

The input to the program should be entered from the keyboard after program prompts as follows. (The program should assume that the number of atoms entered will be ten or less)

```
Input the atomic symbol of atom 1 (0 to exit):        H
Input the atomic weight of atom 1:                    1
Input the atomic symbol of atom 2 (0 to exit):        O
Input the atomic weight of atom 2:                    16
Input the atomic symbol of atom 3:(0 to exit):        0
```

The atoms and atomic weights you entered were

```
H      1
O      16

Input the number of atoms of type 1 in the formula:   2
Input the number of atoms of type 2 in the formula:   1
```

Your program should be organized as follows

```
#include <iostream.h>

class Element {
  public:
    char iElementSymbol[3];
    double iAtomicWeight;
    Element()
};

class MolecularWeightCalculator {
  public:
```

```
    double iMolecularWeight;
    int iNumberOfElements;
    Element iElementArray[20];
    double iFormula[20];
    int iNumberOfFormulaElements;
    double iResult;
  // Copy the information from the function parameters into
  // the internal class variables.
  MolecularWeightCalculator ( Element aElementArray[],
          int aNumberOfElements )

  // Generate the molecular weight from the information in
  // iElementArray and iFormula.
  void calculateMolecularWeight()

  // Print out the final result
  void print()

  // To enter the formula, the element symbol will first have
  // to be compared with the symbols of the existing elements.
  // The element symbol can be read in and the comparison
  // effected as indicated below.
  void getFormula() {
    char symbolBuffer[3] = {0};
    cout << "Input the number of elements in the formula\
        \t";
    cin >> iNumberOfFormulaElements;
    cout << "Enter the element symbol: \t \t \t";
    cin >> symbolBuffer;
    if ( iElementArray[loopInner].iElementSymbol[0] ==
           symbolBuffer[0] &&
         iElementArray[loopInner].iElementSymbol[1] ==
           symbolBuffer[1] )
  }
};

main() {
  Element elementArray[20];
  // place your input routine here (or in a separate function)
  // Note the following procedure for reading in the
  // element symbol.
  cout << "Insert atomic weight for element " << loop <<
        ": \t \t";
  cin >> elementArray[loop - 1].iAtomicWeight;
  MolecularWeightCalculator MW(elementArray,
        numberOfElements);
  MW.getFormula();
  MW.calculateMolecularWeight();
  MW.print() ;
}
```

(12) This problem implements an object-oriented version of sandpile dynamics. The sandpile model is the following. At each time step, a grain is added to each of the points in the sandpile with a probability **addProbability**. If the number of grains of sand,

iGrid, at a grid point, A, then exceeds the number at a neighboring grid point, B, by more than **criticalGradient**, the number of grains at A is decreased by a number **sandUnit** and the number of grains at B is increased by **sandUnit**. The number of grid points of the sand pile is **xDimension**, here given as 200 along the x-direction and **yDimension**, taken as 4, along the y-direction. At the left boundary of the sand pile, we set the number of grains of sand **iGrid[0][loopY] = iGrid[1][loopY]** for all **loopY** so that no sand is transferred through the left boundary point. At the same time, we set **iGrid[xDimension-1][loopY] = 0** so that any sand that reaches the right boundary is removed from the problem (these are boundary conditions). We further apply periodic boundary conditions in the y-direction, so that, for example, the fictitious grid point **iGrid[0][4]** is identified with **iGrid[0][0]**, see the program below. Under these assumptions, the final sand distribution (up to a constant uniform displacement, which depends on how much sand has been added before the equilibrium situation is reached), is independent of the initial sand distribution (neglecting statistical fluctuations). Further when this equilibrium distribution is reached, the average number of grid points per time step at which sand is transferred from one site to another is similarly independent of the initial distribution. Of course, these quantities depend on computational parameters, such as **addProbability**.

In your program employ the following class template. On a two-dimensional grid each site has four neighboring sites and therefore it is possible that the number of grains of sand at a site exceeds the number of grains at two or more neighboring sites by **criticalGradient**. However simplify the problem here by transferring the grains to the first neighboring site that you find for which this condition is fulfilled.

```
#include <fstream.h>
#incflude "dislin.h"

const int xDimension = 200;
const int yDimension = 4;
const int maximumNumberOfSteps = 100000;

class Sandpile {
  double iCriticalGradient;
  double iSandUnit;
  double iAddProbability;
  double iGrid[xDimension][yDimension]; // the sand profile
  int iFlips;       // the number of transfer events per step
  float iFlipsArray[maximumNumberOfSteps],
      xArray [maximumNumberOfSteps];
  public:
     Sandpile(double aCG, double aSU, double iAP)
     void evolve(int aNumberOfSteps) {
       for (int loop = 0; loop < aNumberOfSteps; loop++) {

// Use this code to add grains of sand randomly to the
// sandpile.
        for (int loopX = 1; loopX < xDimension - 1; loopX++) {
          for (int loopY = 0; loopY < yDimension; loopY++) {
```

```
                if (rand()/double(RAND_MAX) < iAddProbability)
                iGrid[loopX][loopY] += 1;
            }
        }

// Implement the left and right boundary conditions here.
// Use the following lines to implement the periodic
// boundary conditions in the y-direction — because
// e.g. -1 % 2 is -1, but you must keep the indices positive!!

      int loopYPlus = (loopY + 1) % yDimension;
      int loopYMinus = (loopY - 1 + yDimension) % yDimension;

// Loop now over *all* grid points and for each grid point
// check if one of its neighbors has iCriticalGradient fewer
// grains of sand. When you find a point for which this is
// the case, immediately transfer iSandUnit from the current
// grid point to this point (use += and -=) and then increment
// iFlips by one. Always apply this sequence of steps to each
// of the four neighboring points.

// Now store iFlips into iFlipsArray[loop] and then
// reinitialize iFlips to zero.
      }

      void createMarginalProfile( aStartingProfile ){

// You should create two possible starting profiles here.
// When aStartingProfile = 0, each grid point should have
// one grain of sand. Otherwise the number of grains of sand
// at a given grid point should be given by
// (iCriticalGradient -1) * (xDimension - loopX);
      }

      void plotFlips(int aNumberOfSteps) {
         qplot(xArray, iFlipsArray, aNumberOfSteps);
      }

      void plotProfile() {
// This function should plot a cross section along the
// x-direction of the sand profile (iGrid) for e.g.
// loopY = 2.
      }
};

main() {
   metafl("XWIN");
   double criticalGradient = 8;
   int numberOfSteps = 100000;
   double sandUnit = 3;
// In the final version you should read addProbability and
// startingProfile from cin.
   double addProbability = 0.05;
   int startingProfile = 0;
   Sandpile S1(criticalGradient, sandUnit, addProbability);
```

```
    S1.createMarginalProfile( startingProfile );
    S1.evolve(numberOfSteps);
    S1.plotFlips(numberOfSteps);
    S1.plotProfile();
}
```

Hand in your program together with the two graphs produced for each of the two starting profiles above. As well, hand in graphs for the constant starting profile when **addProbability** is increased from 0.05 to 0.5.

(13) This is a very simple traffic simulator presented in Phys. Rev. A, vol. 46, N. 10, pp. R6124–6127 (1992). Traffic is modeled as the movement of arrows on a two-dimensional grid. Right arrows move one grid point to the right for even time steps, while up arrows move one grid point upward for odd time steps unless the new grid point is already occupied by either a right or an up arrow. The average velocity of the ensemble of arrows is calculated (in our implementation) by dividing the total number of successful moves of all the arrows in the problem over all time steps by the maximum possible number of moves, which is equal to the number of arrows multiplied by the total number of time steps. Note that a grid point can be occupied by only one type of arrow. Implement periodic boundary conditions (as in the previous problem) in both coordinate directions, so that, for a $n \times n$ point grid, the $(i, n+1)$st grid point is identified with the $(i, 1)$st grid point and, similarly, the grid point $(n+1, i)$ is identified with $(1, i)$ for any i (the 1st and nth grid points are represented by the indices 0 and $n-1$ index in C++). That is, if traffic moves upward from a grid point at the top of the grid, it moves to the point with the same longitudinal (x) coordinate but located at the bottom of the grid in the transverse (y) direction.

You should generate an object-oriented realization of this problem according to the following outline

```
#include <iostream.h>

const int size = 50;

class TrafficSimulator {
   bool iUp[size][size];
   bool iRight[size][size];
   int iArrows;     // This will store the total number of
                    // arrows
   public:
     TrafficSimulator() : iArrows(0) {}   // Default constructor
     void initialize(float aThreshold) {
      iArrows = 0;
                 // Use int(rand() / double (RAND_MAX) +
                 // aThreshold) to place
                 // arrows (which are here represented by the
                 // boolean value true) in either iUp or iRight
                 // with a probability given by aThreshold. Be
                 // sure to place the arrow either in iUp or in
                 // iRight but not in both!
     }
```

```
    double propagate(int aNumberOfSteps) {
      int iMove = 0;
      bool rightSave[size][size], upSave[size][size];
                  // Be careful to save a copy of the iUp
                  // and iRight matrices since otherwise you
                  // will overwrite them during each move! Each
                  // arrow attempts to move rightwards on even
                  // numbered steps and upwards on odd numbered
                  // steps. Implement periodic boundary
                  // conditions as in the last assignment.
                  // The following line returns the fraction
                  // of successful moves.
      return 2 * iMove/(double(iArrows) * aNumberOfSteps);
    }
};

main () {
  srand(time(NULL));
  TrafficSimulator T1;
  for (float probability = 0.05; probability <= 0.9;
       probability = probability + 0.025) {
    T1.initialize(probability);
            // Place the result of propagating 20000 steps for
            // each value of the probability into an array and
            // graph the result using DISLIN
  }
}
```

Hand in your program together with a graph of the output array in **main( )**. Observe the sharp transition from smooth traffic flow to a high-density grid-lock or jammed phase.

Chapter 11
Input and output streams

C++ provides a flexible system for managing input and output operations based on a device-independent approach to reading and writing data. That is, nearly the same interface (i.e. the public internal functions and variables of a class) is used to send data to or receive data from different output devices, such as disk files and memory buffers. This is achieved by grouping all shared functions and variables into a set of base classes. These properties are inherited by a larger set of specialized classes that incorporate the code necessary to manage specific sets of devices by overloading the base class functions and adding a limited additional number of additional features. A typical inheritance diagram for the classes that govern input from and output to various devices is shown in Fig. 11.1.

Figure 11.1 indicates that functions and variables common to all input and output operations are collected in the **ios** base class. Operations involving the standard input device (the keyboard) **cin** are added by the **istream** class. The additional functionality that enables data to be read from files and memory buffers is then incorporated by the **ifstream** and **istrstream** classes, respectively. Because of their location in the inheritance hierarchy, these classes normally encompass the standard input operations associated with **cin** (however, this is not the case for Dev-C++ for which **istream** or **iostream** must be included separately). Output operations are implemented in the same manner by the corresponding output classes. The **iostream** class inherits the internal functions and variables of both the **istream** and **ostream** classes. Finally, the **fstream** and **strstream** classes further specialize the **iostream** class by appending the properties required to read from and write to files and memory buffers.

Following the structure of the inheritance diagram, we first discuss the **iostream** class. Subsequently we discuss the interfaces to files and memory buffers associated with **fstream** and **strstream**.

11.1 The *iostream* class and stream manipulators

The conceptual basis of the C++ I/O subsystem is that variables read into or written from a program initially reside in abstract objects termed streams, examples of which are **cin** and **cout.** These streams may be thought of as smart memory buffers

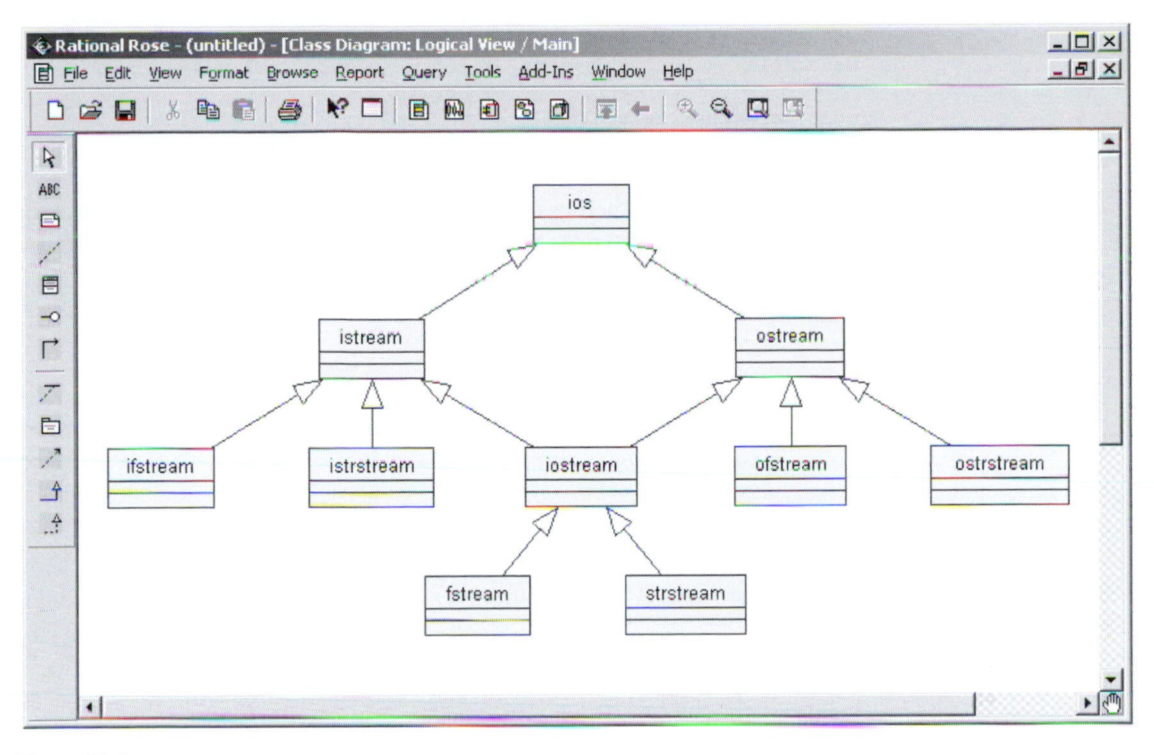

Figure 11.1
Source: Reprinted by permission from Rational Rose Enterprise Edition, © Copyright 2002 by International Business Machines Corporation. All Rights Reserved.

that are enhanced through the inclusion of public, user-accessible functions that can act on the data stored in memory before they are processed by the program or sent to a device. The components that interface the memory buffer with the actual system devices are implemented as private functions and variables of the stream classes, and are invisible to the user.

Stream objects, such as **cout** and **cin**, possess internal member variables labeled flags. Member functions of the stream access the flags, which control its behavior. The flags of **cout** govern in this manner, for example, the notation, precision, and number of spaces employed in printing a value. (In reality, the flag value is typically assigned to specific bit or bits of a binary number, so that one byte can store several flags.) Flags that are common to all input and output classes are defined in the **ios** base class and are addressed through the scope resolution operator **ios::**. Important set member functions for flags and other internal member variables of the **cout** class are

```
cout.precision(n)          // n figures written out in floating
                           // point
cout.setf(ios::fixed)      // retains trailing zeros, turns off
                           // scientific notation
```

```
cout.setf(ios::left)         // left-justifies output
cout.fill('.')               // replaces all blank fields with a
                             // period
cout.setf(ios::scientific)   // scientific notation (floating point)
cout.width(m)                // reserves m spaces for output
```

Generally once the properties of the stream object are changed, it remains in the new state, since the functions that insert data in or remove data from the stream do not alter these properties. The output width, set to **m** above, however reverts to its default value every time a new value is written. Therefore, the test program

```
main () {
        float f = .0001;
        cout.setf(ios::fixed);
        cout.setf(ios::left);
        cout.fill('.');
        cout.width(20);
        cout << f << endl;
        cout.precision(20);
        cout.setf(ios::scientific);
        cout << f << endl;
}
```

yields

```
0.000100............
9.99999747378751636e-05
```

A second way to access the flags defined in **iostream.h** and its base classes is through stream manipulators that are defined in the header file **iomanip.h**. The manipulators change the values of the stream flags when they are directly piped into the input or output stream

```
#include <iomanip.h>

cout << setiosflags(ios::scientific);    // for scientific notation
cout << setiosflags(ios::showpoint);     // for decimal notation
cout << setw(n);                         // set width of output
                                         // field
cout << setf(ios::left);                 // left-justify output
cout << setprecision(n);                 // sets precision to n
                                         // digits after the
                                         // decimal point
```

If desired, output can be sent directly to the printer by replacing all instances of the standard output stream **cout** by the printer stream **cprn**. Other standard streams are **cerr** (the "standard error" stream) and **clog**, both of which are associated by C++ with the terminal, and **caux**, which is bound to the serial communication port and can be employed to send data to a second computer connected through a suitable (null modem) cable attached to the serial port and running appropriate communication software.

11.2 File streams

A file is a (often non-contiguous) region on a storage medium, such as a hard disk, floppy disk, memory card, or CD-ROM. A file resides within a larger group of files called a subdirectory (folder) or directory and is accessed either through its "absolute pathname," which is the full directory name followed by the file name, or by its relative pathname, the path that must be followed to reach the file from the current default directory. An example of an absolute pathname is **X:\Dev-Cpp\include\io.h**, which refers to the Dev-C++ include file **io.h**. The relative pathname to this file from the directory **X:\Dev-Cpp\bin** is in contrast **..\include\iostream.h**, where **..** is a symbol that is interpreted by the operating system as the parent directory, **X:\Dev-Cpp**, immediately above the current directory. (Similarly, a single period is interpreted as the current directory.) The relative path and file name from the parent directory **X:\Dev-Cpp** is simply **include\io.h** (or equivalently **.\include\io.h**).

As stated earlier, by substituting **#include <fstream.h>** (not **ifstream.h** or **ofstream.h)** for **#include <iostream.h>** in a program, additional streams are defined that inherit the properties and behaviors of the **iostream** class but append the additional functionality required by file operations. (However, because of differences in implementation the standard streams, such as **cin** and **cout**, and operations, such as **endl**, are included within **fstream.h** in Borland C++ but not in Dev-C++.) The associated file stream objects inherit the interface of the **iostream** class, which means they operate on files largely in the same manner as **cin** and **cout** interface with the standard devices. However, unlike the standard input and output streams a file stream does not possess a default device such as the keyboard or terminal. Therefore, it must be attached to a specific storage file (opened) before it can be used.

A file stream can be most simply created and attached to a file through the definition statement

```
fstream myFileStream("fileName");
```

Here **myFileStream** can be any legal C++ identifier, while the string **"fileName"** represents either the absolute or the relative file name. Writing e.g. **"file.dat"** in place of **"fileName"** will search for or create a file named **file.dat** in the current directory. If the file is to be used exclusively for input or output, we can write, respectively

```
ifstream myInputFileStream("fileName");
```

or

```
ofstream myOutputFileStream("fileName");
```

Once input or output file streams are opened, they are manipulated in exactly the same manner as **cin** and **cout**, as, for example

```
int r;
myInputFileStream >> r;
myOutputFileStream << setw(8) << r << endl;
```

Assuming that **file.dat** has been attached to the file stream **myFileStream** and that the file contains only numeric values, data can be read up to the end of the file using either of two constructs. The principle of operation of the first procedure, typified by

```
while (myFileStream >> r) {...}
```

is that a null character is automatically placed at the end of a file to demarcate the end of reserved memory in the same way as a null character is employed to demarcate the end of a string. The second procedure

```
while( !myFileStream.eof( ) ) { myFileStream >> r; ... }
```

instead employs the **eof( )** member function of **fstream**, which returns **true** when the null character at the end of the attached file is reached and **false** otherwise.

Each time a file is opened for writing, its previous contents are normally deleted. To append data instead to the end of an existing file, an additional argument **ios::app** must be included in the file stream definition (to employ the **ios** flags, **iostream.h** must be included together with **fstream.h** in Dev-C++)

```
ofstream fou("out1.dat", ios::app);
```

Sometimes the definition of the stream is separated from the process of attaching the stream to the file (opening the file) by use of an **open** statement as in

```
ofstream fou;
fou.open("out1.dat", ios:::out | ios:app);
```

Through a **close** statement, a file stream can be reassigned to another file

```
fou.close();
fou.open("out2.dat");
```

All files are automatically closed by the operating system when a program terminates. The **open( )** and **close( )** functions are particularly useful in examining data generated by a long-running program. If a file is intermittently opened from and written to in append mode and then immediately closed, it can be copied to a second file during program execution.

In general, considerable savings in both memory space and access time can be achieved by storing information in binary as opposed to standard text (ASCII) format. To do this, an **ios::binary** specifier must be included after the file name in either the **open** or the file definition statement. [However, the procedure for actually writing binary information is somewhat complex: for example the **double** variable **a** has to be written to the file with **fout.write((char *) &ff, sizeof (ff));]**. Two examples are

```
fstream mS("myFile.dat", ios::out | ios::binary);
ofstream fou;
fou.open("out1.dat", ios::out | ios::binary);
```

A stream, such as **cin, cout**, or a user-defined file stream, such as **fou** above, can be passed to a function. However, the function must act on the same file stream as that defined in the calling program, since a standard file stream cannot be copied to a second file stream. To avoid copying, a function argument must be an array, reference, or pointer. As we have so far only introduced the first of these variable types, we illustrate the procedure by creating an array of two file streams that is then passed to a function

```
#include <fstream.h>
void print ( ofstream aFstream[ ] ) {
      aFstream[0] << "Hello World" << endl;
}

main () {
      ofstream fout[2];
      fout[0].open("test1.dat");
      fout[1].open("test2.dat");
      fout[0] << "Line 1" << endl ;
      print (fout);
}
```

11.3 The string class and string streams

Finally, many programs are greatly simplified by reading and writing characters from memory buffers. A memory buffer stores data that will be processed further by subsequent statements in the program. A particularly convenient interface to memory buffers is presented by the string stream classes defined in **strstream.h** or the updated header file **sstream.h**. A string stream comprises a stream connected to a memory buffer rather than to a standard output device or a file.

The underlying container into which a string stream places data is a **string** object. This object can be directly accessed through the **str( )** member function of the **strstream** class of **strstream.h** header file or the corresponding **stringstream** class of **sstream.h**. To manipulate individual **string** objects, the header file **string.h** must be included. A blank **string** object can then be defined through either the statement

```
string s;
```

or

```
string s = "";
```

both of which store the one-element array {'**\0**' } in **s**. In general a string can be initialized with any **string** constant such as

```
string s = "A string";
```

The operator + is overloaded for strings, such that if **s1** and **s2** are both string objects, the string **s1+s2** is the concatenation of **s1** and **s2**, where the termination character '**\0**' at the end of **s1** is deleted; for example, **s** + " **:** " + **s** yields the string "A string : A string". Lexicographic comparisons (comparison according to dictionary order) can be performed through the overloaded operators ==, !=, <, <=, >, and >=.

Some important functions that are defined for string objects are

```
s.length()
```

which returns the length of the string **s** *not including* the termination character

```
s.erase(3, 4);                // Result: A sg
```

which erases four characters in the string s, starting at the position to the right of the third character (writing simply **s.erase( );** or **s** = " **";** deletes the entire contents of the string except for the null character). A single element of a string can be accessed through the syntax either **s[i]** or **s.at(i)**, while the entire string can be converted into a standard character array by employing the member function call **s.c_str( )**. Characters can be read directly into or out of a string from a stream through the operators >>, << and the **getline( )** function.

The string class has additional **find( )**, **replace( )**, and **insert( )** functions that with suitable arguments can be respectively employed to find a sequence of characters in a given string, replace a character or a sequence of characters in a string with other values, and insert additional characters into a stream.

To illustrate the application of string streams, suppose that data, represented below by the string "some input data," are to be written to a large number of files labeled **output1.dat**, **output2.dat**, **output3.dat**. These names can be generated automatically from within the program by piping the string "output" followed an integer and finally by ".dat" into a string stream as seen below

```
// For Borland C++ use the following two lines
#include <sstream.h>
#include <fstream.h>

// For Dev-C++ use instead the following five lines
#include <sstream>
#include <fstream>
#include <iostream>
#include <string>
using namespace std;

main () {
        char temp[10] = "output", p[100];
        string s;
```

```
        stringstream inoutStream;
        ofstream fout;
        for (int loop = 1; loop < 4; loop++){
                inoutStream << temp << loop << ".dat" << endl;
        }
        inoutStream >> s;
        cout << s.c_str() << endl;   // Output: output1.dat (c_str()
                                     // is optional here)
        cout << inoutStream.str();   // Output: the entire contents
                                     // of inoutStream
        for (int loop = 1; loop < 3; loop++){
                inoutStream >> p;    // Reads from inoutStream one
                                     // line at a time starting at
                                     // output2.dat
                fout.close();
                fout.open(p);
                fout << "some input data" << endl;
        }
}
```

If the **strstream** header file is employed instead of **sstream**, the keyword **stringstream** must be replaced by **strstream**.

In the above program, four consecutive lines **output1.dat, output1.dat, output2.dat, output3.dat** are written to the string stream **inoutstream**. The first of these is then piped into the string **s**, which is then converted into a character array through the **c_str( )** member function of the **stringstream** class (**cout** processes strings and character arrays identically). This advances the location of the current line in the **stringstream** buffer to the start of the second line. Subsequently, the **inoutStream.str( );** statement converts the entire contents of the buffer including the first line to a **string** object, which is sent to **cout**. The files **output2.dat** and **output3.dat** are generated by piping a line at a time from the string stream starting at the position of the current (second) line into a character array buffer which is then employed as the file name parameter of an **open( )** statement.

For applications requiring **wchar_t** characters, the **wstringstream** class should be employed in place of the **stringstream** class. The **wstring** class must then be used in place of the **string** class and **wcout** and **wcin** must be substituted for **cout** and **cin** (this functionality is not present in the **strstream** classes).

11.4 The *toString( )* class member

A good practice in designing classes is to include a **toString( )** member function in each class that returns a string containing the values of the internal class variables together with appropriate descriptions in a clear, readable form. The advantage of returning a **string** object is that if a class contains one or more user-defined objects as internal member variables, the **toString( )** function of the enclosing class can be easily assembled through calls to the corresponding functions of the member objects. The simple example below demonstrates how string streams can be employed in this context. Note that **ends** is used to terminate an entire string, while **endl** terminates a line within a string

```
// For Borland C++
#include <strstream.h>

// For Dev-C++ employ instead the following four lines
#include <strstream>
#include <iostream>
#include <string>
using namespace std;

class C {
        public:
        int iC;
        C(int aC = 0) : iC(aC){}
        string toString() {
                strstream s;
                s << "iC = " << iC <<ends;
                return s.str();
        }
};

class D {
        public:
        C iC[2];
        int iD;
        D(int aD, C aC[ ]) : iD(aD) {
                iC[0] = aC[0];
                iC[1] = aC[1];
        }
        string toString() {
                strstream s;
                for (int loop = 0; loop < 2; loop++)
                        s << iC[loop].toString() << " ";
                s << endl;
                s << "iD = " << iD;
                s << ends;
                return s.str();
        }
};

main() {
        C C1[2] = {1, 2};
        D D1(2, C1);
        cout << D1.toString();
}
```

The output from **main()** is

```
iC = 1 iC = 2
iD = 2
```

11.5 The *printf ( )* function

Since the C++ language encompasses the C language with a few minor modifications, the standard C input and output operations can be employed in C++ by including the **stdio.h** header file. These or similarly structured functions appear often in older code, DISLIN being one example, and can be convenient when a large amount of output must be formatted into columns. In C, a floating-point number is formatted through the specifier **%i.jf**, where **i** and **j** are the number of integers to the left and right of the decimal point. An integer is represented by **%id** where **i** is the number of positions to be displayed and a string by **%is**, where **i** is the number of characters in the string. A line is written to the terminal through the **printf( )** function, which contains the line formatting information followed by a list of the variables to be written out. In this manner, the program

```
#include <stdio.h>    // C I/O functions

main() {
    int i = 10;
    double x = 1.e2;
    char a[10] = "and ";
    printf("x = %3.4f, %10s i = %10d", x, a, i);
}
```

sends the line

```
x = 100.0000,    and i = 10
```

to the standard output device. A related function **sprintf(buf, "x = %3.4f, %10s, i = %10d", x, a, i);** instead places the output into a memory buffer defined as, for example, **char buf[100];**.

11.6 Assignments

Part I

Locate any errors in the programs below and indicate if they are runtime, compile-time, or link-time errors. If the program will function properly, give its output (if the output is written to a file, specify what the contents of the file will be).

(1)
```
void myFunction ( int k, char l[ ] = "Hello" , int m = 2 ) {
    cout << k << '\t' << l << '\t' << m << endl;
}

main () {
    myFunction(4, "World");
}
```

(2)
```
#include <iomanip.h>

void myFunction ( double &aI ) {
    cout.setf (ios::scientific);
    cout << setw(5) << aI << endl;
```

```
}
main() {
    double a = 12.53;
    myFunction(a);
}
```

(3)
```
main () {
    char c;
    c = 'a' + 256;
    cout << c << endl;
}
```

(4)
```
#include <strstream.h>

void myFunction( strstream aF[ ] ) {
    aF[1] << "Test" << ends;
}

main () {
    char p[100];
    strstream stream[2];
    stream[0] << "This is a " << endl;
    myFunction(stream);
    stream[0] >> p;
    stream[1] >> p;
    cout << p << ends;
}
```

(5)
```
main () {
    int size = 3;
    char a[size] = "cat";
    for (int ix = 1; ix <=size; ++ix)
    cout << (a[ix] = ix + 65) << endl;
}
```

(6)
```
Contents of file.dat: 1 2

#include <fstream.h>

main() {
    fstream myFile("file.dat");
    int i, j;
    myFile >> i;
    myFile << i;
    myFile >> j;
    myFile << endl;
    myFile << i << j << endl;
}
```

(7)
```
Contents of file.dat: 1 2

#include <fstream.h>

main() {
    ofstream myFile("file.dat", ios::app);
```

```
    int i = 3, j = 4;
    myFile << i << endl;
    myFile << j;
}
```

(8) `Contents of file.dat: 1 2`

```
#include <ifstream.h>

main () {
    fstream oF("file.dat", ios::app);
    int i = 3;
    oF << i;
    oF << "end of file";
}
```

(9)
```
main () {
    char A[] = "A test string";
    char *P = A;
    ++P;
    cout << P << endl;
    P += 2;
    cout << P << endl;
    cout << --P << endl;
}
```

(10)
```
main(){
    int b;
    char a = 'c';
    const int c = 'a';
    switch (a) {
        case c: b = 1;
        case c + 1: b = 2;
        case c + 2: b = 3;
        case c + 3: b = 4;
        case c + 4: b = 5;
        default: b = 6;
    }
    cout << b;
}
```

(11)
```
#include <stdio.h>

main() {
    int i = 10;
    double d = 20.0;
    char c[4] = "and";
    printf( "The digits are %4d, %3s, %4.4f", i, c, d );
}
```

Part II

(12) Write a program that reads an $N \times N$ matrix from a file in the form (for the 2×2 case)

2

1 2

3 4

and then overwrites this file by replacing each element by the average of its four adjacent elements (right, left, up, and down). The first line of the file indicates the number of row and column elements of the matrix. If an adjacent element is outside the matrix boundary, its value is taken to be zero. You can close and open the file more than once.

(13) A simple model of the spread of an infection can be generated in much the same manner as the traffic flow model of Section 10.10. In particular, assume that each individual of a population is located at a point on a rectangular grid. The population is randomly inoculated against an infection with a probability p that a given person is inoculated. Subsequently, one person, located at an initially non-inoculated site, is infected. After one time step, the person infects those of his four nearest neighbors that are not inoculated. This proceeds for a large number of steps. Just as the traffic problem demonstrated a sharp transition between steady flow and gridlock, in this case we find that a sharp transition occurs between a localized infection and an infection that spreads throughout the population. This behavior is called percolation and is a general feature of such problems. In the implementation below, you are to write a file containing a rectangular map of all sites in which all uninfected sites appear as 0 and all infected sites appear as 1.

```
# include <fstream.h>

const int numberOfSites = 100;

//sites[0][i][j] - contain inoculation information
//sites[1][i][j] - contain infection information
class Population {
    bool sites[2][numberOfSites+2][numberOfSites+2];
public:
        // Initialize sites.
        Population()
        // Inoculate a random fraction of the population
        void inoculate(double aFraction)
        // Calculate the total number of infected sites
        int total()
        // Infect a particular site.
        bool infect(int aXLocation, int aYLocation)
        // Infects any sites adjacent to an infected site that
        // are not inoculated.
        void advanceTime(int aNumberOfTimeSteps)
        // Writes a 1 for all infected sites, 0 for
        // non-infected sites on a file.
        void writeFile(fstream fout[])
};

main() {
        // Generate an 82 x 82 grid of sites of which 80 x 80
        // sites are actually used for the calculation; the
        // sites at the edges are simply used to implement
        // the absorbing boundary conditions (they can be
        // infected but an infection at the boundary does not
        // spread back to the interior. Ask the user to input
```

```
    // the inoculation probability and the location of a
    // single infected site. If the site is inoculated,
    // ask the user for a new infected site until a
    // non-inoculated site is found; infect this site.
    // Subsequently advance the time 4000 steps, determine
    // the total number of infected sites and write the
    // result to a file. Submit the file for inoculation
    // probabilities of 0.5 and 0.2. Subsequently generate
    // and plot a graph of the total number of infected
    // sites versus the inoculation probability for
    // inoculation probabilities from 0.05 to 0.95. In
    // this case, to find an initial site, automatically
    // search sites in order starting from the center site
    // until one is found that is not inoculated.
}
```

Part II

Numerical analysis

Chapter 12
Numerical error analysis – derivatives

Having introduced the key aspects of C++ programming, we now progress to basic numerical analysis and scientific programming. Our first example is the derivative operator. Numerical approximations to the derivative operator constitute perhaps the simplest non-trivial computational algorithms, yet their error behavior can be used to illustrate the features of far more complex algorithms.

12.1 The derivative operator

The derivative operator is a continuous transformation that, applied to a differentiable function, $f(x)$, yields a second function, $df(x)/dx$, whose value is the slope of $f(x)$ at each point x. However, the slope is the tangent to the function that, except for linear functions, intersects the curve at a single point in the vicinity of x. Since the only numerically accessible information at a point is the function value itself, which is insufficient to construct the tangent, the exact derivative cannot in general be computed. However, a numerical approximation to the derivative can be generated from its original definition as the limit of a discrete expression, namely

$$\frac{df}{dx} = \lim_{\Delta x \to 0} \left(\frac{f(x + \Delta x) - f(x)}{\Delta x} \right) \tag{12.1}$$

This strategy of representing continuous quantities through their underlying definitions as discrete limits can be used to formulate numerical techniques for many problems.

We accordingly define the *discrete forward finite difference operator* associated with a differential element Δx as

$$D^{+}_{\Delta x}(f) \equiv \frac{f(x + \Delta x) - f(x)}{\Delta x} \tag{12.2}$$

This expression maps a function f to a new function $D^{+}_{\Delta x}(f)$ and therefore again represents an operator. Eq. (12.1) can then be rewritten as

$$\frac{df}{dx} = \lim_{\Delta x \to 0} D^{+}_{\Delta x}(f) \tag{12.3}$$

However, in numerical computation Δx remains finite. The left- and right-hand sides of Eq. (12.3) then differ by an *error term* that typically varies as a polynomial function of Δx. If the smallest power of Δx appearing in this polynomial is N, the error decreases as $(\Delta x)^N$ for $\Delta x \to 0$ and the finite difference approximation is said to be $O(\Delta x)^N$ accurate.

To determine N for the derivative approximation of Eq. (12.2), we consider the Taylor series expansion of a continuous and infinitely differentiable function $f(x)$, namely

$$f(x + \Delta x) = f(x) + \Delta x \frac{df(x)}{dx} + \frac{(\Delta x)^2}{2!} \frac{d^2 f(x)}{dx^2} + \cdots \tag{12.4}$$

Inserting Eq. (12.4) into Eq. (12.2), we have

$$\frac{f(x + \Delta x) - f(x)}{\Delta x} = \frac{df(x)}{dx} + \frac{\Delta x}{2} \frac{d^2 f(x)}{dx^2} + \cdots \tag{12.5}$$

which implies, if $d^2 f(x)/dx \neq 0$

$$\frac{df(x)}{dx} = D^+_{\Delta x}(f) + O(\Delta x) \tag{12.6}$$

This equation suggests a simple procedural program for evaluating the derivative of any continuous function. Since the forward difference expression is an operator that acts on a function, $f(x)$, a representation of the operator should have a *function* as one of its arguments. This means that the *type* of one of the parameters should be a function with a prototype, such as **double f(double);** that is, a function of a **double** argument that returns a **double**. This function *must be specified at compile-time* by inserting its name into the operator's parameter list in the source code as below

```
double cube(double aD) {
   return pow(aD, 3);
}

double linear(double aD) {
   return pow(aD, 1);
}

double derivOperator(double f(double),double aX, double aDel) {
   return ( f(aX + aDel) - f(aX) ) / aDel;
}

main () {
   double del = 1.e-1;
   int choice;
   cout << "Choose a function 1 - cube, 2 - square" << endl;
   cin >> choice;
   switch (choice) {
     case 1: cout << derivOperator (cube, 1.0, del);
        break;            // Output: 3.31
```

```
    case 2: cout << derivOperator (linear, 1.0, del);
      break;          // Output: 1
    default: cout << "Incorrect Input — program
      exiting";
  }
}
```

Note that we have assigned the name **del** to Δx in the program. The numerical result, 3.31, obtained for the derivative of x^3 corresponds to the value of $[(1+0.1)^3 - 1]/0.1$, which is as expected of the order of Δx.

12.2 Error dependence

Two general features of the above program deserve immediate comment. First, the finite-difference approximation to the derivative is exact for $f(x) = x$, so that error estimation based on linear or nearly linear functions is completely misleading – specialized cases with small error must be carefully avoided in evaluating the accuracy of nearly *any* numerical method. Further, when Δx is very small, rounding errors in the evaluation of $f(x + \Delta x) - f(x)$ degrade the accuracy of the result. For example, in Borland C++ setting **del = 1.e – 15;** yields 3.330 67 for the derivative of x^3, while **del = 1.e – 17;** instead gives 0. Such large errors occur when the order of magnitude of the change in f over the interval Δx approaches 10^{-14} of the absolute value of f as was discussed in Section 5.14. Since the function values that appear in the finite difference expression are nearly unity, the 14 significant digits of the **double** type are insufficient to distinguish the values of $f(1)$ and $f(1) + \Delta x(df/dx)|_{x=1}$ as Δx approaches 10^{-14}. While a carefully programmed routine for the derivative must therefore insure that for a given function and evaluation point, the choice of Δx yields an acceptably small error, for a wide range of Δx values, the simple program of Section 12.1 is both computationally efficient and accurate.

12.3 Graphical error analysis

Normally, unlike the finite difference case, an analytic expression for the numerical error is not available and the dependence of the error on different input parameters must instead be determined empirically. To illustrate the technique, we determine the variation of the error with step size Δx graphically for our example. Consider the following **main( )** function

```
main () {
  double del = 1.0e - 1;
  float x[10], y[10];              // Array definitions
  for (int l = 0; l < 10; l++) {
    y[l] = derivOperator(cube, 1.0, del);
                                   // Error calculation
    x[l] = del;
    del /= 2;
  }
```

Figure 12.1

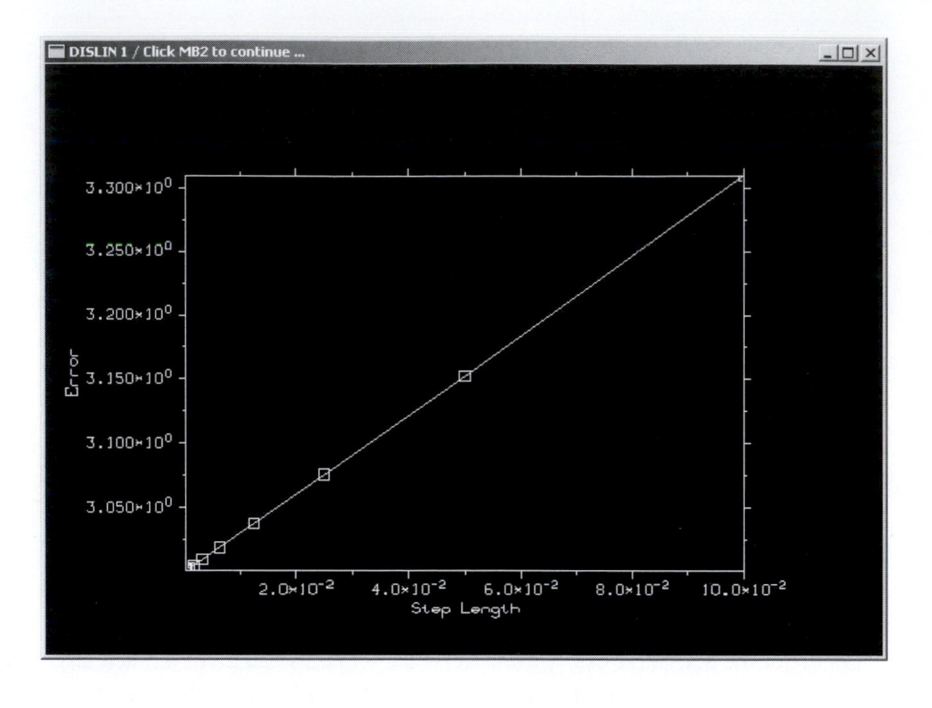


```
    metafl("XWIN");                    // write to terminal
    disini();                          // start plotting program
    name("Step Length", "x");          // x label
    name("Error", "y");                // y label
    labels("EXP","xy");                // exponential format
    incmrk(1);                         // markers at every (1) point
    setscl(x, 10, "X");                // automatically scales x axis.
    setscl(y, 10, "Y");
    //scale("LOG","XY");               // uncomment for log-log plot
    float minX,maxX,minY,maxY,stepX,stepY;
    graf(minX, maxX, minX, stepX, minY, maxY, minY, stepY);
                                       // axes
    curve(x, y, 10);                   // plotcurve
    disfin();                          // terminate plot
}
```

(More information on the plot routines that enter into the above program can be obtained by navigating to the DISGNU or DISBCC icon on the Programs submenu of the Start button and selecting selecting DISHLP or DISMAN or by typing DISHLP at the command line.) The above program yields the graph shown in Fig. 12.1, which clearly demonstrates the linear dependence of the error on the step length. In general, if a numerical method has an $O(\Delta x)^{\alpha}$ dependence on some adjustable input parameter x, the result plotted as a function of $(\Delta x)^{\alpha}$ will be a straight line at small Δx whose y-intercept is a corrected value. That is, the order of the error can be determined by selecting an integer value for α and plotting for different values of Δx the (x, y) pairs composed of $(\Delta x)^{\alpha}$ together with the value of the result at $x + \Delta x$. Different integer values for α are examined until one is found for which the curve near the origin is a straight line.

12.4 Analytic error analysis – higher-order methods

In contrast to the uniqueness of the continuous derivative operator, infinitely many discrete finite-difference operators can be defined. These can be classified according to their intrinsic error. For example, to construct a finite-difference operator that approximates the continuous derivative to an accuracy of $O(\Delta x)^2$ unlike Eq. (12.6), which is $O(\Delta x)$ accurate, return to the Taylor series relationships

$$
\begin{aligned}
f(x+\Delta x) &= f(x) + \Delta x \frac{df}{dx} + \frac{(\Delta x)^2}{2!}\frac{d^2 f}{dx^2} + \frac{(\Delta x)^3}{3!}\frac{d^3 f}{dx^3} + \cdots \\
f(x-\Delta x) &= f(x) - \Delta x \frac{df}{dx} + \frac{(\Delta x)^2}{2!}\frac{d^2 f}{dx^2} - \frac{(\Delta x)^3}{3!}\frac{d^3 f}{dx^3} + \cdots
\end{aligned}
\tag{12.7}
$$

Subtracting these formulas yields immediately

$$
\frac{f(x+\Delta x) - f(x-\Delta x)}{2\Delta x} = \frac{df}{dx} + O(\Delta x)^2 \tag{12.8}
$$

Numerically, this is equivalent to replacing the derivative function in our program by

```
double derivOperator(double f(double),double aX, double aDel) {
       return (f(aX + aDel) - f(aX - aDel) )/(2 * aDel);
}
```

Repeating in this manner the calculation of Fig. 12.1 with the *centered finite difference operator* of Eq. (12.8) generates Fig. 12.2.

Evidently, the numerical error is decreased by orders of magnitude over the range shown, while the parabolic dependence of the error on step length is clearly apparent.

12.5 Extrapolation

We have previously observed that extrapolating the graphs of the previous sections to the origin yields a highly accurate estimate of the converged result. However, as we will again illustrate through the derivative example, such an extrapolation can in fact be performed directly without graphical analysis. The underlying strategy consists of determining a linear combination of two or more finite difference representations of the continuous operator that yields a higher-order accurate procedure.

As a first step, observe that an alternate first-order difference operator is obtained by employing a negative step length $-\Delta x$ in place of Δx in the expression for $D^{+}_{\Delta x}(f)$. Since the signs, but not the coefficients, of all odd-order terms in the

Figure 12.2

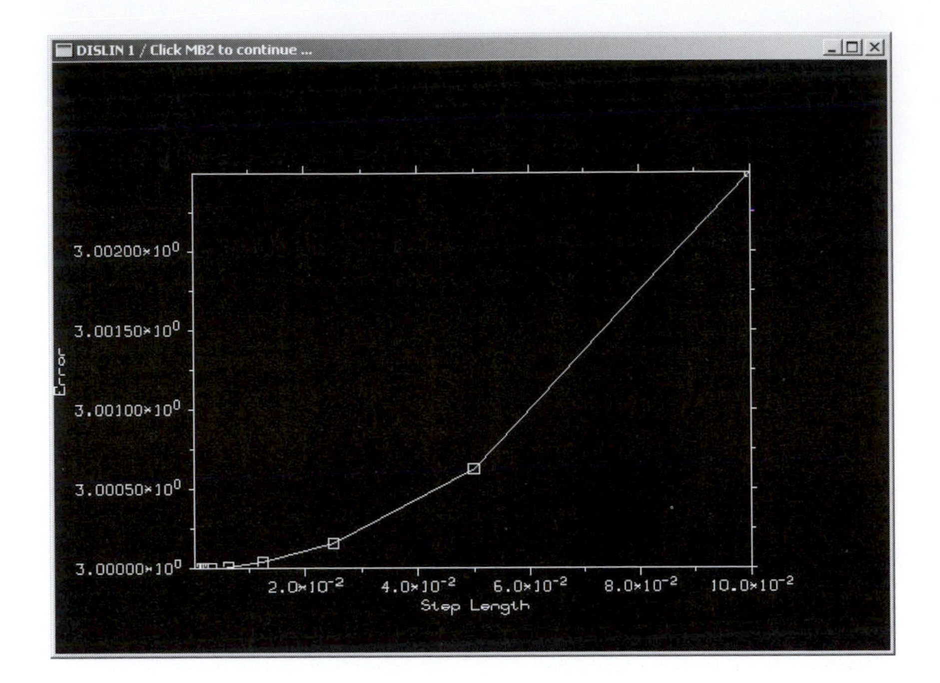

Taylor series expansion are then reversed, we have

$$\begin{aligned} \frac{df}{dx} &= D^{+}_{-\Delta x}(f) + O(-\Delta x) \\ &= \frac{f(x-\Delta x) - f(x)}{-\Delta x} + O(-\Delta x) \\ &\equiv D^{-}_{\Delta x}(f) + O(-\Delta x) \end{aligned} \tag{12.9}$$

where we have introduced the backward difference operator

$$D^{-}_{\Delta x}(f) = \frac{f(x) - f(x-\Delta x)}{\Delta x} \tag{12.10}$$

However, $O(\Delta x) + O\,(-\Delta x) = O(\Delta x)^2$ since the even terms in the Taylor series for the error add in the indicated sum while odd terms cancel. Therefore, adding the expressions for df/dx in Eq. (12.6) and Eq. (12.9) yields the second-order accurate expression of Eq. (12.8). Evidently, such a procedure can be applied to improve the results of first-order accurate numerical calculations of any function that is three-times continuously differentiable. Further, as will be demonstrated in the problem set, by appropriately combining linear combinations of results obtained by replacing **del** by **−del**, **±2*del**, **±3*del**, . . . , arbitrarily high numerical accuracy can in principle be achieved (for sufficiently smooth functions).

12.6 The derivative calculator class

Finally, we develop an object-oriented implementation of the derivative operator. Our paradigm is an abstraction of a handheld calculator in which the user enters the step length and the position at which the derivative will be evaluated. The calculator, which is programmed to store the function that the derivative operator acts on, computes the value of the derivative when its **calculate** button is depressed and subsequently displays the result upon request.

A new feature of this program is that the derivative operator transforms functions rather than variables. Therefore, one of the calculator's internal members should be the function upon which the operator acts. In other words, at compile-time a function with a single **double** parameter and that returns a **double** value should be introduced as one of the calculator's internal variables. To accomplish this, recall from Section 9.1 that a function name is an alias (alternative name) for the memory location of the first instruction of an instruction set. When this instruction set is activated at runtime, the operations associated with the function body are invoked. Therefore, an external function can be accessed from within a class if an internal variable in the class stores the memory address corresponding to the function name. Such variables, which are discussed in detail in Chapter 18, are called pointers. A pointer to an object is declared or defined by preceding a declaration or definition of a variable with the same type as the object by a star and then setting the value of this "pointer" variable to the memory location of the relevant object. Since a pointer to a function then stores the starting memory address and is associated with the function's type, it can be employed in precisely the same manner as the function itself. Accordingly, a program that calculates the input data for an error analysis plot is

```
class DerivativeCalculator {
public:
        void setDx(double aDx) { iDx = aDx; }
        void setX(double aX) { iX = aX; }
        double dx() const { return iDx; }
        double calculateDerivative() const { return (iF(iX + iDx)
            -iF(iX))/iDx; }
        DerivativeCalculator(double aX, double aDx, double
            aF(double)) : iX(aX), iDx(aDx) { iF = aF; }
private:
        double (*iF) (double);
        double iDx;
        double iX;
};

main () {
        DerivativeCalculator DC1(1.0, 0.1, cube);
        float x[10], y[10];
        for (int l = 0; l < 10; l++) {
                y[l] = DC1.calculateDerivative() - 3;
                x[l] = DC1.dx();
                DC1.setDx(DC1.dx()/1.3);
        }
        qplot(x, y, 10);
}
```

12.7 Assignments

(1) A fourth-order accurate finite-difference formula for the derivative operator is

$$\frac{df}{dx} = \frac{f(x - 2\Delta x) - 8f(x - \Delta x) + 8f(x + \Delta x) - f(x + 2\Delta x)}{12\Delta x} + O(\Delta x)^4$$

Apply this formula to the function $f(x) = x^6$ at $x = 1$. This must be done by modifying the **derivativeCalculator** class and corresponding **main( )** program presented in the slides. By plotting your results for the derivative first against the third and then against the fourth power of the step length Δx, prove that the assertion of fourth-order accuracy is correct. To obtain a good graph, take the variable **del**, as defined in the slides in the derivative handout, to be 0.1 and divide **del** by 1.3 for each successive iteration. Now graph the log of the error $df/dx - 6$ as a function of log(x) adapting the graphics routines of Section 12.3 and demonstrate that the slope of the curve near the origin is 4 using the procedure outlined in the text. Uncommenting the **scale** command yields a log–log graph, but this is inconvenient since the exponents of the axis labels rather than the labels themselves are then the desired logarithms. Hand in your graphs as well as your program.

(2) Extending the procedure introduced in Section 12.5, find the interpolation method that produces the fourth-order result of the previous problem and verify this by repeating your calculation of the error against the fourth power of the step length. Hand in your analytic calculations and the graphical result.

(3) Apply the program that you developed in (1) or (2) to the function $f(x) = x^4$ and explain the result.

(4) Using the Taylor series expansion of a function

$$f(x + h) = f(x) + \frac{df(x)}{dx}h + \frac{1}{2!}\frac{d^2 f(x)}{dx^2}h^2 + \cdots \tag{12.11}$$

generate an $O(\Delta x)^2$ accurate asymmetric discrete approximation to the continuous derivative operator using the values of $f(x)$ at $x,\ x + \Delta x,\ x + 2\Delta x, \ldots$ but *not* the values of $f(x)$ at $x - \Delta x,\ x - 2\Delta x, \ldots$. Verify your expression with a plot of the error against the second power of the step length and hand in your calculations and the graph.

Chapter 13
Integration

This chapter presents a short introduction to numerical integration. As many aspects of the error analysis are similar to those encountered in our discussion of derivatives, we concentrate principally on numerical programming practice.

13.1 Discretization procedures

A numerical procedure for evaluating integrals can be derived in the same manner as for the derivative operator. We will view the definite integral

$$I^L f(x) = \int_L^x f(x')dx' \tag{13.1}$$

as an operator, I^L, that acts on an integrand function $f(x)$ to yield a second function $I^L f(x)$ given by the integral from a certain specified lower endpoint L to x. As in the derivative case, to derive a numerical approximation for a continuous operator, we return to its basic definition as a limit of a discrete expression. Here we recall that the integral is defined as the limit as $n \to \infty$ of the discrete summation $I_n^L f(x)$ with

$$I_n^L f(x) = \sum_{k=0}^{n-1} \Delta x \, a_k \tag{13.2}$$

where $\Delta x = (x - L)/n$ and the a_k are suitably chosen values of the function $f(x)$ within the interval $[x + k\Delta x, x + (k+1)\Delta x]$. Note that the limits in the summation are from 0 to $n-1$, corresponding to n intervals of length Δx. A common programming error is to set the upper limit to n. This yields a small $O(\Delta x)$ numerical error that can be mistaken for the intrinsic error of the integration procedure. Unless the error of the integration method is known beforehand to be of higher order, identifying such a mistake can be difficult.

In calculus textbooks, two possible choices for the coefficients a_k are commonly cited. The first of these

$$a_k = f(L + k\Delta x) \tag{13.3}$$

evaluates the function at the left endpoint of each interval and is termed the rectangular rule. If the function values are instead determined at the center of the interval

$$a_k = f(L + (k + 0.5)\Delta x) \tag{13.4}$$

the resulting procedure is called the midpoint rule. While both techniques yield the same integral value in the continuous limit ($\Delta x \rightarrow 0$), their discrete error properties differ considerably.

13.2 Implementation

While the structure of the object-oriented code for integration does not differ substantially from the derivative example, an examination of the steps involved in programming the integration routine illustrates the process of coding numerical problems.

Problem statement: In the initial phase of program development, the desired interaction of the program with the user is described without reference to the actual software implementation. Such a problem statement could take the following form:

> The program will compute an integral of a given function as a discrete sum between any two endpoints. Upon request, the user will enter the input data consisting of the left and right limits of the integration region, the desired accuracy and an initial number of integration intervals. Subsequently, the integral will be evaluated to within the given accuracy and the result will be displayed.

The first iteration of the code, however, should implement as simple a version of the problem as possible. Accordingly, we first simply calculate the discrete integral value for a fixed, user-specified, number of intervals.

Designing the user interface: In the process of developing the problem statement, the user interface should be constructed. In our simplified example, the requested input comprises the lower and upper integration limits, a and b, the number of initial intervals n, and the maximum acceptable error, *eps*. A single value will be output from the program, namely the value of the integral, *res*. For a procedural program, the **main( )** function could be

```
main() {
   double leftLimit, rightLimit, eps, result;
   int numberOfIntervals;
   cout << "Input lower and upper limits, the initial number \
     of intervals and the maximum permissible error : ";
   cin >> leftLimit >> rightLimit >> numberOfIntervals >> eps;
```

```
    cout << "left limit = " << leftLimit <<
         "\nright limit = " << rightLimit <<
         "\nnumber of intervals = " << numberOfIntervals <<
         "\nerror = " << eps << "\nIntegral is " << result;
}
```

where we have employed the new line character **\n** to delimit successive lines in the input and the continuation character \ to write a single string on two lines of the source file.

Requirements specification: Once the purpose of the program is established together with the form of its input and output, a full problem statement that details the required resources and numerical methods should be supplied. This might take the form:

> The program will compute an integral of a given function as a discrete sum between any two endpoints. Upon request, the user will enter the input data consisting of the left and right limits of the integration region, the desired accuracy and an initial number of integration intervals. After computing the integral using the rectangular rule with the user-specified initial number of evaluation points, the number of evaluation points is doubled and the sum recomputed. If the *absolute value* of the difference between the old result and the new result is greater then the convergence parameter, **eps**, the new result for the integral is copied into a variable **oldResult** and the procedure is repeated. (The variable **oldResult** will be initialized to an unrealistically large value at the beginning of the program to avoid additional control statements.) This procedure is repeated a maximum of **maximumIterations** times; if convergence is not reached, the last value will be returned. Subsequently, the result will be displayed.

Object discovery: In an object-oriented implementation, the concrete and abstract objects of the problem should now be identified. While the actual coding is left to the problem set, the integration routine can, as in the derivative case, be represented as a function of an **IntegralCalculator** object. In this view, the object also handles the acquisition of the input data and display of the output data, leading to a class definition such as

```
class IntegralCalculator {
   public:
   IntegralCalculator (double aIntegrandFunction(double));
   void setNumberOfIntervals(const int aNumberOfIntervals);
   int numberOfIntervals();
   void integrate();
   void readInputData();
   void print() const;
```

```
    private:
    int iNumberOfIntervals;
    double iLeftLimit, iRightLimit, iResult;
    double (*iIntegrandFunction) (double);
};
```

Programming and testing functions: After the program is structured, code for the functions should be introduced. Each function should be individually tested before the next function is attempted. Subsequently, each time a single non-trivial change is made to the function body, the function should be rechecked. Otherwise, if an error is incurred while several changes are made in a program, its source can be difficult to locate, since the new lines may interact in unforeseen ways. The procedural implementation of the rectangular rule takes the form

```
double square(double aX) {
    return aX*aX;
}
double integrate(double aLeftLimit, double aRightLimit,
         int aNumberOfIntervals, double integrand(double)) {
    double dx = (aRightLimit - aLeftLimit) / aNumberOfIntervals;
    double x = 0;                    // Rectangular rule
    double sum = 0;
    for (int loop = 0; loop < aNumberOfIntervals; loop++) {
         sum += integrand(x);     // Function evaluation
         x += dx;                 // Update endpoint
    }
    return sum * dx;
}
```

After the function is coded, it should be tested independently by a test driver such as

```
main() {
    cout << integrate(0.0, 1.0, 1000, square);
}
```

that yields the result 0.332 834, close to the expected value of 1/3. In a class implementation each member function should be similarly tested; however, this generally requires first writing a constructor and other required input and output routines. For example, an object-oriented version of the integration program requires a **main( )** function such as

```
main() {
    IntegralCalculator IC1(square);
    IC1.readInputData( );
    IC1.integrate( );
    IC1.print( );
}
```

Accordingly, the constructor, input and output functions must therefore be introduced before the **integrate** function can be written and tested.

Completing the main( ) program: After the functions are individually tested, the full **main( )** program can be constructed. This yields the following code (the **math.h** header file must also be included)

```
main() {
        double leftLimit, rightLimit, eps, result;
        int numberOfIntervals;
        cout << "Input lower and upper limits, the initial number \
          of intervals and the maximum permissible error : ";
        cin >> leftLimit >> rightLimit >> numberOfIntervals >> eps;

        double oldResult = 1.e30;
        int maximumIterations = 25;
        for (int loop = 0; loop < maximumIterations; loop++) {
              result = integrate (leftLimit, rightLimit,
                 numberOfIntervals, square);
              if (fabs(result - oldResult) < eps) break;
              oldResult = result;
              numberOfIntervals *= 2;
        }

        cout << "left limit = " << leftLimit
          << "\nright limit = " << rightLimit <<
          "\nnumber of intervals = " << numberOfIntervals <<
          "\nerror = " << eps << "\nIntegral is " << result;
}
```

Two very common errors are to employ the **abs( )** function in place of **fabs( )**, as discussed in Section 9.4, and to copy the new result into the variable **oldResult** before performing the convergence test.

Testing the final program: Once the program is completed, test cases should be compared with analytic results. In our example, the known answer is $(\mathbf{R}^3 - \mathbf{L}^3)/3$, where **R** is the value of **rightLimit** and **L** is the value of **leftLimit**. Subsequently, different extreme cases on the so-called problem boundaries should be tested to see if the program functions or terminates correctly. These might include testing the program when convergence is attained in a single step, when convergence is not reached after 25 steps, and when the right limit is chosen equal to or less than the left limit. Finally, an error analysis should be undertaken as described in the following section.

After the program is written, additional comment lines should be supplied as necessary. Further, the structure of the program should be reviewed and simplified wherever possible, as long as clarity is not sacrificed for the sake of brevity. The program should then be retested before the next simplification is introduced. The final code should be simple to maintain and modify at a future time by both the author and other programmers.

13.3 Discretization error

The determination of the intrinsic accuracy of a program and, if possible, its comparison with the expected precision of the numerical method constitute an often omitted but critically important step in numerical programming. For the integral operator, an analytic expression can again easily be obtained for the error. We perform this analytic study first for the rectangular rule and then for the midpoint rule.

Rectangular rule: To calculate the error associated with the rectangular rule, we observe from Eqs. (13.1)–(13.3) that the difference between the rectangular rule approximation and the exact result for the integral is given by

$$E_{rect} = \sum_{k=0}^{n-1} \int_{x_k}^{x_k+\Delta x} [f(x_k) - f(x)]\, dx \tag{13.5}$$

where $f(x_k)$ is a constant given by the value of the function at the left endpoint x_k throughout the interval. If $f(x)$ is continuous in the interval from L to R, it can be expanded in the kth interval in a Taylor series in $(x_k - x)$. This yields

$$E_{rect} = \sum_{k=0}^{n-1} \int_{x_k}^{x_k+\Delta x} f'(x_k)(x_k - x)dx + O(\Delta x)^2 \tag{13.6}$$

However, if the magnitude of the first derivative of $f(x)$ is bounded on the interval $x \in [L, R]$ such that over this entire interval $|f'(x)| < |M_1|$

$$|E_{rect}| < \frac{M_1 n(\Delta x)^2}{2} = \frac{M_1(R-L)\Delta x}{2} \tag{13.7}$$

since $n\Delta x$ is the total length of the integration interval. We therefore conclude that unless an exceptional cancellation occurs between positive and negative contributions to the total error, the error varies linearly with the length of Δx; note that the error in each discrete interval is $O(\Delta x)^2$, but the errors add over $O(1/\Delta x)$ intervals. This behavior is confirmed by our numerical results in Fig. 13.1 for the integral as a function of step length.

Midpoint rule: While the rectangular rule yields an $O(\Delta x)$ integration procedure, applying the midpoint rule reduces the order of the error to $O(\Delta x^2)$. To verify this analytically, we write

$$f(x) = f(\bar{x}_{k+1/2}) + f'(\bar{x}_{k+1/2})(\bar{x}_{k+1/2} - x) + \frac{f''(\bar{x}_{k+1/2})}{2!}(\bar{x}_{k+1/2} - x)^2 + \cdots \tag{13.8}$$

where $\bar{x}_{k+1/2}$ represents the value of x at the midpoint of the kth integration interval. The expression corresponding to (13.6) for the error then becomes

$$E_{mid} = \sum_{k=0}^{n-1} \int_{x_k}^{x_{k+1}} \left[f'(\bar{x}_{k+1/2})(\bar{x}_{k+1/2} - x) + \frac{f''(\bar{x}_{k+1/2})}{2!}(\bar{x}_{k+1/2} - x)^2 + \cdots \right] dx \tag{13.9}$$

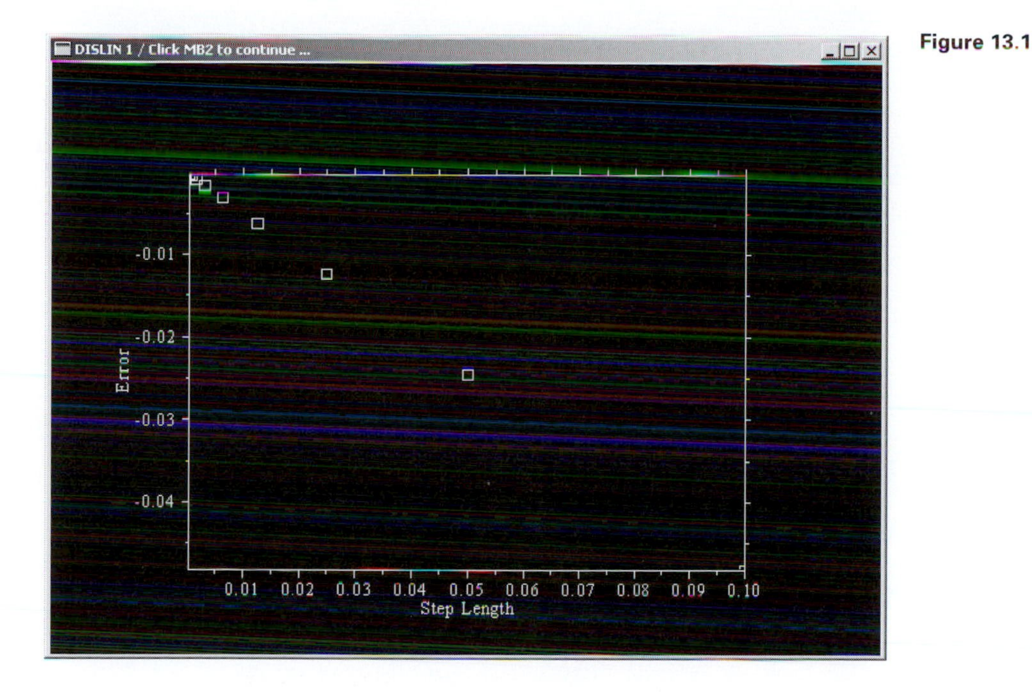

Figure 13.1

Since $f'(\bar{x}_{k+1/2})$ is a constant the first term averages to zero, leaving

$$|E_{mid}| < \frac{M_2 n(\Delta x)^3}{24} = \frac{M_2(b-a)(\Delta x)^2}{24} \tag{13.10}$$

where $|f''(x)| < M_2$ for $x \in [a, b]$. The quadratic convergence dramatically decreases computation times, as verified in the subsequent assignment.

13.4 Assignments

(1) Complete the coding of the **IntegratorCalculator** class introduced in Section 13.2 using the midpoint rule. With your program, verify the convergence predicted in Eq. (13.10). In particular, plot graphs of the error against the inverse of the number of points, n, for the integral of x^3 from 0 to 1 for both the midpoint rule and the rectangle (left-point) rule.

(2) Replace the midpoint rule with Simpson's rule given by

$$\int_L^R f(x)dx = \frac{\Delta x}{6}\sum_{i=0}^{n-1}\left[f(x_i) + 4f\left(\frac{x_i + x_{i+1}}{2}\right) + f(x_{i+1})\right]$$

and determine graphically the order of error of this procedure.

(3) Generate a new convergence curve for the midpoint rule by instead evaluating the integral of the step function

$$f(x) = \begin{cases} 1 & x < 1 \\ 0 & x \geq 1 \end{cases}$$

from 0 to 2. Why does the result not agree with the analytic derivation of Section 13.3?

(4) Introduce an error in the computation of the midpoint integration in problem 1 above by evaluating the sum between 0 and n instead of 0 and $n-1$. Graph out the resulting error curve. Do you see why this is a particularly dangerous error?

(5) Write a program to take the integral of the (numerically evaluated) derivative function of x^4 and show that you recover the original function.

Chapter 14
Root-finding procedures

Another common problem in numerical calculations is that of determining the real solutions over an interval $[L,R]$ of a nonlinear equation $f(x) = 0$. In this chapter, we examine two simple solution techniques for this problem. The first procedure, known as the bisection method, requires only that the function $f(x)$ be continuous over the interval. If the derivative is also continuous near the zero, Newton's method can instead be applied, leading to dramatically shorter computation times. Both the bisection and Newton's method can be formulated in terms of recursive or non-recursive functions.

14.1 Bisection method

If $f(x)$ is continuous over the interval $[L,R]$ and $f(L)f(R) < 0$, it has at least one root in $[L,R]$.Thus, to find all roots in an interval, $f(x)$ is first evaluated on a sufficiently large number of equally spaced points over the interval to insure that no more than one root occurs between each set of adjacent points. We then consider in turn each interval on which the function changes sign. We label the current interval $[a,b]$, and its bisector c. If $f(b)f(c)$ is ≤ 0 so that the root is between b and c including possibly the endpoints of the interval, we obtain a new interval $[a,b]$ with $a = c$. Otherwise, we set instead $b = c$. This process is then iterated until the change in the endpoint is less than a prescribed value, *error*.

Input/output section: As in the previous example, once a problem definition statement is formulated, the user interface should be constructed

```
main() {
        float leftLimit, rightLimit, error, root;
        cout << "input left limit, right limit, error" << endl;
        cin >> leftLimit >> rightLimit >> error;
        cout << "left limit = " << leftLimit << " right limit = "
            << rightLimit << " error = " << error;
        // call to root function to be inserted here
        cout << " root = " << root;
}
```

We assume that only a single root is present between the right and left limits as the coding of the initial search of an interval for subintervals that contain a single root is left to the assignment.

Standard implementation: The bisection procedure is programmed below within a **while** loop that terminates when the convergence criterion is satisfied, although **do . . . while** or **for** statements can equally well be employed. Note that the function returns the average of the left and right endpoints of the interval, which statistically better estimates the root than either endpoint

```
#include <math.h>

double mySquare(double aD) { return aD * aD - 4; }

double bisect ( double aFunction(double), double aLeftLimit,
          double aRightLimit, double aError ) {
      while (fabs(aLeftLimit - aRightLimit) > aError) {
                                                  // fabs NOT abs()!
          double midpoint = (aLeftLimit + aRightLimit) / 2;
          if (aFunction(midpoint) * aFunction(aRightLimit) <= 0)
                aLeftLimit = midpoint;
          else aRightLimit = midpoint;
          }
      return (aLeftLimit + aRightLimit) / 2;
}

main() { ...root = bisect (mySquare, leftLimit, rightLimit,
          error); ...}
```

Recursive implementation: Since the root-finding procedure repeatedly performs the same set of manipulations on intervals of steadily decreasing size, the control loop above can be recast as a recursive function. While such an implementation is far less efficient, the resulting code is more transparent as evidenced below

```
#include <math.h>

double mySquare(double aD) { return aD * aD - 4; }

double bisect ( double aFunction(double), double aLeftLimit,
          double aRightLimit, double aError ) {
      double midpoint = (aleftLimit + arightLimit) / 2;
      if (fabs(aLeftLimit - aRightLimit) < aError) return
          midpoint;
      if (aFunction(midpoint) * aFunction(aRightLimit) <= 0)
          aLeftLimit = midpoint;
      else aRightLimit = midpoint;
      return bisect(aFunction, aLeftLimit, aRightLimit, aError);
}

main() { ... root = bisect (mySquare, leftLimit, rightLimit,
          error); ...}
```

Here the **bisect** function calls itself repeatedly, each time with half the previous interval length until the left and right limits of the new interval differ by less than **aError**. The average of the limits is then passed to the previous function, which passes this result back to the next previous function and so on until the result finally reaches the calling program.

14.2 Newton's method

The bisection method is guaranteed to find a root with a maximum attainable accuracy given by the roundoff error if $f(x)$ is continuous and the initial interval $[a,b]$ contains a single root. However, each iteration only halves the size of the interval so that the computation time required to obtain adequate precision can be long.

The Newton method is an alternative, far more efficient root-finding procedure. However, the method can fail unless the function has a continuous *derivative* in the vicinity of the root and a sufficiently accurate initial estimate of the root, x_1, is available. To derive the Newton method for finding a root of $f(x)$ near x_1, we approximate $f(x)$ by the first two terms in its Taylor series expansion, namely

$$f_T(x) = f(x_1) + \frac{\Delta f(x_1)}{\Delta x}(x - x_1) \tag{14.1}$$

where we have expressed the derivative as $\Delta f / \Delta x$ to emphasize that this quantity will be determined numerically. The zero of the above linear approximation to $f(x)$

$$x_2 = x_1 - \frac{f(x_1)}{\left(\frac{\Delta f(x_1)}{\Delta x}\right)} \tag{14.2}$$

is generally located far closer to the true root value than x_1. Equation (14.2) is iterated until $|x_{x+1} - x_i| < \varepsilon$. Equivalently, from the perspective of similar triangles the ratio of $f(x_1)$ to the distance $x_1 - x_2$ equals the slope of $f(x)$ evaluated at x_1. In coding (14.2) numerically, attention must be paid to the minus sign, which is easily omitted.

A recursive implementation of Newton's method is

```
#include <math.h>

double mySquare(double aD) { return aD * aD - 4; }

double derivative(double aFunction(double), double aX) {
      double delta = 1.e-4;
      return (aFunction(aX + delta) - aFunction(aX - delta))
           / (2. * delta);
}
```

```
double newton(double aFunction(double), double aEstimate,
          double aError ) {
     double del = - aFunction(aEstimate) / derivative(aFunction,
          aEstimate);
     aEstimate += del;
     if (fabs(del) < aError) return aEstimate;
     else return newton(aFunction, aEstimate, aError);
}

main() { ...root = newton(mySquare, estimate, error); ... }
```

The above method has two major drawbacks. First, once the numerical estimate is close to the root, the $O(\Delta x)^2$ error in the centered approximation to the derivative yields an error of the same order to the root value. Therefore, a high-order approximation for the derivative operator and a small value of **delta** are desirable, although too small a value or high an order degrades the accuracy of the derivative through roundoff errors. Secondly, if the initial estimate, x_1, is not sufficiently close to the actual root value, the sign of the derivative of $f(x)$ can be the opposite of its sign near the root. In this case, x_2 is positioned further from the root than x_1, and successive iterations generally diverge. To avoid this problem, the Newton and bisection method can be combined. Here an interval $[a, b]$ containing the root is selected where $x_1 = a$. If x_2 in Newton's method falls within the interval it is employed in the next iteration, otherwise the program applies the bisection method to determine a better estimate for the root. The implementation of this technique is left to the exercises.

14.3 Assignments

(1) As noted in the chapter, to use either the Newton or bisection procedures, the approximate locations of the desired root or roots must first be determined. This is normally done by evaluating the function on a grid of points inside a large interval. Using 173 points on an initial interval $[-20, 20]$ to determine the starting values, print out a table of all the roots of the cosine function within this interval accurate to four decimal places as determined using the bisection method.

(2) Generate a combined bisection and Newton's method program using the procedure outlined in Section 14.2. Using this program together with a scan such as that of the previous problem, print out a table of all the roots of the function $\sin(x^2)$ in the interval $[35, 40]$, starting as in the previous problem from a 111-point grid of initial values on the interval.

(3) Newton's method can take a particularly long time or even diverge for functions with complicated analytic behavior near the root. As an example, compare the number of iteration steps required to attain an accuracy of 10^{-4} for the functions $\sin(x)$ and $x\sqrt{|x|}$ at $x = 0$ using Newton's method for a starting value of $x = 0.1$. Note that the computation times for the bisection method are the same in both cases.

(4) Newton's method may not converge properly if the interval length Δx in the derivative is chosen to be too large or inconsistent with the desired accuracy. Verify this by generating values for the root of x^3 around $x = 0$ for a starting value of $x = 0.1$ and values of Δx of 0.1, $0.1/10^{1/2}$, 0.01, $0.01/10^{1/2}$ ... 0.00001 and graph your values using a base 10 logarithmic scale for the x-axis (see Section 12.3 for the appropriate graphic commands).

(5) The accuracy and convergence of Newton's method can also be compromised if the function is nearly constant near its root. Verify this by repeating your analysis for the value of the root as a function of interval length for the root of x^{11} around $x = 0$ for a starting value of 0.1.

Chapter 15
Differential equations

Perhaps the most useful mathematical construct from a scientific perspective is the differential equation, which enables the global behavior of a physical quantity to be extrapolated from relations describing its local behavior. For example, the local relationship between the force and the acceleration on a point particle is described by a second-order differential equation that can be solved once initial values of the position and velocity (boundary conditions) are specified. This differential equation gives a relationship between the particle's position and velocity at a given time to these values at an infinitesimally later (or earlier) time. The equation can be solved analytically through continuous integration formulas that sum the effect on the particle of the infinitesimal force increments that accrue over a finite time interval.

Even when analytic procedures are prohibitively complex, numerical approaches can generally still be implemented relatively simply. To demonstrate the basic principles, we here examine the motion of a point particle. Discretizing time and space transforms the differential equation into a finite difference equation. The position and velocity are then advanced repeatedly over small time intervals to generate the particle's trajectory at any later time.

15.1 Euler's method

Euler's method signifies a procedure for evolving physical quantities (the dependent variables) described by an ordinary differential equation stepwise in an independent (propagation) variable. Although the procedure is applicable to differential equations of arbitrary order, for simplicity we restrict our discussion to a single massive particle with mass m attached to a spring with force constant k. In this case, the underlying differential equation is

$$a = \frac{d^2x}{dt^2} = -\frac{k}{m}x \tag{15.1}$$

An nth-order ordinary differential equation can always be replaced by a system of n first-order equations that require n independent initial or boundary conditions to specify a unique solution. In mechanics, this transformation is implemented through the introduction of the velocity variable

$$
\begin{aligned}
v &= \frac{dx}{dt} \\
a &= \frac{dv}{dt} = -\frac{k}{m}x
\end{aligned}
\tag{15.2}
$$

In a similar fashion the third-order differential equation

$$\frac{d^3x}{dt^3} = -kv \tag{15.3}$$

becomes

$$
\begin{aligned}
v &= \frac{dx}{dt} \\
a &= \frac{dv}{dt} \\
\frac{da}{dt} &= -kv
\end{aligned}
\tag{15.4}
$$

Euler's method is in each case applied to the resulting first-order equation system.

In the Euler method, the first-order derivatives are replaced by their forward finite difference approximations $D^+_{\Delta t}$ to yield

$$
\begin{aligned}
\frac{x(t+\Delta t) - x(t)}{\Delta t} &= v(t) + O(\Delta t) \\
\frac{v(t+\Delta t) - v(t)}{\Delta t} &= -\frac{k}{m}x(t) + O(\Delta t)
\end{aligned}
\tag{15.5}
$$

or equivalently

$$
\begin{aligned}
x(t+\Delta t) &= x(t) + \Delta t\, v(t) + O(\Delta t)^2 \\
v(t+\Delta t) &= v(t) - k\Delta t\, x(t)/m + O(\Delta t)^2
\end{aligned}
\tag{15.6}
$$

Note that multiplying the error terms represented by $O(\Delta t)$ in Eq. (15.5) by Δt generates $O(\Delta t)^2$ terms in the above equation. Motivated by the simple form of the above equations, we generate code that as closely as possible resembles the algebraic expressions. For the values $k = m = 1$ and Δt =0.06 this yields

```
main() {
        double x[100], v[100], k=1, m=1, dt=0.06;
        x[0] = -1;
        v[0] = 0;     // 2nd-order equation → 2 boundary conditions
        for (int i = 1; i < 100; i++) {
              x[i] = x[i - 1] + dt * v[i - 1];
              v[i] = v[i - 1] - k * dt * x[i - 1] / m;
        }
}
```

Note that if we apply the Euler method to the differential equation $dx/dt = f(t)$, each evaluation of $x(t + \Delta t)$ adds an increment given by $f(t)\Delta t$ to the previous value of $x(t)$. This corresponds to the rectangular rule of the previous chapter for evaluating the integral of $f(t)$. Thus, the Euler procedure can be viewed as a generalized integration technique that can be applied directly to any differential equation.

The second derivative term in Newton's equations can also be programmed directly without first generating a system of first-order equations. This requires the second finite difference operator approximation to the continuous second derivative operator d^2/dt^2 that is defined as follows

$$\begin{aligned}(D_{\Delta t})^2 f(t) &\equiv \frac{1}{\Delta t}\left(D_{\Delta t}\Big|_{t+\Delta t/2} - D_{\Delta t}\Big|_{t-\Delta t/2}\right) f(t) \\ &= \frac{1}{(\Delta t)^2}\left([f(t+\Delta t) - f(t)] - [f(t) - f(t-\Delta t)]\right) \\ &= \frac{1}{(\Delta t)^2}\left(f(t+\Delta t) - 2f(t) + f(t-\Delta t)\right) \end{aligned} \tag{15.7}$$

Inserting this expression directly into Eq. (15.1) yields a solution algorithm for $x(t+\Delta t)$ in terms of the two initial conditions $x(t)$ and $x(t-\Delta t)$.

15.2 Error analysis

While Eq. (15.6) yields a result for $x(t+\Delta t)$ that is accurate to $O(\Delta t)^2$, typically the particle position is computed at a propagation time T, requiring $T/\Delta t$ evolution steps. If the error per step is $O(\Delta t)^2$, the total error is therefore $O(\Delta t)^2(T/\Delta t)$ or $O(\Delta t)$. This reduction of the error order of course precisely parallels that arising in numerical integration for which the error is one order lower than that associated with each integration interval.

A major drawback of the Euler method is that the numerical solutions of non-dissipative problems generally diverge as the propagation time $T \to \infty$, although the rate of divergence decreases rapidly with the time step size Δt. Such a divergence is evident in Fig. 15.1, obtained from Eq. (15.6), of the velocity v against the position x for the boundary conditions $v(0) = 1, x(0) = 0$ for 80 steps of size $\Delta t = 2\pi/80$.

The spring evidently gains energy after each oscillation. This behavior is explained physically as follows. As the particle initially propagates from $x = 0$ to $x = 1$, the position at which the force is evaluated, $x(t_i)$, is closer to the origin than the average displacement, $x(t_i + \Delta t/2)$, of the spring during the propagation interval. Similarly, the value of the velocity variable in the first line of Eq. (15.5) or of Eq. (15.6) that drives the change in the displacement is larger than its properly averaged effective value over the time interval. Accordingly, the restoring force is fictitiously weakened but the velocity is increased, so that a weakened force acts on the particle a shorter amount of time, causing the calculated displacement to overshoot its true maximum value of 1.0. In contrast, as the particle returns from its point of maximum excursion back to zero displacement, the returning force is instead overestimated and the magnitude of the negative driving velocity

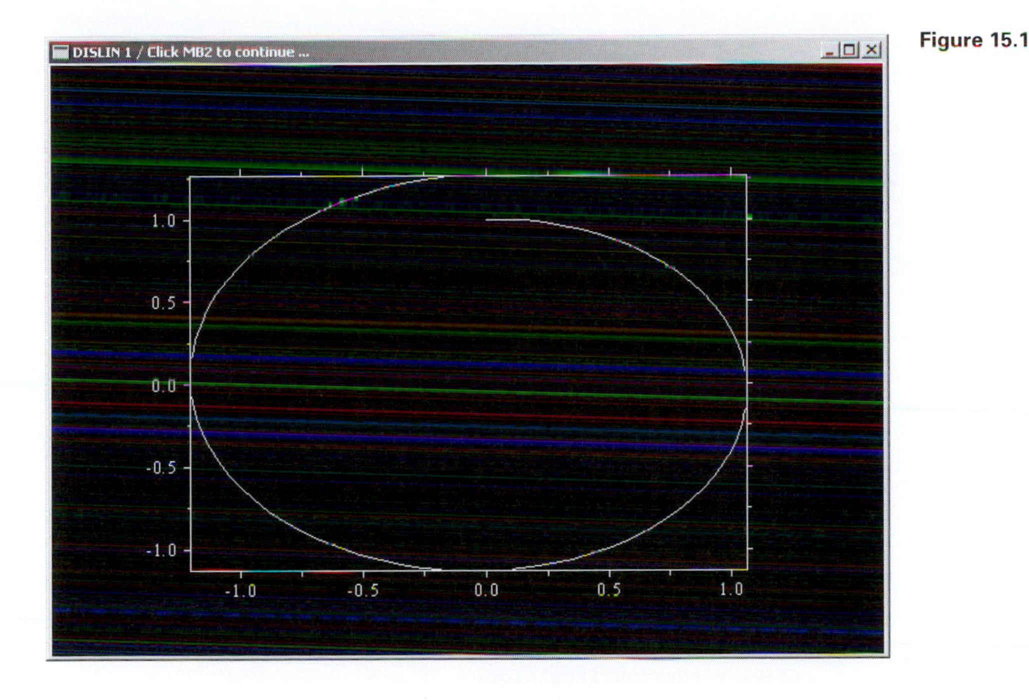

Figure 15.1

term underestimated. Therefore, over this quarter-cycle a stronger restoring force acts over a longer time, causing an overshoot in the magnitude of the negative velocity after a half period, as again evidenced in Fig. 15.1. These errors generate a fictitious numerical driving force in perfect resonance with the motion, yielding the observed energy divergence.

A procedure that eliminates this divergence balances the systematic error in the force in the second line of Eq. (15.6) against that of the velocity term in the first line. This is accomplished simply by replacing $v(t)$ in the first line by $v(t + \Delta t)$, so that these two errors have opposite signs. This yields the modified equation system

$$\begin{aligned} x(t + \Delta t) &= x(t) + \Delta t\, v(t + \Delta t) + O(\Delta t)^2 \\ v(t + \Delta t) &= v(t) - k \Delta t\, x(t)/m + O(\Delta t)^2 \end{aligned} \tag{15.8}$$

which can be shown through algebraic manipulations to be energy-conserving.

15.3 The spring class

Finally, we develop an object-oriented implementation of the differential equation solver. We first introduce a **Spring class** that encapsulates the relevant physical properties of a physical spring. We select the spring constant and the mass of the attached particle along with its position and velocity as the internal **Spring** class variables, although as noted in Chapter 6, other choices are possible

```
class Spring {
        public:
        float iK, iM, iPosition, iVelocity;
        Spring(float aK, float aM, float aPosition, float
          aVelocity) : iK(aK), iM(aM), iPosition(aPosition),
          iVelocity(aVelocity) {}
};
```

We also define a separate **Trajectory** class that records and manages the position and velocity data associated with the motion of a **Spring** object. This class also is responsible for plotting the particle path. By overloading the **propagate( )** function, the **Trajectory** class could as well be used to store and process the trajectories of different types of physical objects

```
class Trajectory{
        float iX[100], iV[100], iSteps;
public:
        void propagate (Spring aSpring, int aSteps, float
            aDt) {
        iSteps = aSteps;
        iX[0] = aSpring.iPosition;
        iV[0] = aSpring.iVelocity;
        for (int i = 1; i < aSteps; i++) {
                iX[i] = aSpring.iPosition +
                        aSpring.iVelocity * aDt;
                iV[i] = aSpring.iVelocity -
                        aSpring.iK / aSpring.iM *
                        aSpring.iPosition *aDt;
                aSpring.iPosition = iX[i];
                aSpring.iVelocity = iV[i];
            }
        }
        void plot() {
                metafl("XWIN");
                qplot(iX ,iV ,iSteps);
        }
};
```

The **main( )** function for this **Spring** implementation is then

```
#include <math.h>

main() {
        Spring S1(1.0, 1.0, 1.0, 0.0);
        Trajectory T1;
        T1.propagate(S1, 80, 2*M_PI/80);
        T1.plot( );
}
```

15.4 Assignments

(1) Demonstrate by regenerating the graph of Fig. (15.1) that the modified Euler equations of Eq. (15.8) are energy conserving.

(2) Program the differential equation solver associated with Eq. (15.7). Using the analytic formula for the motion of a spring to determine its amplitude at time $-\Delta t$, regenerate the graph of Fig. 15.1 using this alternative procedure (using the same input data).

(3) Apply the Euler method to a ball thrown vertically upward in the presence of air resistance such that the force on the ball is given in newtons by

$$\vec{F} = (-9.8m - 0.1v)\hat{z}$$

where m is the mass of the ball and v is the velocity. Assuming that $m = 1$, $v_0 =$ 100m/s, $z_0 = 0$, apply Euler's method and DISLIN to generate a curve of the vertical position of the ball as a function of time from $t = 0$ to $t = 40$ s. Submit your program and two graphs, the second graph should be obtained using half the time step of the first graph in order to demonstrate that your result has converged (find an appropriate time step that does this). In your program first use $v[i-1]$ in the line of code that determines $z[i]$ in your program and then replace $v[i-1]$ in this line by $v[i]$ and compare the accuracy of the two methods by writing out the difference in the value of z at the last point for both methods when the time step is halved.

(4) Write a program based on the Euler method that calculates x as a function of t for a (highly unphysical) system described by the differential equation (note the *third* derivative operator)

$$\frac{d^3x}{dt^3} = -3x - 0.1a(t) + 0.02\cos(t)$$

where $a(t)$ is the acceleration together with specified initial conditions at time zero (give an example of one possible set of values for these initial conditions in your program). You should input a time step length of 0.001 and the total number of time steps (1000) from the keyboard and output the position at the final time, $t = 1$, to the terminal.

(5) The Duffing oscillator is a damped, nonlinear and periodically forced pendulum described by the differential equation

$$\frac{d^2x}{dt^2} = -\left(ax^3 + bx + c\frac{dx}{dt}\right) + F\cos(t)$$

Using the Euler method presented in class, write a program that yields a plot of the velocity versus the position for a Duffing oscillator with $a = 1, b = 0, c = 0.3, F = 10$ for 40,000 time steps of length $\Delta t = 0.03$ and the starting values (initial conditions) $x_0 = 1$ and $v_0 = 6$. (See what happens if you change these values.) In your program first use $x[i-1]$ in the line of code that determines $v[i]$ in your program (that is the line corresponding to the last line in the program on p. 197) and then replace all instances of $x[i-1]$ in this line by $x[i]$ and indicate which of these two procedures runs correctly.

To generate an acceptable plot you must plot the values that you obtain as small points rather than lines. This can be done with the following code. Again submit your program together with the output

```
float minX,maxX,minY,maxY,stepX,stepY;
metafl("XWIN");
disini();
setscl(x, numberOfSteps, "X");
setscl(v, numberOfSteps, "Y");
incmrk(-1);             // This plots markers at each point and
                        // suppresses lines between points
marker(21);             // The marker type (here filled circles)
hsymbl(2);              // The marker size (very small)
graf(minX, maxX, minX, stepX, minY, maxY, minY, stepY);
curve(x,v,numberOfSteps);
endgrf;
disfin();
```

Chapter 16
Linear algebra

Non-trivial matrix problems typically arise in scientific applications when modeling the evolution of a system with several degrees of freedom or in calculating the states of a system that is subjected to a global constraint or set of constraints. This chapter considers two fundamental matrix operations that occur in such problems, namely linear equation and eigenvalue solvers. Since matrix calculations are, however, complex and subject to subtle errors, established library routines are widely employed. Our discussion, therefore, focuses on underlying principles and on the most transparent and rapid techniques.

16.1 Linear equation solvers

We first consider the matrix equation $\mathbf{Ac} = \mathbf{b}$ and summarize the simplest solution method, namely Gaussian elimination. Labeling the elements of the matrix $\mathbf{A}$ by A_{ij} and the elements of the input and output column vectors $\mathbf{b}$ and $\mathbf{c}$ by b_i and c_i, for an $N \times N$ matrix $\mathbf{A}$, the matrix equation represents N linear equations for the N variables, c_i

$$\sum_{j=0}^{N-1} A_{ij} c_j = b_i \tag{16.1}$$

where the A_{ij} and the b_i are given. The Gaussian elimination procedure outlined below subtracts pairs of scaled equations from each other in order to recast this original set of equations as a triangular equation system. The last equation in this new set is a trivial expression of the form $A^{(N-1)}_{N-1,N-1} c_{N-1} = b^{(N-1)}_{N-1}$ that is immediately solved for c_{N-1}. Subsequently in a "back substitution" step the preceding equation, which contains only the two dependent variables c_{N-1} and c_{N-2}, is solved for c_{N-2} and so on.

A simple practical example serves to illustrate the procedure. Consider the problem

$$\begin{pmatrix} 1 & 2 \\ 3 & 1 \end{pmatrix} \begin{pmatrix} c_0 \\ c_1 \end{pmatrix} = \begin{pmatrix} 1 \\ 2 \end{pmatrix} \tag{16.2}$$

which corresponds to the set of equations

$$\begin{aligned} c_0 + 2c_1 &= 1 \\ 3c_0 + c_1 &= 2 \end{aligned} \tag{16.3}$$

Multiplying the first equation in this set by 3 and subtracting it from the second equation yields the triangular system

$$\begin{aligned} c_0 + 2c_1 &= 1 \\ -5c_1 &= -1 \end{aligned} \tag{16.4}$$

In the back substitution step, the second of these equations is solved for c_1 and the result inserted into the first equation to obtain c_0.

In the general case, we subtract A_{i0}/A_{00} times the first row of Eq. (16.1) from row i. This generates a new equation system for the c_j with elements

$$A_{ij}^{(1)} = A_{ij} - A_{0j}\frac{A_{i0}}{A_{00}}, \quad b_i^{(1)} = b_i - b_0\frac{A_{i0}}{A_{00}} \tag{16.5}$$

Repeating for each row $i \neq 0$ eliminates the first element of each row but the first. We then repeat this procedure within the submatrix formed by excluding the first row from the linear equation system. After $N-1$ iterations, we arrive at the triangular equation system

$$\begin{aligned} A_{00}c_0 + A_{01}c_1 + \ldots + A_{0N-1}c_{N-1} &= b_0 \\ A_{11}^{(1)}c_1 + \ldots + A_{1N-1}^{(1)}c_{N-1} &= b_1^{(1)} \\ &\vdots \\ A_{N-1,N-1}^{(N-1)}c_{N-1} &= b_{N-1}^{(N-1)} \end{aligned} \tag{16.6}$$

To implement back substitution the last, trivial, equation is solved and the result for c_{N-1} is inserted into the previous equation; subsequently the values for both c_{N-1} and c_{N-2} are inserted in the third from the last equation and the process is repeated until the first equation is reached. This yields, for example

$$\begin{aligned} C_{N-1} &= b_{N-1}^{(N-1)} \Big/ A_{N-1,N-1}^{(N-1)} \\ C_{N-2} &= \left(b_{N-2}^{(N-2)} - A_{N-2,N-1}^{(N-2)}c_{N-1}\right) \Big/ A_{N-2,N-2}^{(N-2)} \\ C_{N-3} &= \left(b_{N-3}^{(N-3)} - A_{N-3,N-2}^{(N-3)}c_{N-2} - A_{N-3,N-1}^{(N-3)}c_{N-1}\right) \Big/ A_{N-3,N-3}^{(N-3)} \end{aligned} \tag{16.7}$$

The code for Gaussian elimination is presented below

```
#include <stdio.h>
const int n = 2;

// Note : both aA and aB are overwritten
void gauss(double aA[ ][n], double aC[ ], double aB[ ]) {
   // Forward elimination
   for ( int i = 0; i < n; i ++ ) {
      if ( !aA[i][i] ) exit(0);
```

```
    for ( int j = i + 1; j < n; j++ ) {
        double d = aA[j][i]/aA[i][i];
        for ( int k = i + 1; k < n; k++ ) aA[j][k]
                -= d*aA[i][k];
        aB[j] -= d*aB[i];
    }
  }

  if ( !aA[n-1][n-1] ) exit(0);

  // Back substitution
  for ( int i = n - 1; i >= 0; i-- ) {
        aC[i] = aB[i];
        for ( int j = i + 1; j < n; j++ )
                aC[i] -= aA[i][j] * aC[j];
        aC[i] /= aA[i][i];
  }
}

main() {
  double a[n][n] = {{1, 2}, {3, 8}};
  double b[n] = {2, 5};
  double c[n];
  gauss(a, c, b);
  cout << c[0] << '\t' << c[1] << endl;
}
```

Since the second derivative operator, which appears in key physical equations such as the diffusion equation, wave equation, and Schrödinger's equation, is represented by the three-point formula of (15.7), the specialization of the above routine to tridiagonal matrices is particularly useful. To preserve memory space, the diagonal and the upper and lower co-diagonals of the matrices are normally stored as three separate arrays of dimensions N, $N-1$ and $N-1$, respectively. This yields:

```
void tridiagonalSolver(double aLowerCodiagonal[ ], double
        aDiagonal[ ], double aUpperCodiagonal[ ], double
        aInputVector[ ], double aOutputVector[ ], double
        aScratch[ ], int aNumberOfPoints) {
    // Forward elimination
    double bsave = aDiagonal[0];
    if ( !bsave ) exit(0);
    aOutputVector[0] = aInputVector[0] / bsave;
    for ( int loop = 1; loop < aNumberOfPoints; loop++ ) {
        aScratch[loop] = aUpperCodiagonal[loop - 1] / bsave;
        bsave = aDiagonal[loop] - aScratch[loop] *
                aLowerCodiagonal[loop - 1];
        if ( !bsave ) exit(0);
        aOutputVector[loop] = (aInputVector[loop] -
                aLowerCodiagonal[loop - 1] *
                aOutputVector [loop - 1]) / bsave;
    }
```

```
    // Back substitution
    for ( int loop = aNumberOfPoints - 2; loop > -1; loop-- )
        aOutputVector[loop] -= aScratch[loop + 1] *
            aOutputVector [loop + 1];
}

main() {
    double diagonal[2] = {1, 8};
    double upperCodiagonal[1] = {2};
    double lowerCodiagonal[1] = {3};
    double inputVector[2] = {2, 5};
    double outputVector[2];
    double scratch[2];
    tridiagonalSolver(lowerCodiagonal, diagonal, upperCodiagonal,
        inputVector, outputVector, scratch, 2);
    cout << outputVector[0] << '\t' << outputVector[1] << endl;
}
```

Often the diagonal elements of the tridiagonal matrix are known to be non-zero and the check for non-zero values of **bsave** can therefore be omitted. If the upper and lower co-diagonals are identical, equal to a fixed value, or unchanged when the program is run multiple times, the above code can be further modified to improve execution speed.

16.2 Errors and condition numbers

Numerical errors in Gaussian elimination principally arise from two sources. The first of these occurs when, for example, $|A_{00}| << |A_{i0}|$ in Eq. (16.5) so that the number of significant digits retained from A_{ij} is small after $A_{0j}A_{i0}/A_{00}$ is subtracted. This error, which can arise at each forward step of the algorithm, can be eliminated by pivoting. This refers to interchanging the order of rows in the equation system at each forward step to minimize the magnitude of the subtracted values.

A second, far more problematic source of error results if the equation system contains equations that are nearly linearly dependent; that is, if multiplying a number of equations by different coefficients and summing yields an approximation to another equation in the set. In this case, small changes in the input vector, **b**, yield large variations in the solution vector, **c**. This can be understood geometrically for $N = 2$, as two nearly identical equations describe almost parallel lines. A small change in the equation describing one of these lines resulting from a minor variation in **b** substantially displaces the intersection point of the two lines and therefore leads to a markedly different solution vector, **c**. In the N-dimensional case the solution, which corresponds to the single point of intersection of N hyperplanes with $N - 1$ dimensions, behaves identically.

A test of whether two or more equations in a system of N equations are nearly dependent is performed by computing the *condition number* of the equation system, defined as the ratio of the largest to the smallest of the matrix eigenvalues

(eigenvalues are discussed in the next section). A matrix is singular if its condition number is infinite and ill-conditioned if the condition number is of the order of 10^6 in single precision and 10^{12} in double precision. Routines for evaluating condition numbers can be found in standard numerical libraries and are generally incorporated into matrix solvers.

16.3 Eigenvalues and iterative eigenvalue solvers

In this section, we discuss an iterative procedure for evaluating matrix eigenvalues based on the linear equation solvers of the previous section. Recall first that each solution Θ_i to a matrix equation $\mathbf{A}\Theta_i = \lambda_i \Theta_i$ is called an eigenvector, while the corresponding constants λ_i are termed the eigenvalues. If $\mathbf{A}$ is an $N \times N$ non-singular real-valued matrix, it possesses N linearly independent real or complex eigenvalues and eigenvectors. The eigenvectors then span the N-dimensional space of vectors $\mathbf{x}$; that is, any N-component vector can be written as a linear combination of the eigenvectors.

While the computation of the exact eigenvalues of a matrix is somewhat complex, an iterative procedure is fast and simple to program if only a small number of eigenvectors and eigenvalues are desired. In particular, we repeatedly solve the equation system

$$(\mathbf{A} - \lambda^{(i)}\mathbf{I})\mathbf{x}^{(i+1)} = \mathbf{x}^{(i)} \tag{16.8}$$

where $\mathbf{I}$ is the identity matrix and $\lambda^{(i)}$and $\mathbf{x}^{(i)}$ are the ith estimate for the eigenvalue and its corresponding eigenvector. The vector components of the initial estimate, $\mathbf{x}^{(0)}$, can generally be chosen randomly, although $\lambda^{(0)}$ must be closer to the desired eigenvalue than to any other eigenvalue. If we expand the vectors $\mathbf{x}^{(i)}$ and $\mathbf{x}^{(i+1)}$ as a linear combination of the eigenvectors, Θ_k, of $\mathbf{A}$, such that, for example

$$\mathbf{x}^{(i)} = \sum_{k=1}^{N} Y_k^{(i)} \Theta_k \tag{16.9}$$

where $i = 0$ in the first step of the procedure we have

$$\sum_{k=1}^{N} (\lambda_k - \lambda^{(i)}) Y_k^{(i+1)} \Theta_k = \sum_{k=1}^{N} Y_k^{(i)} \Theta_k \tag{16.10}$$

However, since the eigenvectors, Θ_k,are linearly independent, the coefficients of each eigenvector on both sides of the equation must be separately equal. Therefore

$$Y_k^{(i+1)} = \frac{1}{\lambda_k - \lambda^{(i)}} Y_k^{(i)} \tag{16.11}$$

and the relative amplitude of the eigenvector in $\mathbf{Y}^{(1)}$ that is closest to the estimated eigenvalue will be enhanced.

The new approximation to the eigenvector can be employed to generate an improved estimate of the eigenvalue. Returning to (16.8) and taking the product of both sides with the transpose of the column vector $\mathbf{x}^{(i+1)}$, we obtain

$$(\mathbf{x}^{(i+1)})^{\mathrm{T}}\mathbf{A}\mathbf{x}^{(i+1)} - \lambda^{(i)}(\mathbf{x}^{(i+1)})^{\mathrm{T}}\mathbf{x}^{(i+1)} = (\mathbf{x}^{(i+1)})^{\mathrm{T}}\mathbf{x}^{(i)} \tag{16.12}$$

Since $\mathbf{x}^{(i+1)}$ is increasingly dominated by the eigenvector closest to the estimate $\lambda^{(i)}$ and, therefore, the product $\mathbf{A}\mathbf{x}^{(i+1)}$ is approximated by the corresponding eigenvalue multiplied by $\mathbf{x}^{(i+1)}$, we can establish our subsequent estimate for the eigenvalue by setting $(\mathbf{x}^{(i+1)})^{\mathrm{T}}\mathbf{A}\mathbf{x}^{(i+1)}$ to $\lambda^{(i+1)}(\mathbf{x}^{(i+1)})^{\mathrm{T}}\mathbf{x}^{(i+1)}$ This yields

$$\lambda^{(i+1)} = \lambda^{(i)} + \frac{(\mathbf{x}^{(i+1)})^{\mathrm{T}}\mathbf{x}^{(i)}}{(\mathbf{x}^{(i+1)})^{\mathrm{T}}\mathbf{x}^{(i+1)}} \tag{16.13}$$

In all such iterative methods the amplitude of $\mathbf{x}^{(i)}$ will grow or decay exponentially with the number of iteration steps, i, and therefore must be periodically renormalized to, for example, unity. In the sample program below this operation is performed after every iteration step

```
const int matrixSize = 2;

void gauss(double aA[][matrixSize], double aC[], double aB[]) {
// ...insert rest of gauss routine
}

// Renormalize vector to unit magnitude
void normalize(double aVector[ ]) {
  double normalizationConstant = 0;
  for ( int loop = 0; loop < matrixSize; loop++ )
    normalizationConstant += aVector[loop] * aVector[loop];
  normalizationConstant = sqrt(fabs(normalizationConstant)) *
    normalizationConstant / fabs(normalizationConstant);
  for ( int loop = 0; loop < matrixSize; loop++ )
    aVector[loop] /= normalizationConstant;
}

double eigenvalueSolver (double aMatrix[ ][matrixSize],
    double aEstimate, double aResult[ ],
    double aInitialEigenvector[ ], double aError) {
  double scratchMatrix[matrixSize][matrixSize];
  double scratchEigen[matrixSize];
  double eigenvector[matrixSize];

  for ( int loop = 0; loop < matrixSize; loop++ )
    eigenvector[loop] = aInitialEigenvector[loop];

  // A matrix, scratchMatrix, equal to A - \lambda_i is defined
  // and the unnormalized eigenvector estimate, eigenvector, is
  // copied to scratchEigen. The number of iterations = 25.
  for ( int loop = 0; loop < 25; loop++ ) {
    for ( int loopInner1 = 0; loopInner1 < matrixSize;
       loopInner1++ ) {
      for ( int loopInner2 = 0; loopInner2 < matrixSize;
       loopInner2++ )
```

```
      scratchMatrix[loopInner1][loopInner2] =
          aMatrix[loopInner1][loopInner2];
      scratchMatrix[loopInner1][loopInner1] =
          aMatrix[loopInner1][loopInner1]- aEstimate;
      scratchEigen[loopInner1] = eigenvector[loopInner1];
    }

    gauss(scratchMatrix, aResult, scratchEigen);

    // The vector products required in Eq. (16.13) are computed
    // here.
    double innerProduct = 0, norm = 0;
    for ( int loopInner = 0; loopInner < matrixSize; loopInner++ ) {
      innerProduct += aResult[loopInner] * eigenvector[loopInner];
      norm += aResult[loopInner] * aResult[loopInner];
    // The new, unnormalized eigenvector approximation is stored
    // in eigenvector.
      eigenvector[loopInner] = aResult[loopInner];
    }
    // Eq. (16.13) is implemented to find the new eigenvalue
    // estimate. This estimate is placed in aEstimate
    double change = innerProduct / norm;
    aEstimate += change;
    normalize(aResult);
    if ( fabs(change/aEstimate) < aError ) break;
  }
  return aEstimate;
}

main( ) {
  double matrix[matrixSize][matrixSize] = {{1, 2}, {3, 8}};
  double initialEstimate = 0.8;
  double eigenvector[matrixSize] = {.1, -0.8};
  double resultVector[matrixSize];
  double error = 1.e-4;
  double result = eigenvalueSolver(matrix, initialEstimate,
      resultVector, eigenvector, error);
  cout << "The eigenvalue is: " << result << endl;
  cout << "The eigenvector is: " << resultVector[0] << '\t' <<
      resultVector[1];
}
```

16.4 Assignments

(1) Often in the analysis of experimental data, an analytic expression that depends on one or more parameters must be fit to a set of experimental data points. These data, however, are invariably affected by measurement error and, almost never coincide with the assumed expression. The parameters must, therefore, be chosen to minimize the integrated deviation of the assumed curve from the actual data points. Perhaps the most straightforward and effective procedure is the least squares curve fitting method

that can be employed to fit a linear combination of n functions $f_j(t)$ of the form

$$u(t) = \sum_{j=1}^{n} c_j f_j(t)$$

to m observations $u_1, \ldots, u_m$ at times $t_1, \ldots, t_m$ with $m > n$. To estimate the n coefficients $c_1, \ldots, c_n$ a local minimum of the least squares error defined as

$$I(c_1, \ldots, c_n) = \sum_{k=1}^{m} \left(\sum_{j=1}^{n} c_j f_j(t_k) - u_k \right)^2$$

is obtained by requiring that I be unaffected to first order if any parameter c_i is changed. This implies that

$$\partial I / \partial c_i = 0$$

for each c_i which yields the matrix equation **Ac**=**b** with

$$A_{ij} = \sum_{k=1}^{m} f_i(t_k) f_j(t_k), \quad b_j = \sum_{k=1}^{m} f_j(t_k) u_k$$

The problem is accordingly to employ the Gaussian equation solver presented in the text to implement the least squares method to fit the equation

$$u(t) = c_1 + c_2 t + c_3 t^2$$

to the data:

0.0	0.50395
0.3	1.18914
0.8	1.71725
1.1	2.68171
1.4	3.79318
1.7	5.17096
1.9	5.76176

Hand in your program together with the values that you obtained for the fitting parameters and a curve of your result with the data points superimposed (you can use two calls to **curve ( )** in graphics code such as that of Section 12.3).

(2) Solve the linear equation system

$$\begin{pmatrix} 1 & 0 & i \\ -i & 1 & -i \\ -1 & 0 & i \end{pmatrix} \mathbf{X} = \begin{pmatrix} i \\ 0 \\ -i \end{pmatrix}$$

You must rewrite the function **gauss** for complex quantities. To do this, include the **complex.h** header file and replace **double** by **complex <double>** where appropriate. You will also have to define a global variable **CI** = i through the statement **const**

complex<double> CI = complex<double>(0.0, 1.0); and to employ the **abs( )** function, which automatically generates the correct magnitude when applied to complex numbers. Submit your program together with the program output.

(3) Generate an object-oriented implementation of the iterative eigenvalue finding procedure. Your program should give the same output as the program in the text but take the following form

```
main() {
  double matrix[matrixSize][matrixSize] = {{1, 2}, {3, 8}};
  double initialEstimate = 0.8;
  double eigenvector[matrixSize] = {.1, -0.8};
  double resultVector[matrixSize];
  double error = 1.e-4;
  EigenvalueSolver ES(matrix, initialEstimate, resultVector,
       eigenvector, error);
  ES.compute( );
  ES.normalizeEigenvector( );
  cout << ES.toString( );
}
```

(4) Consider a system of pipes linking three input pipes to three output pipes such that the fluid flowing out of the three output pipes is given in terms of the fluid entering the three input pipes in the following manner

$$
\begin{aligned}
f_1^{out} &= \frac{1}{6}f_1^{in} + \frac{1}{3}f_2^{in} + \frac{1}{2}f_3^{in} \\
f_2^{out} &= \frac{1}{3}f_1^{in} + \frac{1}{3}f_2^{in} + \frac{1}{2}f_3^{in} \\
f_3^{out} &= \frac{1}{2}f_1^{in} + \frac{1}{3}f_2^{in}
\end{aligned}
$$

That is, one-sixth of the water flowing into the first input pipe emerges from the first output pipe as does one-third of the water flowing into the second input pipe and half of the water flowing into the third input pipe.

Now suppose that 4/3 liters/second of water is observed to emerge from the first output pipe, 3/2 liters/second from the second output pipe, and 7/6 liters/second from the third output pipe. Using the Gaussian elimination solver, determine the rate of water flow into each of the input pipes.

(5) Kirchhoff's laws in a resistor network state that the sum of the voltage drops around any closed loop is zero and that the sum of the currents flowing into a junction is zero. Consider the following set of equations that arise from Kirchhoff's laws, where the unit of the first 5 is volts, the coefficients of the currents are in ohms, and the currents are given in amperes

$$
\begin{aligned}
5 - 3i_1 - 6(i_1 - i_2) &= 0 \\
6(i_1 - i_2) - 10i_2 - 5(i_2 + i_3) &= 0 \\
-5(i_2 + i_3) - 3i_3 &= 0
\end{aligned}
$$

Draw the electrical circuit and indicate the location and direction of the currents i_1, i_2, and i_3 that give rise to the above equations and then solve them using the Gaussian matrix solution procedure. Hand in the program, your circuit diagram, and your output.

(6) Linear equation solvers can be used to balance chemical reaction formulas. As a example, consider the decomposition reaction, where the four coefficients $\alpha, \ldots, \delta$ are unknown

$$\alpha\, O_2 + \beta\, C_4H_9NH_2 \rightarrow \gamma\, CO_2 + \delta\, H_2O + N_2$$

Since the number of atoms of each species must be the same on both sides of the formula, the above expression maps into the set of four linear equations

$$\begin{aligned} &\text{O:} \quad 2\alpha = 2\gamma + \delta \\ &\text{C:} \quad 4\beta = \gamma \\ &\text{H:} \quad 11\beta = 2\delta \\ &\text{N:} \quad \beta = 2 \end{aligned}$$

which in turn produce the linear equation system

$$\begin{pmatrix} 2 & 0 & -2 & -1 \\ 0 & 1 & 0 & 0 \\ 0 & 4 & -1 & 0 \\ 0 & 11 & 0 & -2 \end{pmatrix} \begin{pmatrix} \alpha \\ \beta \\ \gamma \\ \delta \end{pmatrix} = \begin{pmatrix} 0 \\ 2 \\ 0 \\ 0 \end{pmatrix}$$

We solve this system for the variables $\alpha, \ldots, \delta$ and insert the values into the formula for the reaction (the first equation of the problem). Finally, this formula is multiplied by the smallest integer that yields integer values for all five coefficients (recall that to balance a chemical formula, the coefficient of one of the reactants is set to unity and the final results for the coefficients are multiplied by the smallest constant that makes all coefficients integers).

In this problem, first use the **gauss( )** routine to balance the chemical reaction following the series of steps described above. Next, generate an object-oriented implementation of the procedure. Here the input will take the form of vectors such that for this example the program should write prompts to and then read data from the keyboard as follows

```
Input the number of different atoms present: 4
Input the number of input reactants:         2
Input the number of output reactants:        3
Input the formula for input reactant 1:      2, 0, 0, 0
Input the formula for input reactant 2:      0, 4, 11, 1
Input the formula for output reactant 1:     2, 1, 0, 0
...
```

The main( ) program should contain the lines

```
EquationSolver ES(inputMatrix, numberOfInputReactants,
   outputMatrix, numberOfOutputReactants, numberOfAtoms);
ES.assembleEquationMatrix( );
```

```
ES.solveLinearEquation( );
ES.multiplyResultByConstant( );
ES.printResult( );
```

In reserving space for the input and output matrices assume that the largest possible problem that you need to consider has 10 atoms and 10 input and 10 output reactants. Hand in your program and a screenshot of the input and output.

Part III

Advanced object-oriented programming

Chapter 17
References

Reference and pointer variables differ fundamentally from the standard variable types in that they store memory locations rather than values. In the case of a reference, the memory location is fixed to that of a preexisting variable so that the reference variable and the variable corresponding to the stored memory address are effectively identical. The address stored in a pointer however can be changed arbitrarily at runtime. Since pointers and references present considerable conceptual and practical challenges to the programmer, this and the subsequent chapter examine these topics individually and in depth.

17.1 Basic properties

A *reference* variable provides an *alias*, which means an alternative name or a nickname, for an existing variable. Numerous examples of alternative names can be found in high-level computer applications, for example the cell "A1" of a spreadsheet can be assigned a second name such as "grade." Setting grade = 5 then has the same effect as writing A1 = 5. The C++ implementation of a reference variable takes the form

```
int A1 = 0;
int & grade = A1;
grade = 5;
cout << A1 << endl;        // Output: 5
```

Since a reference variable is simply a new name for the variable to which it is assigned, the two variables must possess identical properties. Therefore a reference *cannot be reassigned to a new variable* and *must be initialized to a variable of exactly the type stated in its definition*. For example, declaring **double &grade = A1;** yields a compiler error in the code above. A further implication is that, since a reference is an alternative name for a *variable*, it cannot be initialized to a constant.

Memorizing the above statements will prevent countless programming errors.

Since a reference variable is simply an alternative name for a compiler-allocated variable that occupies a fixed memory location, a reference, once defined, can be employed in the same manner as an ordinary variable.

17.2 References as function arguments

The obvious question is why is a reference a valuable concept? Primarily, references are employed to prevent the automatic copying of function parameters into new memory space. Recall that when a function in C++ is invoked, memory space is allocated by the operating system for its arguments that is separate from the memory space assigned to the parameter variables in the calling statement. The function then copies the values of the parameters into the newly reserved locations for the function arguments and operates on these copies. Thus, if the values of function arguments are changed in the function body, the corresponding function parameters in the calling block remain constant.

If a function could only change variables in the calling block through its single return value it could not simply implement operations, such as switching the values of two variables. As we have already encountered in discussing array parameters, however, C++ circumvents this limitation for certain variable types by copying the *address* of the variable's memory location in the calling block to the function memory space rather than the *contents* of the memory location itself. Once the function receives this information, it can access the value stored at the address and thus modify the original variable in the calling block. If the memory location in the variable passed to the function can be reassigned, the variable that holds this information is termed a pointer. If, however, the memory location information is fixed in the same manner as the memory location of a standard variable, the associated variable is, as noted above, a reference (or an array member).

An example of a reference function argument is:

```
void zero( int &aI ) { aI = 0; }

main() {
	int i = 4;
	zero( i );
	cout << i << endl; }		// Output: 0
```

Recall that through the act of reserving memory space for its arguments and copying the parameter values into this space, when a function is called, it first allocates (defines) and initializes the variables in its argument list. Therefore the initial statement that is implicitly executed by the function is

```
int &aI = i;
```

Otherwise, the reference variable **aI** would be illegally created without having been initialized. Accordingly, *a reference parameter in a function cannot have a* *constant or a variable of a different type as a calling argument*, and the following statements yield compile-time errors

```
double d = 4.0;
zero ( d ); // Error: an int reference cannot be set to a double
zero ( 4 ); // Error: a reference variable cannot be set to a
            // constant
```

A useful application of the feature that a reference must be initialized to a variable of the same type occurs in passing arrays to functions. If the function signature contains a reference to an array with, for example, 20 elements as in:

```
void print( int (&aI)[20] ) { cout << aI[6] << endl; }
```

the function can only be called with a parameter of matching type, namely a 20 element array of integers. Therefore, the size of the array in **print( )** is predetermined and does not have to be passed as a separate function argument.

17.3 Reference member variables

If two physical objects share a common, independently accessible component, the variable corresponding to the component can be typed as a reference member variable. In the same manner as a **const** internal variable, cf. Section 7.5, reference variables must be initialized in the initializer list prior to the body of the class constructors. As an example, suppose that several **PositionSensor** objects share a single **Printer** subcomponent so that a change to the **Printer** object settings affects the behavior of all **PositionSensor** objects identically. This can be modeled by creating a single **Printer** object in, for example, the **main( )** function and including this object by reference in each **PositionSensor**. Schematically, we have

```
class Printer {
        public:
        int iPrinterSetting;
};

class PositionSensor {
        public:
                Printer& iPrinter;
                PositionSensor ( Printer &aPrinter ) :
                        iPrinter(aPrinter){}
};

main() {
        Printer P1 = {3};
        PositionSensor PS1( P1 );
        PositionSensor PS2( P1 );
        P1.iPrinterSetting = 5;
        cout << PS1.iPrinter.iPrinterSetting << " " <<
```

```
        PS2.iPrinter.iPrinterSetting << endl; // Output: 5 5
}
```

A change to the properties of the common **Printer** reference variable **P1** then simultaneously propagates to all **Printer** objects.

17.4 *const* reference variables

A reference variable may be declared **const**, in which case the value of the variable that it is defined to equal cannot be changed by assigning a new value to the *reference* variable. This definition permits a **const** reference to be assigned to a non-**const** variable. Seen from the viewpoint of our spreadsheet example, such a construct corresponds to assigning a new name to a spreadsheet shell that permits read-only access; that is the cell contents can only be viewed but not modified through the name. On the other hand, a non-**const** reference generally cannot access a **const** variable since this would negate the functionality of the original **const** variable definition. These considerations are illustrated below, where the first reference definition leads to a compile-time error or, in some compilers such as Borland C++, a compile-time warning, while the second is correct

```
const int i = 4;
int k;
int &j = i;                    // Compile-time error or warning
const int &l = k;              // Valid
```

A non-**const** reference, such as **j**, can, however, always be assigned to a **const** variable through a **const_cast** operation, such as **int &j = const_cast<int &> (i);**.

That a **const** reference can be assigned to a non-**const** variable enables the programmer to avoid the overhead of copying function arguments, while still preventing changes to these arguments from within the function body. A **const** variable in the calling block cannot, however, at least in principle be passed to a function with a non-**const** argument – again, since the first action taken when the function is invoked is to define and initialize the non-**const** parameter with the **const** reference argument. Accordingly, the first function call in the **main( )** program below is valid, while the second potentially yields a compile-time error (if only a warning message is generated the output is of course 3 3)

```
void pr (const int &aI) { cout << aI << " ";}

void pr2 (int &aI) { cout << aI << " "; }

main () {
      const int i = 3;
      pr(i);  //Output: 3
      pr2(i); //Potential error: should not cast const int to int
}
```

17.5 Reference return values

A difficult but useful C++ construct occurs when a function returns a reference variable; that is, when the return type is a reference such as **int & myFunction( )**. In this case, the memory location associated with the return variable *inside* the function will be identical to the memory location of the variable to which it is assigned *outside* the function in the calling program. The function can then be either an rvalue or an lvalue but in the latter case a *variable* of the same *type* as the return variable must appear on the right side of the assignment operator. As an example, we construct a function that can be used to set elements of two arrays, **a** and **x**, equal at the same time as the array index is shifted by one (so that the starting array elements are accessed as **shift(1,a)** and **shift(1,x)**

```
double& shift(int aI, double aV[ ]) {
        return aV[aI - 1];
}                                          // shift(1, a) is a[0]

main() {
        double target[2];
        double source[2] = {0, 1};
        shift(2, target) = shift(2, source);  // Sets target[1] =
                                              // source[1]
        double result = shift(2, target);
        cout << result << endl;               // Output: 1
}
```

To understand the program, observe that since **target** is an array parameter in the statement **shift(2, target) = shift(2, source);** the argument **aV** in the body of **shift(2, target)** occupies the same memory location as the array variable **target** in **main( )**. Similarly, **aV** in **shift(2, source)** coincides with **source**. However, as **aV[1]** in **shift(2, target)** is returned as a reference variable, it is an alternate name for the return value, **aV[1]**, in the function **shift(2, source)**. Since these **aV[1]** elements also represent **target[1]** and **source[1]**, the program has, as desired, equated the second elements of the arrays **source** and **target**.

That a reference return variable enables a function to return the same object as is passed to it through its argument can considerably simplify program syntax. For example, the nested chain of function calls below writes different messages to a single file. The **fstream** object is passed both into and out of the function as a reference variable, which insures that the input and the output refer to the same object

```
#include <fstream.h>

fstream& print( fstream& aStream, char aMessage[80] ) {
        aStream << aMessage << endl;
        return aStream;
}

main () {
```

```
fstream myStream("output.dat");
print( print (myStream, "This is the first line"),
        "This is the second line"); }
```

In the next chapter, we will demonstrate that the above construct can be further enhanced in object-oriented programs through the application of **this** pointer.

17.6 Assignments

Part I

If the programs below contain errors, find these and identify each one as run-type, compile-type, or link-type. If the program will run correctly, give its output.

(1)
```
int& function( int &k, int &j ) {
    int l;
    if ( k < j ) l = k ;
    else l = j;
    return l;
}

main () {
    int k = 2;
    int l = 3;
    cout << function(k, l) << endl;
}
```

(2)
```
int& function( int &j , int &k ) {
    if ( j > k ) return j;
    else return k;
}

main () {
    int k = 4, j = 5;
    function (k, j) = 6;
    cout << k << '\t' << j << endl;
}
```

(3)
```
#include <iostream.h>

void myFunction(int b[ ][3]) {
int (&c)[ ] = b[1];
for (int loop = 0; loop <= 3; loop++)
    cout << c[loop] << '\t';
}

main() {
    int b[3][3];
    int step = 0;
    for (int loopOuter = 0; loopOuter < 3;
                loopOuter++) {
        for (int loopInner = 0; loopInner < 3;
                loopInner++)
            b[loopInner][loopOuter] = step++;
    }
```

```
   myFunction (b);
}
```

(4)
```
main() {
int i = 10;
{
   int i = 15;
   int &j = i;
}
cout << j << endl;
}
```

(5)
```
ostream & f( ) {return cout;}
main () {
   f( ) << "This";
}
```

(6)
```
int j = 0;

class C {
   int iI;
   public:
   int i( ) {return iI; }
   C(int & aK) {iI = aK++ + ++j ;}
 };

main() {
   int i = 1, k = 2;
   C C1(i);
   C C2(i);
   C C3(k);
   cout << C2.i( ) << '\t' << j << endl;
}
```

(7)
```
void myFunction (int &aI ) {
   aI = 4;
}

main () {
   const int a = 3;
   myFunction(a);
   cout << a;
}
```

(8)
```
double myFun(const int &l) {
   int& k = l;
   int m = k;
   return m;
}

int k = 3;

main(){
   float k = 4.5;
   cout << myFun(k);
}
```

(9)
```
int& myFun(int& i){
    int&j = i;
    return j;
}

main(){
    int i = 5;
    int j = 4;
    myFun(i) = j;
    cout << i;
}
```

(10)
```
void f(const int& aI) {
    aI = 3;
}

main () {
    int i = 4;
    f(i);
    cout << i << endl;
}
```

Part II

(11) By using reference return variables, generate a function**, fMin,** of two unequal **double** reference arguments, that sets the smaller of the two arguments to the value of a **double** variable **a** (equal to 10 below) through the following statement

```
double a = 10;
fMin(x, y) = a;
```

(12) Generate a recursive factorial function using a function with the signature

```
int factorial( int & i)
```

Do you get a warning message when compiling your function and if so what is the meaning of this message? How can it be eliminated? Replace the argument int &i with **const int &i**. Does your function still run properly? Explain your result.

(13) Rewrite the **change** function in Section 9.8 using reference variables. Comment on why this is a safer programming construct.

(14) In this problem, you will implement an elementary version of the *simplex* method for finding the extremum of a function. While this procedure is here applied to maximize a function, **parabola**, of only two variables, the advantage of the procedure is that it can be immediately generalized to a function of multiple arguments. Our procedure can be further made far more accurate through several minor modifications that are discussed in numerous textbooks on computer algorithms.

The simplex procedure is constructed as follows, assuming that we seek the maximum of a function. First three noncollinear points called below **iLeast**, **iMiddle**, and **iHighest** are specified; the triangle formed by these points is called a simplex. Next the points are sorted such that **parabola(iLeast) < parabola(iMiddle) < parabola(iHighest)**.

We wish next to substitute a new point for **iLeast** that yields a larger value of **parabola( )**. To do this, we reflect the point **iLeast** through the line segment joining **iMiddle**, and **iHighest** using the formula **iLeast = 2*iCenter – iLeast**, where the variables are treated as vector quantities and **iCenter** is the midpoint of the side of the simplex opposite the point **iLeast**. Repeating the sorting and reflection steps a sufficient number of times yields a value that is close to the point corresponding to the minimum of **parabola( )**. You should program this algorithm by completing the implementation of the program below; hand in both the completed program and the first and last ten lines of your output.

```
#include <fstream.h>
#include <string.h>
#include <strstream.h>

const int numberOfDimensions = 2;

class Point {
public:
        double iCoordinate[numberOfDimensions];
        double coordinate(int aIndex) {
                return iCoordinate[aIndex];
        };
        string toString( );
};
// This is the (global) function to be maximized.
double parabola(Point aPoint) {
        double xShift = aPoint.iCoordinate[0] - 1;
        double yShift = aPoint.iCoordinate[1] - 2;
        return -(xShift * xShift + yShift * yShift);
}

// This global function orders the points aP1 and aP2 so that
// aF(aP1) < aF(aP2).
void sortPoint(double (*aF) (Point), Point& aP1, Point& aP2);

class Simplex {
private:
        double iDelta;
        Point iLeast;
        Point iMiddle;
        Point iLargest;
        Point iCenter;
        double (*iF)(Point);
// The next three functions are helper functions (sortPoint could
// be included here as well)
// Sorts the three points such that iF(iLeast) < iF(iCenter) <
// iF(iLargest)
        void sort( )
// The next function determines the midpoint of the line extending
// between the points iMiddle and iLargest and places this result
// in iCenter.
        void oppositeCenter( );
```

```
// Generates a new estimate of iLeast according to iLeast =
// 2*iCenter - iLeast, where these quantities are taken as
// vectors.
        void reflect( );
public:
// For a number of times given by aMaximumSteps, sort the points,
// find the center of the simplex segment opposite iLeast and
// reflect iLeast through this opposing simplex segment. For
// debugging purposes, you may wish to print out the value of iF
// (iLeast) here.
        void findMaximum(int aMaximumSteps);
// Construct the intial simplex using iLeast = aStart, iMiddle =
// aStart + Delta * xu and iLargst = aStart + yu, where xu and yu
// are unit vectors in the x and y direction respectively.
        Simplex( double (*aF)(Point), double aDelta ,
            double aStart[ ]);
};

main( ) {
        double startingPoint[numberOfDimensions] = {0, 0};
        Simplex S(parabola, 0.1, startingPoint);
        S.findMaximum(100);
}
```

Chapter 18
Pointers and dynamic memory allocation

Pointer variables are a central feature of C and C++ as they enable the contents of any accessible memory location to be addressed directly. To accomplish this, a pointer variable stores a memory address, but, unlike a reference variable, the address can be arbitrarily assigned or reassigned. Further, through a simple operation on the pointer, the value stored at the address can be arbitrarily manipulated.

Unfortunately, introducing variable types with enhanced functionality inevitably leads to new categories of errors. For example, pointers, like reference or array variables, can be employed to implement call-by-reference semantics in function arguments. However, such a construct, by identifying variables in two disjoint blocks, compromises the intent of block structure. That is, if the variable in the function body is assigned an unintended value, the corresponding variable in the calling block, which can be physically far removed from the function block, will change as well, leading to an error that is difficult to detect.

That pointers can be reassigned to new memory locations enables numerous advanced procedures, such as memory allocation at runtime under program control. However, such an operation is far more error prone than pass by reference. Briefly, the program issues a request to the operating system for the allocation of new memory for a variable. If successful, the operating system returns the address of the starting location of the assigned memory. Because the address stored in a pointer can be altered, the running program can then set the value of a preexisting pointer variable to this new memory address. However, if the pointer variable is defined within a block, it will be automatically deallocated when the block terminates. The address to the new memory is then lost leading to memory that cannot later be deallocated. Alternatively, memory can be addressed by two pointers and deallocated through an operation on one of these. The second pointer then contains the address of memory that can be arbitrarily overwritten.

Evidently, the topic of pointers must be learned sequentially and methodically. The student is therefore advised to understand thoroughly the contents of each section in this chapter before proceeding to the next section.

18.1 Introduction to pointers

As noted earlier, a type declaration or definition in C++ establishes the amount of memory space required for a variable and the manner in which the bit pattern stored at this memory location is to be interpreted. A variable of pointer type stores a value interpreted as the *starting memory address of a variable of a specified type*, in other words, *the value of a pointer is a memory address.* That is, a **double** pointer that holds a value of, for example, 80000 interprets the eight-byte region from physical memory location 80000 to location 80007 as the storage location of a **double** variable. The amount of memory space reserved by any pointer variable equals the number of bits required to store a hardware memory address – on a 32-bit machine, an address requires four bytes so that *applying the sizeof( ) function to any pointer yields 4.*

To designate the type of a pointer to a variable of, for example, type **int**, the notation **int** * is employed, where a pointer type is denoted by a star. When multiple pointer variables are defined on the same line, a star must precede each pointer. As a result, the following statement defines **j** to be an **int**, while **i** and **k** are pointers to integers

```
int *i, j, *k;
```

18.2 Initializing pointer variables

Although **i** and **k** are defined by the statement above, they are not yet initialized. Therefore, the 4-byte memory space allocated for these variables (on a 32-bit machine) will contain a random bit pattern and the variables are said to point to random memory locations. The pointers must be set to the memory addresses of actual **int** variables before they can be meaningfully used.

The most conceptually straightforward technique for setting the *value* of the pointer **i** above to a relevant memory location casts a constant value representing the starting memory address of the desired location to an integer pointer. Since the standard in C++ is to represent memory addresses by hexadecimal constants, which are distinguished by preceding the constant value by a **0x** or **0X**, the syntax for this operation is

```
int *i = (int *) 0x0012FF88;                    // 0x = hexadecimal
```

A hexadecimal value can also be read in from the keyboard (although in this case it is not preceded by **0x**) using the following syntax (here the **hex** manipulator can be omitted and the memory address read in as an integer)

```
int j, *i;
cin >> hex >> j;                                // Input: 0012FF88
i = (int *) j;
```

Such a procedure is of limited value since most memory locations of interest in a program are fixed by the operating system at runtime; the only useful static memory locations are those associated with physical devices, such as a sound or video card.

18.3 The address-of and dereferencing operators

Address-of operator: The memory location of a variable allocated when a program is run can be obtained by applying the *address of* operator **&** to the variable (this use of the ampersand symbol has a completely different meaning from the ampersand that appears in reference declarations and definitions). To illustrate, the lines

```
int j;
cout << &j << endl;                    // Output: e.g. 0012FJ4A
```

display the address of the variable **j** on the terminal in standard (hexadecimal) format. Since the **&** operator returns a memory address, it can be employed to initialize a pointer to the address of a second variable

```
int *k = &j;                          // k now contains 0012FJ4A
```

The pointer variable **k** now contains the address of an area of memory storage that the compiler has assigned to hold the **int** variable **j**. *Unlike a reference, the memory address held in a pointer can be later reassigned to the address of a different variable* as in the lines below

```
int l;
k = &l;
```

Dereferencing and pointer-to-member operators: Once a meaningful memory address is stored in a pointer, we generally wish to manipulate the *value stored at this address* through the pointer variable. This value is accessed by applying the dereferencing operator, *, as follows

```
int i = 4;
int *j = &i;
*j = 5;
cout << i << ' ' << *j;                          // Output: 5 5
```

That is, ***j** is an *alias* for the *variable* **i**; it is the **int** variable located at the value of **j**. Evidently, the address of and dereferencing operators are inverse operators; that is, ***(&i)** is the same as **i** itself. For future reference, another procedure for dereferencing the variable **j** above is to write **j[0]**, which is identical to ***j**.

A special form of the dereferencing operator exists for pointers to objects. In particular, consider an object of a class **A** with a member variable **b** and a member function **void f( )**

```
class C {
        public:
        int b;
        void f( ) { cout << "test" << endl; }
};
```

Now suppose that in **main( )** we create a pointer **pC** to a variable of type **C** as follows

```
main() {
      C C1 = { 1 };
      C *pC = &C1;
      cout << (*pC).b << endl;       // Output 1: parenthesis required!
      pC -> f( );                    // Output: test
}
```

If we wish to access the public class member **b** through this pointer, we first dereference **pC** to obtain an alias for the underlying variable (object) **C1** that **pC** points to and then use the member-of operator, **.**, to address the public internal variable **b** of the object. This sequence of operations yields the expression **(*pC).b** appearing in the third line of the **main( )** function body. Here the parentheses are *required* since the precedence of the member-of operator is higher than that of the dereferencing operator.

Since large objects are often manipulated through pointers in order to avoid undesired copy operations, a special operator is assigned to the combination of dereferencing and member selection. This *pointer-to-member* operator appears on the fourth line of the **main( )** program above. The syntax **pC -> f( );** is precisely equivalent to **(*pC).f( )**, but prevents the frequent error of omitting the parentheses around the dereferenced object in the latter expression.

18.4 Uninitialized pointer errors

As noted in the introduction to this chapter, the ability of pointers to address any memory location permitted by the operating system creates new categories of errors. A particularly common runtime error results when a value is assigned to an uninitialized pointer as in

```
int *i;                              // i can point anywhere in memory
*i = 20;                             // Typically forbidden by O/S
```

Here the pointer variable **i** contains the address of a random memory location that from statistical considerations is almost certainly located outside the memory space allocated to the program by the operating system. Placing a value into this location, therefore, raises a general protection fault. On a practical level, this means that the operating system displays an error message stating that the program has performed an illegal instruction.

18.5 NULL and void pointers

An extremely helpful but often neglected technique for preventing values from being assigned to an uninitialized pointer is to set every pointer to address **0** (which has the alias **NULL**) when defined or when detached from a meaningful memory location. A logical test can then be employed to insure that a pointer is properly initialized

```
int *p = NULL;                  // or int *p = 0;
cout << p << endl;              // Output: 00000000
...additional lines such as int l; p = &l; ...
if ( p ) *p = 3;                // Insures p is initialized
```

Here **int** *__p__ = **0;** can if desired be used in place of **int** *__p__ = **NULL;**.

Another construct that, while not recommended in C++, still appears in many programs, is the **void** pointer. A **void** pointer can be set to a pointer of any type. A cast is then employed to convert the **void** pointer to the appropriate type. As an example

```
void *v;

int iv = 10;
int *jv = &iv;
double dv = 20;
double *jdv = &dv;

v = jv;
cout << (*static_cast<int *>(v)) << endl;           // Output: 10
v = jdv;
cout << *((double *)(v)) << endl;                   // Output: 20
```

where we illustrate the two effectively identical procedures for the type cast. Note that **void** pointers violate type-safety since a **void** pointer can be incorrectly cast to a pointer of a completely inconsistent type with the type of the pointer to which it was originally assigned.

18.6 Dangling pointers

Assigning values to uninitialized pointers is common but not particularly troublesome since the error is normally trapped by the operating system. A far more severe problem results from dangling pointers, which are pointers to deallocated memory space as in the example below

```
int *i;
{
        int j;
        i = &j;
}                               // j is now destroyed
*i = 3;                         // Error: i's memory is deallocated!
```

Recall that deallocated memory is memory that the program has released and that has therefore been reclaimed by the operating system. However, such memory generally initially remains accessible to the running program and the value stored in the memory persists until the memory location is either overwritten by a new definition in the program or allocated to another running process. If a statement in the program attempts to read or modify the value of a deallocated variable, often the program will still function properly but, on occasion, large errors will appear or the program will terminate unexpectedly. Locating dangling pointers efficiently accordingly often requires specialized software tools.

18.7 Pointers in function blocks

Dangling pointers can potentially arise whenever a block that contains pointer definitions terminates. Frequently this occurs when pointers are passed as arguments to a function as in the example below

```
void test( int *(&aI), int *(&aJ ) {
        int k = 4;
        *aI = k;                // OK: address stored in aI unchanged
        aJ = &k;
}                               // Error: k's memory deallocated

main() {
        int l = 1, m = 2;
        int *pI = &l;
        int *pJ = &m;
        test( pI, pJ );
        cout << l << endl;      // Output: 4
        cout << *pJ << endl;    // Output: maybe 4 but unpredictable
}
```

Since **pI** and **pJ** are passed as *references* to pointers, the pointers **aI** and **aJ** inside the function are simply alternative names for these variables. Only the value of the variable pointed to by **aI** is changed inside **test( )**, not the memory location stored in **aI,** so that **pI,** which points to the address space of the variable **l** in the main program, remains correctly allocated. Therefore, **l** has a value of 4 after the function call.

On the other hand, the *value* of the pointer reference variable (that is the address of the variable that the pointer points to) **aJ** in the function is reassigned to the memory address of the local **int** variable **k**. As **k** is subsequently deallocated when the function block terminates, **pJ** at the end of the program points to deallocated memory.

18.8 The const keyword and pointers

A pointer can in general change both its *value*, which is the memory address that it points to, and the value stored at this address through the dereferencing operator. Therefore, there are three possible ways to define a pointer using the keyword **const**.

Constant pointers: The first application of **const**, typified by

```
int j = 1;
int * const i = &j;
```

indicates that the address that **i** points to is fixed, but applying the dereferencing operation to **i** yields an **int** value that can be changed. This behavior is clearly alluded to by the notation; reading the definition from right to left, **i** stores a memory address, which is declared **const** through **const i**, while dereferencing this address through the dereferencing operator * yields a non-constant **int** variable. Since the address of the pointer cannot be changed, **i** *must* be initialized to a sensible address when defined. For the above definition, we have

```
int k = 2;
*i = 30;                // OK, the int value can be changed
i = &k;                 // Error: the address in i is constant.
```

Note the similarity of this construct to a reference variable. In fact, since a constant pointer cannot be detached from the variable to which it points, the only meaningful operations that can be performed on the pointer require dereferencing to access the underlying non-constant variable value. Therefore, a reference is a constant pointer that is automatically dereferenced, transforming its syntax into that of an ordinary variable.

*Pointer to a **const***: A second use of const as applied to pointers is

```
const int *i;
```

or equivalently

```
int const *i;
```

Here the value ***i** located at the address held in **i** is a **const int** in the sense that it cannot be changed *through* **i**, but the value of **i** (the memory location stored in the pointer) can be altered. That is, we have

```
const int *i;           // i holds a random value
int l;
int *k = &l;
i = k;                  // OK: address held in i can be chaged
*k = 3;                 // OK: k is not a constant pointer
*i = 4;                 // Error: *i = l cannot be changed through i
```

*Constant pointer to a **const***: A final construct is

```
const int * const i = &j;
```

In this case, not only is the memory location stored in the pointer variable fixed but the value stored at this location also cannot be changed through the pointer variable. Such a pointer is closely related to a **const** reference variable.

18.9 Pointer arithmetic

Suppose that the memory allocated to a variable of type **variableType, sizeof(variableType),** is **n** bytes, where **variableType** can be any built-in or user-defined type (such as **int**, **double** or **Spring**). Then if, for example, **variableType *i = (variableType *) 00000040;**, ***(i + 1)** is the value of the variable of type **variableType** stored at starting address 40 + **n**, ***(i + 2)** the value stored at 40 + 2***n** and so on. Note that this implies that ***(i + k)** – the value of the variable of type **variableType** at starting address **i** + **n*****k** – yields the same result as **i[k]**. Although rarely used except for the prefix and postfix operators ++ and −−, other arithmetic operators are defined to operate on pointers in the same manner.

18.10 Pointers and arrays

An array variable resembles a constant pointer as it evaluates to a constant memory address. Consequently, an array variable can be assigned to a pointer. However, an array definition and a pointer definition are fundamentally different since **int a[10];** allocates memory for 10 **int** variables, while **int * const a;** simply allocates memory for a single pointer variable that stores the memory address of an **int**. These concepts are illustrated by the following code

```
int a[3] = { 1, 2, 3 };
cout << a << end;                               // Possibly: 001H2B00
cout << a + 2 << '\t' << *(a + 2) << endl;      // Output: 001H2B08 3
   int *p = a;
   cout << p[2];                                // Output: 3
```

18.11 Pointer comparisons

A major source of error occurs when attempting to compare two values through pointers to variables containing these values. In this case, to compare the values the comparison operator must be applied to the *dereferenced* pointers, not to the pointers themselves. That is, in the code

```
int *a = &i, *b = &j;
if (a == b) cout << "same addresses";
if (*a == *b) cout << "same values";
```

the first **if** statement compares the *values* of the pointer variables **a** and **b**, which are the *addresses* of the integer variables that **a** and **b** point to, and thus determines if **a** and **b** point to the same variable. While this may in fact be the intention, generally the objective is instead to compare the values of the *variables* that **a** and **b** point to. The pointers must then be dereferenced before the comparison is performed as in the second **if** statement above. Recall that a similar error occurs when the index operator, [ ], is omitted in comparing two array variables. However, since arrays can never occupy the same memory location, the logical comparison will then always return a **false** value.

18.12 Pointers to pointers and matrices

Consider a pointer variable that contains the memory address of a second pointer, which we take for concreteness to be a pointer to an **int**. Just as the type of a pointer to an **int** is denoted **int** *, the type of a pointer to an **int** pointer is **int** **. Again this definition is interpreted from right to left: in the statement **int** ****j;**, applying the rightmost dereferencing operator * to **j** yields a variable of type **int** *; dereferencing this variable one more time is equivalent to dereferencing **j** twice and yields an **int** value. A concrete example is

```
int i = 4;
   int *p = &i;
   int **pp = &p;
   cout << **pp << ' ' << *p << endl;                    // Output: 4 4
```

Recall that **p[0]** is equivalent to ***p**. Similarly, **pp[0]** is equivalent to ***pp** which is the pointer **p**. By extension, **(pp[0])[0]**, which can also be written **pp[0][0]**, is equivalent to ****pp**. A two-component matrix can therefore be constructed from a two-dimensional array of two pointers as follows

```
int a[2] = {1, 2};
int b[2] = {3, 4};
int *p = a;
int *q = b;
int *pp[2];
pp[0] = p;
pp[1] = q;
```

In this case, writing, for example, **pp[0][1]** yields the value 2. Note that this procedure applies equally well to two-dimensional arrays with rows of differing lengths.

18.13 String manipulation

Recall that a string is an array of characters terminated by a null character. C++ functions and operators are generally programmed to read a string until the null character is reached. A string can be generated at the point of definition in either of two ways

```
char s1[6] = "Hello";
char *s2 = "World";
```

An advantage of the second procedure is that space is automatically reserved for the null character that terminates the string "World." However, the length of the character array (6 bytes) allocated to **s2** must be established if a new string with a different length is later assigned to **s2**.

The standard insertion and extraction operators generally suffice to read a character string from the keyboard, or to write a string to the terminal unless the input line contains whitespaces. Otherwise, the **getline(s1, i1)** member function of any input stream can be employed to read the first **i1 – 1** characters of an entire input line up to a carriage return, including whitespace characters. The function then appends a null character and stores the result in the array **s1**, which therefore must be at least **i1** characters long. A second form of the function, **getline(s1, i1, 'c'),** reads **i1 – 1** characters from the input stream but only up to the termination character, **c**. That is, the output from the program

```
char s[10];
cin.getline(s, 10, '|');
cout << s;
```

is, for example, "**A Test**" if the input line is

```
A Test|
```

C++ provides native (that is, not contained in **#include** statements) string handling functions that act on either character arrays or pointers to character arrays. For the strings **s1** and **s2**, we can obtain the length, copy the first $n = 2$ letters in the string **s2** into **s1** or compare the first $n = 2$ letters through the **strlen( )**, **strncpy ( )**, and **strncmp( )** functions as follows

```
int n = 2;
cout << strlen(s1);                    // Output: 5
cout << strncpy(s1, s2, n);            // Output: Wollo
cout << strncmp(s1, s2, n);            // Output: 0 (s1 is now Wollo)
```

(the functions **strcpy( )** and **strcmp( )** function identically but do not take a third argument.) The output of **strncmp( )** is 0 if the first **n** characters are the same, −1 if **s2** is after **s1** in dictionary ordering and +1 otherwise. In **strncpy( )**, which copies the first **n** characters of **s2** into the string **s1**, the string **s2** must be the same size or smaller than **s1** to prevent a runtime memory allocation error.

Similar operations can be performed on the corresponding **string** class objects introduced in Section 11.3; however, the member functions are rather different in nature. The code for **string** objects corresponding to the above example is

```
#include <string.h>
main() {
    string s1 = "Hello";
    string s2 = "World";
    cout << s1.length( ) << endl;                  // Output: 5
    cout << s1.replace(0, 2, s2.c_str( ), 2);      // Output: Wollo
    cout << s1.compare(0, 2, s2) << endl;          // Output: 0
}
```

Recall here that a **string** class object is converted to a string represented as a character array through the **c_str( )** member function of the **string** class.

The **string** class also implements **begin( )** and **end( )** iterators that return pointers to the initial and final elements of a **string** object. This enables iterations over the characters in a string as in the following example

```
string s = "a string";
for (char* g = s.begin( ); g != s.end( ); g++)
cout << *g ;
```

which simply writes the string **s** back to the terminal.

18.14 Static and dynamic memory allocation

Static memory allocation: Memory for variables and arrays that is allocated by the *compiler* is termed statically allocated memory. When the compiler processes definitions, instructions are generated that can eventually lead to memory allocation at runtime. The compiler also issues further instructions to deallocate memory at the point that a variable will no longer be used (usually when the block containing the definition is terminated). This can occur in two ways, typified by the statements

```
float a[10];
void fun( ) { int b[10]; }
void main( ) { int c[10]; { int d[10]; } }
```

The *lifetime* of the *global* array **a** is the period during which it reserves memory, namely from the beginning to the end of the program, while the lifetime of the *local* arrays **b, c, d** extends from their definition statements to the end of the block in which the statement is located.

Dynamic memory allocation: While a compiler allocates and deallocates memory efficiently, the amount of memory required by a program often depends on the outcome of a user interaction or a logical condition and is therefore not apparent until runtime. In such a case, a program should allocate the minimum amount of required memory so that it can exploit the resources of large computers but still run successfully on small machines. A paradigm for this type of program is a computer game that reserves new memory every time a menu item is selected and a new player icon is created. Such a game can be played on nearly any hardware platform but the number of icons that can be created depends on the memory size.

Runtime memory requests from the operating system for a variable of a certain type are implemented in C++ through the **new** operator. This operator returns a temporary *pointer* whose value is the starting memory address of the new, dynamically allocated space and whose type is that of a pointer to the specified type. To save the starting memory address so that the dynamically allocated space can be accessed later by the program, it must be placed into a preexisting compatible pointer variable that has been allocated by the compiler through an appropriate pointer definition. That is

```
int *i;         // Compiler allocated pointer to an int
i = new int;    // Memory is allocated for an int and its address is
                // stored in i
```

These two steps are commonly combined into a single line as follows

```
int *i = new int;
```

A dynamically allocated built-in atomic variable, such as a **double** or **int**, can be initialized when defined by writing, for example

```
double *d = new double(100.0);
```

The syntax for dynamic allocation of an array of ten doubles is instead

```
double *d = new double[10];
```

Accordingly, *parentheses for initialization should not be confused with square brackets for array allocation*. The maximum amount of memory that can be allocated is determined by the operating system; if the system is unable to service a request, the new operator either throws an exception in newer C++ compilers, which leads to program termination with an error message or returns the null pointer, which a pointer with a value of 0. (The latter behavior can be forced in modern compilers by including the **new.h** header file, followed by the statement **new_handler set_new_handler(0);**.) In the latter case, to determine if an allocation request is correctly fulfilled requires a simple logical comparison test

```
int *i = new int;
if (i == NULL) exit(0);
```

As explained above, since dynamic memory is allocated by the operating system at runtime, it can depend on the values of runtime variables as in the example below

```
int *iP;
cin >> n;
if (n > 0) {
   iP = new int [n];
   if (iP == 0) cout << "out of memory";
}
```

The lifetime of memory allocated through a call to **new** *extends until the end of the program unless it is deallocated with a* **delete** *or* **delete [ ]** *statement. The second of these statements must be employed whenever an array is allocated as illustrated below*

```
int *i = new int;
delete i;
i = 0;
if ( !i ) i = new int[10];
delete [ ] i;
i = 0;
delete i;
```

The square brackets after **delete** in the second case insure that the memory associated with the entire array will be deleted. *If the brackets are omitted, only the memory at the position of the first array element is certain to be deleted*, so that a 'memory leak' can result. Further, the **delete** *statement can always be applied to a null pointer, in which case no action is taken at runtime. As well,* **delete [ ]** *can be applied to both non-array variables and the null pointer.*

Note that the compiler-allocated pointer **i** is still allocated until the block in which it is defined terminates – the **delete** statement *frees the memory that* ***i*** *points to but does not delete* ***i*** *itself*. Accordingly, following the **delete** statement the value of **i** can still be set to the memory address of a different variable or to a new dynamically allocated memory address. Whenever a dynamically allocated variable is deleted, any pointers to this variable should be set to the null pointer so that any further attempt to assign a value to the dereferenced pointers will generate a runtime error.

18.15 Memory leaks and dangling pointers

Memory leaks: We have remarked that the **new** operator returns an address to dynamically allocated memory. To manipulate the contents of the memory or to deallocate the memory, this address must subsequently be available to the program. This could be done in the extreme case by writing the memory location to the terminal and then entering this value from the keyboard each time that the memory needs to be accessed; however, the address is of course normally stored in one or more compiler-assigned pointer variables. If all these pointers to the dynamically allocated memory are destroyed or reassigned, the memory address is lost. Therefore, the memory, while still reserved by the operating system for program operation, becomes inaccessible (garbage) and cannot be further used or deleted.

A characteristic error leading to a memory leak is the unintended reassignment of a pointer to dynamically allocated memory

```
char *ptr = new char('a');
ptr = 0;                              // Address to new memory lost
```

While the above statement only causes the loss of a single byte, if located, for example, in a **for** loop, the effect can be severe. For example, all memory available to the running program will be quickly exhausted though the following statement

```
for ( ; ; ; ) int *garbage = new int[1000];
```

which can lead to the computer slowing down and then ceasing to function until rebooted.

Dangling pointers: A closely related, but intrinsically more serious problem occurs when values are mistakenly written to and read from deallocated memory. In this case, the program will often continue to run, in most cases without producing an error. However, occasionally the deallocated memory will be used by the program for a different purpose, such as a function call or a new dynamic memory assignment, or the memory may be reallocated by the operating system to another running program. The program is therefore only sporadically incorrect. Further, the error does not appear until the deallocated memory is accessed, which can be long after deallocation occurs so that the source of the difficulty can be extremely difficult to identify.

The simplest example of a dangling pointer (a pointer to deallocated memory) is

```
int *p = new int [20];
delete [ ] p;                  // The memory p points to is deleted
```

Subsequently writing

```
*p = 30;                       // Bad error: uncertain result
```

places the value 30 into the unsafe memory location originally assigned to **p[0]**. Dereferencing **p** later in the program still yields 30 unless the contents of this location were in the meantime overwritten or **p** was assigned to a different memory address.

A highly recommended procedure for eliminating dangling pointers is to assign the null (0) address to all pointers to deallocated memory. In this case, we would write

```
p = NULL;
```

immediately after **delete [ ] p;**. Attempting to read or write from the memory pointed to by **p** as in

```
*p = 30;                       // Good error: program crash
```

is certain to lead to a runtime error, since the operating system always prevents memory access at address 0.

Dangling pointer pairs: A more subtle form of a dangling pointer error can occur if two pointers point to the same memory location. If the memory assigned to one of these pointers is deleted and its address is set to **NULL**, the second pointer variable points to deallocated memory.

The simplest manifestation of this problem is accordingly

```
int *p = new int[20];
int *p2 = p;
delete [ ] p;
p = NULL;
*p2 = 30;                                // Error: the memory at p2's
                                         // address is destroyed!
```

A dangling pointer can also result from the deallocation of *compiler* allocated memory at the end of a block as in

```
double *r;
{
    double d = 3;
    r = &d;
}                                        // Error: r's memory deallocated
```

This problem does not, however, occur with dynamically allocated memory, which persists until an explicit **delete** statement is issued

```
double *r;
{
    double *d = new double[10];
    r = d;
}
r[10] = 10;                              // OK: pointer variable d
                                         // destroyed but not its memory
```

The reason in fact that the memory associated with a pointer **d** cannot be automatically deallocated once **d** is destroyed at runtime is that no facility exists to determine if another pointer has been reassigned to point to **d**'s memory as in the code above.

18.16 Dynamic memory allocation within functions

Memory can be allocated to an external pointer from within a function as follows

```
void allocate( int *(&aI) ) {
    aI = new int[10];
}

main() {
    int *i;
    allocate( i );
    i[5] = 3;
}
```

Although the pointer variable **aI** inside the function is destroyed when the function terminates, the memory assigned to this variable persists and therefore remains assigned to the variable, **i**, used as a parameter in the calling program. However, the function argument must be a reference to a pointer variable (or an array of pointers, in which case **i** in the main program must also be defined to be an array of pointers). Otherwise, as in Section 18.7 the pointer variable **aI** inside the function will be a *copy* of the pointer, **i,** employed as the function parameter in the calling program. Setting the *value* of **aI** inside the function to the location of the newly accessed memory will not then affect the *value* of **i**, resulting in a memory leak when the function terminates.

Recall as well from Section 18.7 that performing the same manipulation using compiler-assigned memory always generates a dangling pointer pair, since the lifetime of the compiler-assigned memory only extends to the end of the function block as in

```
void allocateError(int *(&aI) ) {
    int j[10];
    aI = &j;                          // Error: j's memory deleted
}
```

Similarly, adding the statement **delete[ ] aI;** to the end of the block of **allocate ()** in the first program of this section creates a dangling pointer pair so that the pointer **i** in the calling program points to deallocated memory.

A second procedure for assigning memory from within a function passes the pointer to the assigned memory back to the calling program through a pointer return value

```
float *assign() {
    float *a = new float [20];
    return a;
}                                     // OK: memory persists
```

Again, the pointer cannot point to compiler-assigned memory that is destroyed at the end of the function block

```
float *assignError() {
    float a[20];
    float *p = a;
    return p;
}                                     // Error: memory destroyed
```

18.17 Dynamically allocated matrices

Although the dynamic allocation of matrices is a common feature of scientific programming, efficient allocation strategies are relatively infrequently discussed. We accordingly examine this topic in some detail below.

Standard memory allocation: A dynamically allocated matrix (two-dimensional array) in C++ is implemented as an *array of dynamically allocated pointer variables* that is stored in a variable of type pointer to a pointer (cf. Section 18.12). In the standard procedure for memory allocation, the value of each pointer in this dynamically allocated array is then set to the address of a further dynamically allocated array. Following this procedure yields an $N \times M$ matrix

```
double **A = new double* [N];
for (int i = 0; i < N; i++) A[i] = new double [M];
```

To deallocate the memory assigned to the matrix (but not **A,** the compiler-assigned pointer variable), the dynamically allocated array pointed to by each element of **A** must be deleted *before* the dynamically allocated pointer array assigned to **A** is deleted

```
for (int i = 0; i < N; i++) delete [ ] A[i];
delete [ ] A;
```

If only the second of these lines is present, the addresses to the arrays containing the actual matrix elements are lost to the program and their memory therefore becomes garbage.

Optimal memory allocation: A significant drawback with the above dynamic allocation strategy is that memory for each matrix row is allocated through a separate call to **new**. If the rows are long, they can be located by the operating system in widely separated areas in memory. Numerous page faults are then likely to occur when iterating through all matrix elements. To keep the allocated memory of an $N \times M$ matrix contiguous a single array of $N \times M$ elements should first be allocated. Then the pointers of a dynamically allocated array of pointers can be set to point to every Mth element of the contiguous array. For a 2×3 matrix of doubles

```
double *B = new double[6];                          // Contiguous
double **A = new double*[2];
for (int loop = 0; loop < 2; loop++) A[loop] = B + 3 * loop;
```

To deallocate the dynamically assigned memory, the memory associated with the underlying array **B** should be deleted after the pointer array **A**

```
delete [ ] A;
delete [ ] B;
```

If **B** is instead deallocated first, the matrix **A** will temporarily point to unallocated memory, potentially leading to incorrect memory access (note that the memory that is allocated last in a structure is as a general rule deallocated first).

18.18 Dynamically allocated matrices as function arguments and parameters

A dynamically allocated matrix can be passed to a function through an argument of pointer to pointer type

```
void print (int **aA, int n, int m) {
   for (int i = 0; i < n; i ++) {
      for (int j = 0; j < m; j ++)
         cout << aA[i][j] << ' ';
      cout << endl;
   }
}

main () {
   int **A = new int *[2];
   A[0] = new int[2];
   A[1] = new int[2];
   A[0][0] = A[0][1] = A[1][0] = A[1][1] = 6;
   print (A, 2, 2);                          // Output: 6 6
}                                            //            6 6
```

However, the above **print** function *cannot* be used for the statically allocated matrix

```
int B[2][2] = {1, 2, 3, 4};
print (B, 2, 2);                             // Error!
```

since, in this case, the compiler requires that the second matrix dimension be specified in the function parameter list (for example, **void print(double *A[2], int n, int m)**) as the type of **B** is that of an array of two-component integer arrays.

In a similar fashion, a dynamically allocated, but not a statically allocated, matrix can be assigned to a variable of type pointer to a pointer

```
int A[2][2] = {1, 2, 3, 4};
int **B = new int*[2];
B[0] = new int[2];
B[1] = new int[2];
int **D = B;                                 // OK
int (*E)[2] = A;                             // OK
D = A;                                       // Error!
```

Clearly, the manipulation of complex types, especially in conjunction with dynamic memory allocation, requires an understanding of their underlying structure.

18.19 Pointer data structures and linked lists

Pointers in conjunction with dynamically allocated memory allocation yield far more flexible procedures for manipulating data than statically allocated variables and arrays, since the size of the structure into which the data are placed can expand or contract in a logical fashion as elements are added or deleted. Further, the properties of this structure can reflect the underlying relationships among different data elements.

In this section, we will consider two basic examples of such data structures, namely linked lists and trees. A linked list is a collection of data with the property that each element in the list contains both data and a system for navigating to the succeeding element in the list. A tree, on the other hand is formed from groups of data elements, each of which possess methods for reaching one or more unique elements that are directly below it in the tree.

We introduce lists and trees below. Additional detail on the implementation and application of these data structures can be found in the assignments and references.

Lists: A *Linked list* is a group of data ordered such that each element of data has a subsequent data element. Consequently, data can be read starting from the initial data element, called the root, forward to the final data element. In a software implementation, functions can be provided, for example, to print data, insert new data at a given position in the list, delete an element from the list, and so on.

As an example of an application, consider a one-dimensional array with only a few non-zero elements. Considerable savings in computation time can be achieved if only these elements are stored, together with their indices.

A single component of a list is called a node. The simplest node stores a single data value together with a pointer whose value is the address of the subsequent array element

```
class Node {
public:
        Node *iNext;
        double iValue;
};
```

The "head" of the list is simply a pointer of type **Node*** to the first list element, while the value of the pointer contained in the last element of the list is set to **0** or **NULL**. The latter choice enables a logical test for the end of the list when iterating through the list elements in the same manner as the zero character facilitates the determination of the end of a string. Hence, a three-element list can be generated through code such as (note the order of definitions)

```
Node NLast = {0, 3};
Node NMiddle = {&NLast, 1};
Node NFirst = {&NMiddle, 8};
Node *Head = &NFirst;
```

Because of the null pointer at the tail of the list, a **for** loop can directly iterate through the list items as in the following **printList( )** function that writes out the data in the list

```
void printList(Node * const aHead) {
    for (Node *N = aHead; N != NULL; N = N -> iNext)
        cout << (N -> iValue) << endl;
}
```

Since a **Node** element can only be reached by navigating through each intervening element starting from the head, variations of the above construct are required to achieve more complicated functionality. For example, to add a **Node** element to the end of a non-empty list, the elements must be iterated until the **NULL** pointer is detected. The **next** pointer of this final element must then be reassigned to the new node, whose **next** pointer must be set to **0**

```
void addNodeAsLast(Node *aHead, Node *aNew) {
    Node *N;
    if (aHead == 0) exit(0);
    for (N = aHead; N -> iNext; N = N -> iNext);
    aNew -> iNext = 0;
    N -> iNext = aNew;
}
```

Our first list example can then be rewritten as

```
main() {
    Node NFirst = {0, 8};
    Node NLast = {0, 3};
    Node NMiddle = {0, 1};
    Node *Head = &NFirst;
    addNodeAsLast(Head, &NMiddle);
    addNodeAsLast(Head, &NLast);
    printList(Head);
}
```

The **Node** class can be combined with the functions that manipulate lists to generate a **List** class; further, dynamic memory allocation can be employed to generate and remove nodes. These issues will be examined in the assignment problems.

Trees: Trees arise in numerous contexts including sorting algorithms, combinatorial procedures, program compilation, and data encryption. While some of these applications are introduced in the chapter assignments, the interested reader should consult a text on computer algorithms for a more thorough discussion.

To introduce the concept of a tree, note first that the unidirectional list of the previous subsection can be slightly generalized by including a second pointer in the **Node** class that allows the list to be traversed in either the forward or the reverse direction

```
class BiNode {
public:
    BiNode *iPrevious;
    BiNode *iNext;
    double iValue;
};

void printList(BiNode * const aHead) {
    for (BiNode *N = aHead; N != NULL; N = N -> iNext)
```

```
        cout << (N -> iValue) << endl;
}

main() {
    BiNode NLast = {0, 0, 3};
    BiNode NSecond = {0, &NLast, 1};
    BiNode NFirst = {0, &NSecond, 8};
    BiNode *Head = &NFirst;
    NFirst.iPrevious = Head;
    NSecond.iPrevious = &NFirst;
    NLast.iPrevious = &NSecond;
    printList(Head);                    // Output: 8 (\n) 1 (\n) 3
}
```

A binary tree is now obtained if each of the pointers **iPrevious** and **iNext** of a given node point either to **NULL** or to an independent node that is not pointed to by any other node. That is each node may point to zero, one or two other nodes but is only pointed to by a single node. In the case that the node points to zero nodes, the node is called a leaf. The level of a node is the number of pointers separating it from the root node.

Accordingly, a sample binary tree with six nodes that contains two leaves to the left of the root and one leaf to the right can be constructed as follows

```
#include <iostream.h>

class BiNode {
public:
    BiNode *iPrevious;
    BiNode *iNext;
    int iValue;
};

void printTree(BiNode *aRoot) {
    if (aRoot -> iPrevious != 0) printTree(aRoot -> iPrevious );
    if (aRoot -> iNext != 0) printTree( aRoot -> iNext );
    cout << aRoot->iValue << endl;
}

main() {
    BiNode Six = {0, 0, 14};
    BiNode Five = {0, 0, 3};
    BiNode Four = {0, 0, 9};
    BiNode Three = {&Six, 0, 7};
    BiNode Two = {&Four, &Five, 12};
    BiNode Root = {&Two, &Three, 4};
    printTree( &Root );
}
/*Output:
9
3
12
14
7
4
*/
```

Operations on tree structures are typically implemented through recursion. As an example, to print out all the tree elements the **printTree( )** function above recursively traverses the left node and then the right node.

18.20 Assignments

Part I

Give the output of the code if the program will function properly. If there is an error or if a potential error condition, such as overflow, underflow, incorrect memory access, etc., arises, so that the program result or runtime state is unpredictable, indicate instead where the error is and what type of error has been made

(1)
```
int *myFunction(int *aI, int *aJ) {
    *aI = *aI * (*aJ);
    return aI;
}

main () {
    int *aI;
    int *aJ;
    *aI = 2;
    *aJ = 3;
    cout << (*myFunction (aI, aJ)) << endl;
}
```

(2)
```
void function( int *aI, int *aJ ) {
    *aJ = 20;
    aI = aJ;
}

main () {
    int j = 10;
    int k = 15;
    function (&j, &k);
    cout << j << '\t' << k << endl;
}
```

(3)
```
ostream* function (ostream *aOut) {
    *(aOut) << "Hello" << endl;
    return aOut;
}

main () {
    function(function(&cout));
}
```

(4)
```
class myStruct {
public:
    int firstElement;
    myStruct *secondElement;
};
```

```
main () {
    myStruct S3 = {4, NULL};
    myStruct S2 = {2, &S3};
    myStruct S1 = {1, &S2};
    cout << (S1.secondElement)->firstElement << " "
                << S1.firstElement << endl;
}
```

(5)
```
int *myFunction ( ) {
    int *aV = new int [3];
    for (int loop = 0; loop < 3; loop++) aV[loop] = loop;
    return aV;
}

main () {
    int *v;
    v = myFunction ( ) ;
    cout << v[1] << endl;
}
```

(6)
```
main () {
    int *a = new int[9];
    for (int loop = 0; loop < 9; loop++) a[loop] = loop;
    int **b = new int*[3];
    for (int loop = 0; loop < 3; loop++) b[loop] = a + 3*loop;
        for (int loopO = 0; loopO < 3; loopO++) {
            for (int loopI = 0; loopI < 3; loopI++)
                b[loopI][loopO] = loopO;
    }
    delete [ ] a;
    cout << b[2][2];
    delete [ ] b;
}
```

(7)
```
main () {
    const int j[4] = {1, 2, 3, 4};
    int *k = &j;
    k[2] = 10;
    cout << j[2] << endl;
}
```

(8)
```
main () {
    int a[][2] = {{1,2},{3,4}};
    cout << *(1[a]+1) << endl;
}
```

(9)
```
void print(int aArray[][2]) {
    cout << *aArray[1] << endl;
}

main () {
    int B[3][2] = { 1, 2, 3, 4, 5, 6 };
    print(B);
}
```

(10)
```
class Person{
public:
    Person* Boss;
    Person* Underling;
    Person() : Name('\0') {};
    Person(char* aName) : Name(aName){}
    const char* Name;
    int i;
};

main() {
    Person TopLevel("Tom");
    Person MiddleLevel("John");
    Person LowerLevel("Norm");
    TopLevel.Boss = NULL;
    TopLevel.Underling = &MiddleLevel;
    MiddleLevel.Boss = &TopLevel;
    MiddleLevel.Underling = &LowerLevel;
    LowerLevel.Boss = &MiddleLevel;
    LowerLevel.Underling = NULL;
    TopLevel.i = 3;
    MiddleLevel.Boss->i = 4;
    TopLevel.i++;
    cout << LowerLevel.Boss->Boss->Name << endl;
    cout << TopLevel.Underling->Name << endl;
    cout << LowerLevel.Boss->Boss->i << endl;
}
```

(11)
```
int l = 5;
int *lp = &l;
int myFunction(int*(&p)) {
    p = lp;
    return *p;
}
main () {
    l = 4;
    int *p = &l;
    myFunction(p);
    cout << *p;
}
```

(12)
```
main(){
    int** m = new int*[2];
    m[0] = new int [3];
    m[1] = new int [2];
    for (int i = 0; i < 3;i++) m[0][i] = i;
         for (int i = 0; i < 2; i++) m[1][i] = i+3;
           cout << (*m)[1];
}
```

(13)
```
main () {
    char A[ ] = "Test";
    char *P = A;
    ++P;
```

```
    cout << P << endl;
    P += 2;
    cout << P << endl;
    cout << --P << endl;
}
```

(14) Construct a linked list class starting from the sample program given in the chapter. Employ dynamically allocated nodes in the **main( )** program. The structure of your program should be

```
#include <iostream.h>
// include stringstream class and possibly string class
// (depending on the compiler)
class Node {
    public:
        Node( int );
// public get and set member functions
        void setData(const int aData)
        int data() const
        void setNext( Node *aNode )
        Node *next() const
    private:
        double iData;
        Node *iNext;
};

Node::Node (int aData) : iData(aData) {
    iNext = 0;
}

class List {
    public:
        void addNode(Node *, int);
        string asString();
        List();
    private:
        void addNodeAsFirst(Node *);
        void addNodeAsLast(Node *);
        Node *iFirst;                    // the head node
};

List::List() {
    iFirst = 0;
}

// This function makes the argument of the function, aNew,
// the new head (iFirst) of the list.
void List::addNodeAsFirst(Node *aNew) {
}

// Traverse the list until you reach the last node and
// then set the subsequent node to aNew and further make
// aNew point at the null pointer.
void List::addNodeAsLast(Node *aNew) {
```

```
    // Use the following lines to traverse the list:
    Node *N;
    for (N = iFirst; N -> next( ); N = N -> next( ));
}

// This function adds a node at a specified position aN.
// aN = 0: the node becomes the new head.
// The node becomes the last node if aN is any value larger
// than the current length of the list.
void List::addNode(Node *aNew, int aN) {

    // Case 1: The list is empty. Set aNew to the head node
    // and set its pointer to the null pointer.
    if ( !iFirst ) {
          iFirst = aNew;
          iFirst -> setNext(0);
          return;
    }

    // Case 2: aN = 0. Use the addNodeAsFirst function.

    // Case 3: aN > 0. Traverse aN - 1 elements through
    // the list: if the list terminates before aN nodes,
    // use the addNodeAsLast function to add aNew.
    // Otherwise, set the iNext pointer of node aN - 1 to
    // point to aNew and the iNext pointer of aNew to point
    // to the previous aNth element.
}

// Write all the elements of the list to a string using
// a stringstream.
string List::asString() {
}

main() {
    List L;
    Node* N1 = new Node(1);
    Node* N2 = new Node(3);
    Node* N3 = new Node(6);
    L.addNode(N1,0);
    L.addNode(N2,0);
    L.addNode(N3,6);
    Node* NewNode = new Node(2);
    L.addNode(NewNode, 2);
    cout << L.asString();
}
```

Hand in the program and your output.

(15) In a sparse matrix, most matrix elements are zero. Such a matrix can be represented as a linked list. One procedure is for each element of the list to contain the row and column of the element together with its value. Write a function that takes a matrix **int M[10][10]** as an argument (take for concreteness **M[1][5] = 6, M[2][7] = 10** and **M[8][9] = 5)** and returns a pointer to the linked list representation of the matrix,

where the list is an object of the **List** class defined in the previous problem. That is, your program should have the form

```
#include <iostream.h>
#include <sstream.h>

const int DIMENSION = 10;

class Node {
    public:
        int iRow;
        int iColumn;
        int iValue;
        Node *iNext;
        void setNext(Node *aNext)
        Node *next()
        Node(int aRow, int aColumn, int aValue)
};

// List class (as in the previous problem) except for the
// body of

string List::asString() {
}

List *generateSparse(int aMatrix[ ][DIMENSION]) {
    // Dynamically allocate a List using 'new List ...' and
    // each successive node.
    // To add the nodes using your previous List class, use
    L1 -> addNode(mNode, INT_MAX);
}

main() {
    int M[DIMENSION][DIMENSION] = {0};
    M[1][5] = 6;
    M[2][7] = 10;
    M[8][9] = 5;
    List* L1 = generateSparse(M);
    cout << L1 -> asString( );
}
```

and should yield the output:

```
row 1 column 5 value 6
row 2 column 7 value 10
row 8 column 9 value 5
```

(16) In analogy to the linked list class presented above, generate code for a binary Tree class. In this problem, the **Tree** class contains only a *single pointer* to a root **TreeNode**, which similarly contain pointers to the right and left subtrees. This establishes an iterative data type, since the structure of the tree at each level is identical. Since an empty tree or subtree is then demarcated through a root pointer value of **NULL**, tree

traversals are greatly facilitated. Model the same binary tree as that of Section 18.9 following the following structure

```
#include <iostream.h>
#include <sstream.h>

class Tree;

class TreeNode {
public:
     Tree *iLeft;
     Tree *iRight;
     TreeNode(int aValue) : iLeft(0), iRight(0) {iValue =
        aValue;}
     TreeNode(Tree *aLeft, Tree *aRight, int aValue)
     int iValue;
};

class Tree {
public:
     Tree() : iRoot(0) { };
     Tree(TreeNode *aTN)
     // Return a pointer to the left node. Note the use of a
     // reference argument here: the pointer in the program
     // must not be a copy of the pointer returned by the
     // function; otherwise we cannot use this function from
     // e.g. main() to set the left node in the Tree object
     // (see the main() function below).
     Tree* &leftNode()
     Tree* &rightNode()
     string asString( );
     TreeNode *root()
private:
     TreeNode *iRoot;
};

     // Note the use of the recursive tree structure here.
     // Three lines are missing in this function: find them.
string Tree::asString() {
        stringstream sout;
     // The next line is necessary in case the function is
     // called on an empty tree.
     if (root()) {
        if (leftNode()) sout << (leftNode() -> asString());
     }
}

main() {
     TreeNode Six(0, 0, 14);
     TreeNode Five(0, 0, 3);
     TreeNode Four(0, 0, 9);
     TreeNode Three(0, 0, 7);
     TreeNode Two(0, 0, 12);
     TreeNode Root(0, 0, 4);
```

```
    Tree TRoot(&Root);
    Tree T2(&Two);
    Tree T3(&Three);
    Tree T4(&Four);
    Tree T5(&Five);
    Tree T6(&Six);
    // Insert here lines that construct the tree corresponding
    // to the example in the textbook (e.g. the root has a
    // left node given by T2 and a right node given by T3).
    cout << TRoot.asString();
}
```

Hand in the program and your output.

(17) Write code that dynamically allocates a three-dimensional matrix corresponding to, for example, the statically allocated matrix **int A[3][2][4]**. In analogy to the discussion in the text, do this in an optimal fashion (one which involves the fewest page faults).

(18) Write code that reads values from a file, stores these in two 2×2 dynamically allocated matrices, multiplies these matrices in the most efficient possible fashion and stores their result in a third 2×2 matrix. Use exclusively pointer arithmetic; that is, refer to **A[1][1]** as *(*(A+1)+1) or as *(*A+4). Use a separate function to dynamically allocate the space for each of the matrices.

(19) Build a simple priority queue structured on a dynamically allocated array that orders a set of nodes such that the node with the smallest value is located first in the array and the node with the largest value is located last. Adding an element to the array increases the size of the queue up to a predetermined maximum value (the size of the array minus one element). Once the array reaches this size, if an additional element is added (pushed on) to the array, a new sort takes place and the largest array element is then removed (popped) from the queue so that the size of the queue does not increase beyond the maximum value. Follow the following format (note the use of the **typedef**, which allows the priority queue in principle to be applied to any compiler-recognized type for which a logical comparison operator is defined).

```
#include <iostream.h>
#include <sstream.h>

typedef int myType;
const int SIZE = 100; // The largest possible size of the
                      // queue.

class PriorityQueue {
public:
// The constructor argument is the maximum number of elements
// that the queue is allowed to store. This must be less than
// SIZE or SIZE - 1 depending on how the push function is
// implemented.
      PriorityQueue( int aMaximumNumberOfElements );
// The push function places a new element into the array and
// simultaneously sorts the elements of the array so
```

```
// that the smallest element in iArray[0], the next smallest
// is iArray[1] ... If the priority queue size would exceed
// iMaximum NumberOfElements the largest element is deleted
// from the array.
        void push( myType aElement );
// The pop function removes the last (largest) element from
// the queue and decreases iNumberOfElements by one (this can
// be done in one line).
        myType pop();
// Return a string representation of the array, using a
// stringstream.
        string toString();
        private:
        myType *iArray;
        int iMaximumNumberOfElements;
        int iNumberOfElements;
};

void main ()
{
        int i = 3;
        int j = 7;
        int k = 5;
        int l = 1;
        PriorityQueue PQ(3);
        PQ.push(i);
        PQ.push(j);
        PQ.push(k);
        PQ.push(l);
        cout << PQ.toString();
}
```

(20) A priority queue data structure facilitates numerous applications. A simple example occurs in coding theory. The Huffman code is a procedure for decreasing the number of bits required to store data by assigning short code words (patterns of zeros and ones) to data that occur frequently and long code words to infrequent data. That is, a letter such as ‘e’ that occurs extremely frequently might be represented simply as 1 while a very infrequent letter such as ‘z’ might instead be mapped into the code word 000001.

Although a Huffman code can be generated through far shorter programs, we present below a transparent procedure that will help conceptualize the practical application of tree structures. In particular, the data symbols, namely the letters of the alphabet, are placed at the end nodes of a tree. If the path to a letter then, for example, consists of a branch to the left from the root node, followed by a branch to the right and then one to the left, the code word that represents the letter will be 101. The coding problem is to obtain the optimal method for placing the data symbols (letters) into the tree, given the known frequency of each symbol (each letter).

The Huffman technique operates as follows. First the symbols are arranged in order of decreasing probability, in our case by placing each symbol into a separate

Tree containing a single **TreeNode** with an internal variable whose value is the symbol frequency. These **Tree** objects are then stored in a priority queue with the **TreeNode** containing the most probable symbol appearing as the first element, the **TreeNode** with the next most probable symbol as the second element, and so on. The two lowest-frequency symbols are then removed from the queue and placed into a new **Tree** such that the more probable symbol is placed to the left of the less probable symbol. The frequency stored in the root node of this **Tree** is then the sum of the frequencies of its right and left nodes. This **Tree** is then placed back into the priority queue, which now has one fewer **Tree** elements, and the process is repeated (in the implementation below, the **PriorityQueue** stores pointers to **Tree** objects instead of the **Tree** objects themselves). Finally, at the end of the process only one **Tree** remains in the **PriorityQueue** and the Huffman code values associated with the various symbols correspond to the order of left (1) and right (0) branches that are encountered in traversing the tree from the root to the edge at which the particular symbol has been placed.

The code framework below implements the above strategy where the **PriorityQueue** is stored in a dynamically allocated array as in the previous problem. Note the use of the **typedef**, which allows the priority queue in principle to be applied to any compiler-recognized type, although the code used to compare two **PriorityQueue** elements would have to be rewritten in most cases

```
#include <sstream.h>

const int DIMENSION = 100;

class Tree;

class TreeNode {
    public:
// (Default) constructor
        TreeNode(int a Frequency = 0, string
            aNodeLabel = " ")
        string iNodeLabel;
        int iFrequency;
        Tree *iLeft;
        Tree *iRight;
};

class Tree {
    public:
        Tree(TreeNode *, int );
        Tree(TreeNode *);
        Tree() : iRoot(0) {}
        string toString();
        int frequency() {return iRoot -> iFrequency;}
        void setFrequency(int aFrequency) {
            iRoot -> iFrequency = aFrequency;
        }
```

```
        string nodeLabel() {return iRoot -> iNodeLabel;}
        void setNodeLabel(string aNodeLabel) {
            iRoot -> iNodeLabel = aNodeLabel;
        }
        void computeCode(string s = "");
        TreeNode* root();
        Tree* &leftNode() {return (iRoot -> iLeft);}
        Tree* &rightNode() {return (iRoot -> iRight); }
    private:
        TreeNode *iRoot;
};

typedef Tree *myType;

class PriorityQueue {
    public:
        PriorityQueue( int aMaximumNumberOfElements
            = DIMENSION - 1 )
// As in the previous example except that the element with the
// *largest* value of totalFrequency is placed into iArray[0].
        void push( myType aElement )
        int numberOfElements()
        myType pop()
        string toString()
// This function starts the computation of the Huffman code
// for the largest frequency tree stored at iArray[0] by
// passing this array element to the computeCode function in
// the Tree. Can only be applied at the end of the
// calculation. Start the procedure from the root of the tree
// as follows:
        void computeCode() {
        iArray[0] -> computeCode();
        }
// Pop the pointers to trees with the smallest values of the
// frequency off the stack and store these in temporary
// variables; these will become subtrees of a new compound
// Tree. Create a new Tree intialized with a new TreeNode,
// set the pointers of the right and left nodes of the Tree to
// the pointers of the subtrees, set the frequency of
// the root of the new TreeNode to the sum of the frequencies
// of the two subtrees and the label of the new tree node to
// the concatenation of the labels of the component nodes:
// that is, if the nodes q and z are combined with q the left
// node, the new label is qz. Finally push this new TreeNode
// back onto the priority queue. Use the set and get functions
// for the node label and frequency.
        void combineTrees() {
            ...
            // You will need this or an equivalent line:
            Tree *NewTree = new Tree(new TreeNode);
            ...
        }
    private:
        myType iArray[DIMENSION];
```

```
        int iMaximumNumberOfElements;
        int iNumberOfElements;
};

// This visits all members of the tree recursively and writes
// iCharacter and then iFrequency for each of the node into a
// stringstream.
string Tree::toString()

// This routine is more involved. Here is one way to
// proceed. Starting from the root, move recursively to the
// left in the tree if there is a non-Null pointer to the
// left. Append a 1 to the string s if this is the case and
// again attempt to move to the left. When it is impossible to
// move to the left, you have one too many ones in the string.
// Delete the last one, being careful to note that erase
// operates to the right of the character position specified
// as its first argment. Now if there is a node to the left,
// append a zero to the string and proceed instead to the
// right. Iterate until you can no longer step to either the
// left or right. You have then come to the end of the tree
// branch, output the character at this end node and the value
// of the string you have built.
void Tree::computeCode(TreeNode *aRoot, string s) {
            ... some code lines
                leftNode() -> computeCode( s );
            ... some more code lines
                rightNode() -> computeCode( s );
            ... some more code lines
                cout << nodeLabel() << s << endl;
}

// get member function
TreeNode* Tree::root()

void main () {
    TreeNode nA(20, "a");
    TreeNode nE(40, "b");
    TreeNode nC(8, "c");
    TreeNode nX(1, "x");
    TreeNode nQ(1, "q");
    TreeNode nU(4, "u");
    Tree tA(&nA);
    Tree tE(&nE);
    Tree tC(&nC);
    Tree tX(&nX);
    Tree tQ(&nQ);
    Tree tU(&nU);
    PriorityQueue PQ;
    PQ.push(&tA);
    PQ.push(&tE);
    PQ.push(&tC);
    PQ.push(&tX);
    PQ.push(&tQ);
    PQ.push(&tU);
```

```
    cout << PQ.toString();
    PQ.combineTrees();
    PQ.combineTrees();
    PQ.combineTrees();
    PQ.combineTrees();
    PQ.combineTrees();
    PQ.computeCode();
    cout << PQ.toString();
}
```

Chapter 19
Advanced memory management

The following two chapters consider advanced or infrequently used features of C++ programming. We first examine the dynamic allocation of memory for an internal pointer class variable through a class constructor. In this case, if an object of the class type is deallocated, its internal variables including the pointer variable are destroyed. However, the memory dynamically assigned to this variable persists, generating a memory leak, unless a special "destructor" function is introduced. Similar problems occur when an object is copied to, or used to initialize a second object. Because such allocation issues are largely invisible to the programmer, they must be understood in detail before undertaking memory allocation within objects.

19.1 The *this* pointer

Before memory management within classes can be analyzed, several additional C++ constructs must be introduced. The first of these is the **this** pointer, which exists in all (non-static) classes and points to the calling object. Dereferencing the **this** pointer generates the calling object itself. That is, we could write for example in any class **A**

```
class A {
        public:
        int iA;
        void print ( ) { cout << (this -> iA) << endl; }
};
```

instead of simply **void print () { cout << iA << endl; }**, since ***this** yields the object that calls the **print( )** function.

Perhaps the most common application of the **this** pointer is demonstrated below

```
#include <iostream.h>

class C {
        public:
        int i;
```

```
    C& print ( ) {
        cout << i << " ";
        return *this;
    }

    C* add(int j) {
        i += j;
        return this;
    }
};

main ( ) {
    C C1 = {2};
    C1.add(3)->print( ).add(3)->print( );        // Output: 5 8
}
```

Since the **.** and **->** functions are left-associative, the last line of the **main( )** program is interpreted as (((**C1.add(3))->print( )).add(3))->print( );**. Consequently the **add( )** function of the object **C1** is applied first, returning a pointer to the calling object, namely **C1**. Next, this pointer is dereferenced through the pointer-to-member operator and the **print( )** function is called, which returns a reference to **C1**. Thus the combined effect of these operations is to return the initial object **C1** with modified internal data members, to which a further series of **add** and **print** functions are applied. While the above program still functions if **C** is employed in place of **C&** as the return type of the **print( )** function, this requires a superfluous and potentially dangerous (through the copy constructor, cf. Section 19.6) copy operation when the **C1** object is returned.

19.2 The *friend* keyword

A class, **A**, may define a particular function or class external to itself as a **friend**. A **friend** can access the private data members of **A** directly without being restricted to calling the member functions of **A**'s public (or protected) interface. That is if we write

```
class A {
    int iJ;
    friend class B;
};
```

all the private data and member functions of **A** appear as public members to functions in class **B** (which can be a subclass of **A**) so that the following code is valid

```
class B {
    A* iA;
    public:
    void printA( ) {cout << iA->iJ << endl;}
```

```
   B(int aJ) {
                 iA = new A;
                 iA->iJ = aJ;
   }
};

main () {
   B B1(5);
   B1.printA( );          // Output: 5
}
```

Similarly, to give a single function **int bfun(int bI)** in class **B** access to the private members of **A**, a line

```
friend int B::bfun( int );
```

should be included in the (public or private) interface of class **A**. As the friend function, **bfun**, is not a member of the class **A**, it does not have an associated **this** pointer to **A**. Although convenient, the **friend** construct violates the object-oriented principles of encapsulation and information hiding.

19.3 Operators

Although any C++ operator with the exception of the scope resolution operator, ::, the dereferencing operator, *, the member-of operator, **.**, the **sizeof** operator, and the if-then-else ternary operator, **?:**, can be overloaded, new operator symbols cannot be introduced, and existing precedence rules cannot be altered.

Note that the latter limitation explains why certain operators have illogical precedence. For example, the C++ stream extraction and insertion operators overload preexisting C operators for bit insertion and extraction that have a relatively high precedence. The C++ operators should rather have a low precedence, so that writing, for example, **cout << i = 3 << endl;** would not yield a compile-time error.

The two procedures for overloading existing operators are presented individually below.

Overloading through friend functions: The conceptually simplest method for overloading an operator implements the operator as a friend function that can access the internal class variables. The procedure is best clarified through an example. Consider a class **Vector** with two private data members **iX** and **iY**. We wish to define an addition operator + such that adding two Vector objects **Vector1** and **Vector2** with internal data members **iX1**, **iY1**, and **iX2**, **iY2** yields a new **Vector** object with data members **iX1 + iX2** and **iY1 + iY2**.

A binary operator expressed as a **friend** constitutes a special two-argument function, denoted by **operator + (V1, V2),** in which the first and second arguments are the expressions to the left and to the right of the + sign, respectively. This yields the class definition

```
class Vector {
public:
   double iX, iY;
   friend Vector operator+(Vector, Vector);
   Vector(double aX,double aY) : iX(aX), iY(aY) { }
   void printVector( ) {cout << iX << "\t" << iY << endl; }
};
```

The friend function is the two-argument global function

```
Vector operator+(Vector aV1, Vector aV2) {
   Vector Temp(0, 0);
   Temp.iX = aV1.iX + aV2.iX;
   Temp.iY = aV1.iY + aV2.iY;
   return Temp;
}
```

Subsequently the operator can be called as in

```
main () {
   Vector V1(1,2), V2(2,3);
   Vector V3(0,0);
   V3 = V1 + V2;
   V3.printVector( );             // Output: 3 5
}
```

The statement **V3 = V1 + V2;** in **main( )** can in fact be replaced with the equivalent statement **V3 = operator+(V1, V2);**

Alternatively, an **operator** function **friend Vector operator+(const Vector&, const Vector&);** can be employed to prevent copying of the input arguments. However

```
Vector& operator+(Vector& aV1, Vector& aV2) {
   Vector* Temp = new Vector(0, 0);
   Temp -> iX = aV1.iX + aV2.iX;
   Temp -> iY = aV1.iY + aV2.iY;
   return *Temp;
}
```

yields a memory leak (except in the case that the operator is called through a statement such as **Vector& V3 = V1 + V2;**) since when the function returns, the address to the dynamically allocated memory is lost (the **Vector** object to which the result of the operator is assigned in the calling program has already been allocated a different memory location). In our previous implementation, the compiler insures that the memory assigned to **Temp** is properly deallocated.

Inclusion in the class definition: The second technique for operator overloading incorporates the operator directly into the class definition. The operation **Vector1 + Vector2** is then interpreted schematically as **Vector1.+(Vector2)**, where + is a member function of the **Vector1** object and **Vector2** is an argument of type **Vector**. In this case, the member variables **iX** and **iY** of **Vector1** are directly accessible to the operator function yielding the code

```
class Vector {
public:
   double iX, iY;
   Vector operator + (Vector);
   Vector(double aX, double aY) : iX(aX), iY(aY) { }
   void printVector( ) {cout << iX << "\t" << iY << endl; }
};

Vector Vector::operator + (Vector aV) {
   Vector Temp(0, 0);
   Temp.iX = iX + aV.iX;
   Temp.iY = iY + aV.iY;
   return Temp;
}
```

Note carefully that here the *binary* operator + has only a *single* argument and can therefore be called through either the standard notation **V3 = V1 + V2;** or the alternate notation **V3 = V1.operator+(V2);**. In general, the **friend** implementation of binary operators is preferable, since both operator arguments are treated symmetrically with respect to implicit or user-defined conversions.

The extraction operator is often overloaded at the global scope in order to print the internal state of an object in a convenient format, as for example

```
ostream& operator<<(ostream &out, Vector V) {
   out << "The vector components are " << V.iX << " and "
       << V.iY << endl;
   return out;
}
```

The statement **cout << V3;** then yields the output **The vector components are 3 and 5**. If the internal class variables **iX** and **iY** are **private**, the operator must be a **friend** of the **Vector** class.

19.4 Destructors

Having introduced operators and friend functions, we can now examine memory management in classes that dynamically allocate memory within constructors. That is, consider the following class

```
class A {
public:
```

```
    A( int aValue = 0, int aSize = 2) : iValue(aValue),
                    iSize(aSize) {
        iArray = new int [aSize];
    }
    ~A() {
        delete [ ] iArray;
    }
    void print () {
        cout << iValue << endl;
    }
    int iSize;
    int *iArray;
    int iValue;
};
```

Unfortunately, this seemingly innocuous class will in general lead to runtime errors unless *three* auxiliary functions, the destructor, assignment operator, and copy constructor, are introduced.

The motivation for the destructor function arises when an object of type **A** goes out of scope as, for example, in the following seemingly harmless code

```
void wrong( int aValue ) {
   double *newArray = new double[10]
   A A1 ( aValue );
   delete [ ] newArray;
}
```

At the end of this function block, the memory dynamically allocated to **newArray** is deallocated by a **delete [ ]** statement. Further, **newArray** and the internal variables of **A1** are compiler assigned and are therefore deallocated. However, the memory that the internal variable **iArray** within **A1** points to has been dynamically assigned at runtime through the **new** statement in the constructor of **A1**. The compiler cannot predict if the user input will lead to a memory request and therefore does not generate instructions for deleting this memory. Therefore the **wrong** function generates a memory leak.

In the example above, since the internal variables are public, a simple solution exists, namely

```
void wrong( int aValue ) {
   double *newArray = new double[10]
   A A1( aValue );
   delete [ ] newArray;
   delete [ ] A1.iArray;
}
```

If **iArray** were instead **private**, a separate **public** member function would have to be introduced into the **A** class definition to deallocate its memory just as a separate **public** member function would be required to initialize **private** member variables in the absence of constructor functions. C++ therefore provides a dedicated "destructor" member function that is automatically called whenever an object of the class type is destroyed in the same manner as the constructor is called when an object is created. Since the destructor cannot be called explicitly and does not return a variable, it does not possess an independent name or return value. Instead, the destructor is denoted in the class definition by the class name preceded by the ∼ character (not followed by the class name is intended as a mnemonic for object destruction). In our example, we have

```
~A() {
   delete [ ] iArray;
}
```

However, when a *pointer* to a dynamically allocated vector object is destroyed, only the compiler-allocated pointer variable is deallocated, while the dynamically allocated memory pointed to by the pointer persists (cf. the discussion at the end of Section 18.15). Instead, *an explicit delete statement must be supplied* as in

```
void Right(int& aSize) {
   A *A1 = new A( );
   delete A1;           // Assumes that a destructor is supplied
                        // for A1
}
```

Explicitly deleting the memory assigned to the pointer **A1** calls **A1**'s destructor.

If memory may or may not be dynamically assigned to a given pointer by the constructor, recall that **delete** or **delete [ ]** operators can always safely be applied to null pointers. Hence, any pointer that will be the target of a **delete** statement in the destructor should be set to the null pointer in the constructor if no dynamic memory is allocated.

When an object is constructed, its base classes are first constructed and initialized in the order they are declared. Within each class, class members are then initialized in order of declaration. The destructor destroys memory in the opposite order to which it is allocated by the constructor.

19.5 Assignment operators

Recall that when an object is assigned to a second non-**const** object of the same type, the values of the compiler-allocated internal variables of the first object are set equal to the values of the corresponding variables of the second object. However, this procedure has unexpected consequences if memory is dynamically allocated through the constructor and assigned to an internal compiler-allocated pointer variable as in the following example

```
main () {
   A A1(3);
   A A2;
   A2 = A1;
   A2.print( );
}                       // Output: 3
```

In this event, the *values* of all *compiler*-assigned variables, namely **iArray, iSize,** and **iValue,** are copied from **A1** to **A2** in the **main( )** function, a procedure termed a "shallow copy." Therefore, in the **main( )** program, the variable **iArray** in **A2** subsequently points to the same memory as **iArray** does to in **A1** and the program can appear to function properly (although in many environments an error message will appear at runtime). However, a severe memory leak is again present. When **A2** is initially constructed through the default constructor in the definition **A A2;**, memory for an array of two integers is dynamically allocated by the operating system and the address of this memory is stored in the object's **iArray** pointer. When the *value* of (the memory address stored in) **A2.iArray** is overwritten by **A1.iArray** through the assignment statement **A2 = A1;**, the memory address of the original dynamically allocated array in **A2** is lost. This memory can therefore not subsequently be reclaimed by the operating system until program termination, even if a destructor function is specified.

An easier to locate (since it generally results in a system error message) but perplexing error that appears in classes with a destructor but not an assignment operator is

```
A A2(2);
{
        A A1(2);
        A2 = A1;
}                              // A1 is destroyed; A2 points to deallocated
                               // memory
A2.iArray[0] = 1;              // Error: illegal memory access
```

Since the memory locations pointed to by **iArray** in both **A1** and **A2** are identical after the assignment statement, once **A1** is destroyed by exiting the block in which it is defined, **A2** points to deallocated memory.

To circumvent the above difficulties, the assignment operator for objects of type **A** must be overloaded. The user-defined assignment operator then prevents the value of the pointer variable, **iArray**, in **A1** from being copied to the pointer variable in **A2** in the statement **A2 = A1;**. Instead, if the length of the dynamically allocated array, **A2.iArray**, is greater than or equal to the length of **A1.iArray**, each *element* of **A1** must be copied into the corresponding element of **A2** (a "deep copy"). If the size of the array in **A2** on the other hand is less than the size of **A1,** the dynamically allocated memory in **A2** should be first deleted and a new array allocated with the same dimension as **A1.iArray** followed by the deep copy operation. As well, the user-defined assignment operator must copy all compiler-assigned variables from **A1** to **A2** since the default assignment operator that normally does this is no longer generated. The signature of the assignment operator is **typeName& operator = (const typeName&)**, where the **const** keyword is optional. Accordingly, for our sample class, the overloaded assignment operator takes the form

```
class A {
// ...same lines as before
        A& operator = (const A& aA) {
// Ensure that the object is not copied onto itself
                if ( this == &aA ) return *this;
// Copy compiler-assigned variables
                iValue = aA.iValue;
                iSize = aA.iSize;

// Assign new memory to the pointer variable if the dimension of
// the preexisting array is smaller than that required by the
// new array.
                if (iSize < aA.iSize) {
                        delete[ ] iArray;
                        iArray = new int[aA.iSize];
                }
// Perform a deep copy of the dynamically allocated elements in aA
// to those of the calling object.
                for (int loop = 0; loop < aA.iSize; loop++)
                        iArray[loop] = aA.iArray[loop];
                return *this;
        }
};
```

The first line of the above function prevents the object's memory from being released in the case that the object is copied onto itself. A simpler procedure for copying the elements of the array **aA.iArray** to **iArray** is provided by the **memcpy** function (the last argument of this function is the number of bytes to copy)

```
memcpy(iArray, aA.iArray, iSize * sizeof(int));
```

The **memcopy** function is a generalization of the **strcpy(s2, s1)** function that copies a string **s1** into a second string **s2** and can be used anywhere in the program to copy both single- and multiple-dimension arrays.

19.6 Copy constructors

The final problem that occurs for classes that dynamically allocate memory through a constructor appears when an object is defined and simultaneously initialized with a second object of the same type. This difficulty, which is essentially identical to that associated with the assignment operator, has four possible origins. One of these is

```
main () {
        A A1;
        {
                A A2(A1);
        }
        A1.iArray = 2;          // Possible runtime error
}
```

Alternatively the definition **A A2(A1);** can be replaced by

```
A A2 = A1;
```

as these two statements are in fact handled identically by the compiler despite the misleading presence of the equality sign in the second version. The last two cases arise when an object of type **A** is passed by value to or returned by value from a function, both of which occur when the function **test( )** is called in the program below

```
A test(A aA) { aA.print( ); return aA; }

main() {
        A A1(1);
        test(A1).print( );       // Output: 1 (\n) 1
}
```

In all the above cases, compiler-assigned variables are copied from one **A** object to a second **A** object through a shallow copy and the address of the dynamically assigned memory in the second object is again lost. These copies do *not* proceed through the assignment operator but rather through the default *copy constructor*, which is a function with the signature **classType (const classType&).** This function must be overloaded to implement a deep copy as illustrated below (the **const** keyword is again optional)

```
A (const A& aA) {
// Copy compiler-assigned variables
        iValue = aA.iValue;
        iSize = aA.iSize;
// Assign new array memory to the pointer member variable
        iArray = new int[aA.iSize];
// Perform a deep copy of the dynamically allocated elements in aA
// to those of the calling object.
        memcpy(iArray, aA.iArray, iSize * sizeof(int));
}
```

The reference parameter in the function signature is required as C++ does not allow a constructor in a class to have a non-reference object of its own class type as a parameter. Since the copy constructor has largely the same purpose as the assignment operator, it can be simply coded through the syntax

```
A(const A& aA) {
        iArray = new int[aA.Size];
        *this = aA;
}
```

To prohibit copying of objects of a given class, a copy constructor without a body

```
A (const A &) { };
```

can be placed in the private part of the class

19.7 Assignments

Part I

Find the errors, if any exist, in the following programs and indicate if they are runtime, link-time, or compile-time errors. If the program will run, describe its output

(1)
```
#include <iostream.h>

class MyClass {
   public:
        MyClass() : nI(2), nJ(3) {
            i = new int[nI];
            j = new int[nJ];
            cout << "In Constructor 1" << endl;
        }

        MyClass(int aNi, int aNj) : nI(aNi), nJ(aNj) {
            i = new int[nI];
            j = new int[nJ];
            cout << "In Constructor 2" << endl;
        }

        ~MyClass() {
            cout << "In destructor" << endl;
            delete[] i;
            delete[] j;
        }

        MyClass(const MyClass& aMyClass) {
            cout << "In Copy constructor" << endl;
            i = new int[aMyClass.nI];
            j = new int[aMyClass.nJ];
            nI = aMyClass.nI;
            nJ = aMyClass.nJ;
            cout << nI << endl << nJ<< endl;
            for (int loop = 0; loop < nI; loop ++) {
             i[loop] = aMyClass.i[loop];
            }
            for (int loop = 0; loop < nJ; loop ++) {
             j[loop] = aMyClass.j[loop];
            }
        }
   private:
        int* i;
        int* j;
```

```
        int nI;
        int nJ;
};

main() {
        MyClass MC1(2, 1);
        MyClass* MC2 = new MyClass[2];
        {
           MyClass& MC3 = MC1;
           MyClass* MC4 = &MC3;
           MyClass MC5(3, 2);
           MyClass MC6 = MC5;
        }
        MyClass MC7(MC1);
        delete [] MC2;
}
```

(2)
```
class MyClass2;

class MyClass1 {
   public:
        MyClass1() : iA(0) {}
        MyClass1(int aIA) : iA(aIA) {}
        MyClass1(MyClass2);
        double getIA(){ return (iA);
        }
   private:
        double iA;
};

class MyClass2 {
   public:
        MyClass2() : iB(0) {}
        MyClass2(int aIB) : iB(aIB) {}
        MyClass2(MyClass1 aMyClass1)
           {iB = aMyClass1. getIA()/2.;}
        double getIB(){ return iB;}
   private:
        double iB;
};

        MyClass1::MyClass1(MyClass2 aMyClass2)
           {iA = 2*aMyClass2.getIB();}

main() {
        MyClass1 MyClass1Object1(3.5);
        MyClass1 MyClass1Object2;
        MyClass2 MyClass2Object;
        MyClass2Object = MyClass1Object1;
        MyClass1Object2 = MyClass2Object;
        cout << MyClass1Object2.getIA() << endl;
}
```

(3)
```
class myClass {
    public:
        int i;
        myClass& square() {i = i*i; return *this;}
};
main(){
        myClass myClass1;
        myClass1.i = 4;
        cout << myClass1.square().square().i;
}
```

(4)
```
class Base {
    public:
        Base operator + (const Base&);
        Base operator + (const Base&) const;
        const char* c() const { return iC; }
        Base (char aC[40]) {for(int i=0;i<40;i++)
            iC[i]=aC[i];}
    private:
        char iC[40];
};

void charMerge(char mergedLine[40], const char char1[40],
            const char char2[40]){
        int i,j;
        for (i=0;char1[i]!=NULL ;i++ ) {
                mergedLine[i]= char1[i];
        } /* endfor */
        for (j=0;char2[j]!=NULL ;j++ ) {
                mergedLine[i+j] = char2[j];
        } /* endfor */
}

Base Base::operator + (const Base& B1) const {
        char pass[40]={""};
        charMerge(pass, B1.c(),iC);
        const Base B3(pass);
        return B3;
}

Base Base::operator + (const Base& B1) {
        char pass[40]={""};
        charMerge(pass, B1.c(),iC);
        char pass2[40]={""};
        char plusArray[40]={"Plus"};
        charMerge(pass2,pass,plusArray);
        Base B3(pass2);
        return B3;
}

main() {
        const Base B1("First ");
        Base B2("\n\t Second");
        const Base B3 = B2 + B1;
```

```
        cout << B3.c();
}
```

(5)
```
class Point {
public:
        Point(int aX, int aY) : iX (aX), iY(iY) {}
        int x() { return iX; }
        void setX(int aX) { iX = aX; }
        int y() { return iY; }
        void setY(int aY) { iY = aY; }
private:
        Point(const Point&);
        int iX;
        int iY;
};

main() {
        Point P1(2,3);
        Point P2(P1);
        cout << P2.x() << endl;
}
```

(6)
```
class Point {
public:
        Point(int aY) : iY (aY), iX(iY) {}
        int x() { return iX; }
        void setX(int aX) { iX = aX; }
        int y() { return iY; }
        void setY(int aY) { iY = aY; }
private:
        int iX;
        int iY;
};

main() {
        Point P1(2);
        Point P2(3);
        P2 = P1;
        cout << P1.x() << endl;
}
```

(7)
```
class C {
public:
        C() {
           i = new int(0);
           cout << "constructed" << endl;
        }
        C(const C& aC) {
           cout << "copy" << endl;
           i = new int;
           *i = *(aC.i);
        }
        const C& operator = (const C& aC) {
           if (this != &aC)
           {
```

```
            cout << "assigned" << endl;
            *i = *(aC.i);
        }
        return *this;
        }
        ~C() {
            cout << "Destroyed" << endl;
            delete i;
        }
        int* i;
};
C f(C aC) {
        C C1(aC);
        C C2;
        C2 = C1;
        return C2;
}
void main ()
{
        C C1;
        C C2 = f(f(C1));
}
```

(8)
```
class B {
public:
        int &fun( );
};

class C {
        int iI;
public:
        friend int& B::fun( );
        void print ( ) {
            cout << iI << endl;
        }
};

int& B::fun( ) {
        C C1;
        return C1.iI;
}

main () {
        C C1;
        B B1;
        B1.fun( ) = 3;
        C1.print( );
}
```

(9)
```
class C {
        int iI;
public:
        friend int& ::fun( );
        void print( ) {
            cout << iI << endl;
        }
```

```
};

int& fun( ) {
        C C1;
        return C1.iI;
}

main () {
        C C1;
        int i;
        fun( ) = 3;
        C1.print( );
}
```

(10)
```
class C {
public:
        int* iI;
        C( ) { iI = new int[2]; }
        void print ( ) { cout << iI[0] << endl; }
        C operator = ( C & aC) {
            iI[0] = aC.iI[0];
            iI[1] = aC.iI[1];
            return *this;
        }
};

main () {
        C C1;
        C1.iI[0] = 3;
        C1.iI[1] = 4;
        C1.print( );
        C C2;
        C2 = C1;
        C2.print( );
}
```

Part II

(11) Provide a destructor, copy constructor, and assignment operator for the linked list class given in problem (14) of the previous chapter. Also, introduce a destructor into the **Node** class so that when the **List** is destroyed, as indicated below, the destructors of all the nodes are called sequentially. Use the following format

```
#include <iostream.h>
#include <sstream.h>

class Node {
        public:
            Node( int );
            void setData(const int aData)
            int data( ) const
            void setNext( Node *aNode )
            Node *next( ) const
```

```
        private:
            int iData;
            Node *iNext;
};

Node::Node (int aData = 0)

class List {
        public:
            void addNode(Node *, int);
            string asString( );
            List( );
            // Destructor: If iFirst is not zero, define a
            // pointer to iFirst, then make iFirst point to
            // the next list element and delete the original
            // iFirst.
            ~List( )
            // Assignment operator:
            // (1) Check that the object is not being
            // assigned to itself
            // (2) Delete all elements in the list to avoid
            // memory leaks.
            // (3) Assign a new node to iFirst.
            // (4) Step through all non-null nodes in aL and
            // assign these to new nodes in the current
            // object.
            // (5) Set the last node to the null pointer.
            List& operator = (const List& aL) {
                  if ( this == &aL ) return *this;
                  ... some lines ...
                  iFirst = new Node;
                  Node *pN = iFirst;
                  Node *aN = aL.iFirst;
                  for ( ; aN ; aN = aN -> next( ),
                        pN = pN -> next( ) ) {
                              ... some lines ...
                  }
                  ... some lines ...
            }

            // copy constructor (you can use the assignment
            // operator as noted in the text)
            List( const List& aL )

        private:
            void addNodeAsFirst(Node *);
            void addNodeAsLast(Node *);
            Node *iFirst;
};

// these are all as in the corresponding program from the
// previous chapter.
List::List( )
void List::addNode(Node *aNew, int aN)
void List::addNodeAsFirst(Node *aNew)
```

```
void List::addNodeAsLast(Node *aNew)
string List::asString( )
main( ) {
        List L;
        Node* N1 = new Node(1);
        Node* N2 = new Node(3);
        Node* N3 = new Node(6);
        L.addNode(N1,0);
        L.addNode(N2,0);
        L.addNode(N3,6);
        Node* NewNode = new Node(2);
        L.addNode(NewNode, 2);
        cout << L.asString( );
        {
            List L2;
            L2 = L;
            L2 = L2;
            cout << L2.asString( );
        }
        List L3(L);
        L3 = L;
        cout << L3.asString( );
}
```

(12) Generate a **Tree** class in which space for the elements is dynamically allocated and supply a constructor, destructor, copy constructor, and assignment operator for the class by completing the program outlined below. Note that, while this code yields a **Tree** structure for integer values, any type of value can be employed by changing the **typedef** statement.

```
#include <iostream.h>
#include <sstream.h>

typedef int ElementType;

class TreeNode {
        friend class Tree;
        // Default constructor (yields an uninitialized value
        // for iValue)
        TreeNode( ) : iLeftTreePtr(0), iRightTreePtr(0)
        // Non-default constructor
        TreeNode(ElementType aValue, Tree *aLeftTreePtr = 0,
           Tree *aRightTreePtr = 0)
        // The destructor should call the destructors of the
        // left and right branches of the tree. This will
        // invoke the destructors recursively until the delete
        // operators act on the zero nodes of the leaves,
        // which produces no result.
        ~TreeNode( )
        ElementType iValue;
        Tree *iLeftTreePtr;
        Tree *iRightTreePtr;
};
```

```
class Tree {
        public:
            // Construct empty tree (root points to zero)
            Tree( )
            // Construct a tree such that the root points to
            // a new TreeNode with iValue = aValue
            Tree(ElementType aValue)
            // Return a pointer to the left tree in the
            // TreeNode pointed to by iRootNodePtr.
            Tree* &LeftTreePtr( ) const
            Tree* &RightTreePtr( ) const
            string asString( );
            // Get member function for iRootNodePtr
            TreeNode *RootNodePtr( ) const
            // Destructor — return an empty tree by
            // calling delete on iRootNodePtr, which then
            // invokes the recursive delete function
            // of the TreeNode class. Set iRootNodePointer
            // to 0 afterwards.
            ~Tree( )
            // Assignment operator — First check if the
            // Tree object is being assigned to itself.
            // Otherwise, delete the current tree and
            // recursively copy the elements of aTree to the
            // current tree using the recursive copyTree
            // function.
            Tree &operator = (const Tree& aTree) {
            ... some lines ...
                // This line is critical since you must have
                // created left and right nodes before you can
                // start to copy the elements of aTree to
                // them!
                iRootNodePtr = new TreeNode
                   (aTree.RootNodePtr( )
                    -> iValue, new Tree, new Tree);
            ... some lines ...
                return *this;
            }
            // Copy constructor - uses above assignment
            // operator.
            Tree(const Tree& aTree) {
                // Create a TreeNode before applying the
                // assignment operator (a value cannot be
                // assigned to a null pointer) as follows:
                iRootNodePtr = new TreeNode;
                ... one line ...
            }

        private:
            TreeNode *iRootNodePtr;
            // Helper function to copy a tree.
            void copyTree(const Tree& aTree) {
                // First handle aTree.iRootNodePtr == 0.
                // Then delete the current tree.
                // Since the current tree now has
                // iRootNodePtr == 0, to
```

```
                        // manipulate it, you must first write
                           iRootNodePtr = new TreeNode
                              (aTree.RootNodePtr( )
                                 -> iValue, new Tree, new Tree);
                        // Now you can use the copyTree function to
                        // copy the elements of aTree into the current
                        // tree recursively. Before copying check that
                        // the elements exist in aTree through
                        // statments such as
                        // 'if (aTree.LeftTreePtr( ))' If no element
                        // exists be sure to set the coresponding node
                        // to zero in the current tree.
                        ... more lines ...
                        }
                }
};

// Use as before a stringstream and its str( ) method
// together with a recursive procedure to print out the
// values stored in the non-zero nodes of the tree.
string Tree::asString( )

main() {
        Tree* TRoot = new Tree(4);
        Tree* T2 = new Tree(12);
        Tree* T3 = new Tree(7);
        Tree* T4 = new Tree(9);
        Tree* T5 = new Tree(3);
        Tree* T6 = new Tree(14);
        T3 -> LeftTreePtr( ) = T6;
        T2 -> LeftTreePtr( ) = T4;
        T2 -> RightTreePtr( ) = T5;
        TRoot -> LeftTreePtr( ) = T2;
        TRoot -> RightTreePtr( ) = T3;
        cout << TRoot -> asString( ) << endl;
        Tree TNew1(*TRoot);
        cout << TNew1.asString( ) << endl;
        Tree TNew2;
        TNew2 = *TRoot;
        cout << TNew2.asString( ) << endl;
}
```

(13) Any periodic waveform can be decomposed into a discrete sum of sine and cosine functions called a Fourier series. For an antisymmetric (with respect to the transformation $t \to -t$) rectangular wave formed by periodically extending the function

$$f(t) = \begin{cases} 1 & 0 < t < T/2 \\ -1 & -T/2 < t < 0 \end{cases}$$

throughout all time, we obtain the formula

$$f(t) = \sum_{n=1}^{\infty} c_n \sin\left(\frac{2n\pi t}{T}\right)$$

where the Fourier coefficients c_n are given by

$$c_n = \frac{2}{n\pi}(1 - \cos(n\pi))$$

In this problem, you are to program and graph the function generated by superimposing the first 30 Fourier coefficients of a square wave according to the last two above equations. The program should have the following form

```
#include <iostream.h>
#include "dislin.h"
#include <math.h>

class FourierCoefficients {
public:
        FourierCoefficients(const int aNumberOfCoefficients);
        double coefficients(int iIndex);
// Supply as well a destructor, copy constructor and
// assignment operator!
        int numberOfCoefficients( );
private:
        int iNumberOfCoefficients;
        double* iCoefficients;
};

class Wave {
public:
        Wave(const int aNumberOfPoints, FourierCoefficients
          *aFourierCoefficients);
// Supply as well a destructor, copy constructor and
// assignment operator! Employ the fact that the
// FourierCoefficient class has its own copy constructor,
// assignment operator and destructor.
// Use qplot here.
        void draw();
        double time(int aIndex);
// Computes the waveform from the Fourier coefficients.
        void computeWaveForm( );
private:
        float* iWaveform;
        float* iTime;
        int iNumberOfPoints;
        FourierCoefficients *iFourierCoefficients;
};

main ( ) {
         const int numberOfCoefficients = 30;
         const int numberOfPoints = 500;
         FourierCoefficients FC1(numberOfCoefficients);
         FourierCoefficients FC(FC1); // To check your copy
                                      // constructor
         Wave W1(numberOfPoints, &FC);
         Wave W(W1);
```

```
        W.computeWaveForm( );
        W.draw( );
}
```

Note that, once calculated, the Fourier coefficients are included in the **Wave** class through a *pointer* variable. This type of inclusion is very useful when the Fourier coefficients are independent quantities that may be used in other classes (and would consequently be included through a pointer or reference variable there). Hand in your code and graphical output.

(14) A neural network is a simple numerical model that attempts to capture the structure of the brain. In the brain, around 10^{11} cells called neurons are joined through an even larger (10^{14}) number of interconnections, or synapses. As similar patterns are presented to the brain's network, the states of the synapses change slowly with time so that successive patterns of the same type are accepted (remembered) with larger probability.

To form a simple numerical model of the above process, consider a time series of values, which we here take simply to be either +1 or −1. We define a set of a given number, **numberOfNeurons**, of neurons, **neurons[numberOfNeurons]**, each of which has a synaptic weight, **weight**[**numberOfNeurons**]. The values of the weights and pattern variables are initially set to zero. Assume that, at each time, the states of the neurons are given by the last **numberOfNeurons** values of this time series. The task of the neural network model is then, from a series of **numberOfNeurons** synaptic weights, to extrapolate the value of the next value in the time series. Each prediction should incorporate information from all the weights as well as the states of all neurons. Thus, we predict that the subsequent element in the time series (the future element) will be given by the sign of the inner product of all the weights with the values of all the neurons, that is **weight[0]*neurons[0] + weight[1]*neurons[1] + ... + weight[numberOfNeurons − 1] * neurons[numberOfNeurons − 1]**

To generate the pattern of neural impulses (the values of *neurons*), a series or time sequence of characters is read, one character at a time, from either the keyboard or an input file. The input '**y**' is interpreted as a 1, while any other input except '**e**', which is used to exit the program, is interpreted as a −1. To implement memory, the system is trained by adding the quantity **input** * **neurons[loop] / numberOfNeurons;** to the value of **weight[loop]**, where **loop** = **0, 1, . . . , numberOfNeurons −1** whenever its estimate of the next weight is inaccurate. Notice the effect of this term: whenever the neuron and the input have the same sign so that the neuron is "predicting" correctly, the weight is increased, while, if the signs are different so that the neuron is "wrong," the weight is decreased. Subsequently, we shift the weight vector such that **weight[loop – 1]** → **weight[loop]** and the actual new observation is inserted into the first position of the weight vector. Observe that if, for example, every value of the stimulus is +1, **weight[loop]** will equal 1 for **loop = dimensions**, thus the weights, at least in this case, "remember" the neural impulses. A prediction for the future value of **neurons[0]** is obtained at each step from the sign of the inner product of the stimulus vector with the weight vector; this involves information from all the neurons and weights. In the program below, which assumes terminal input, the number of correct predictions is recorded and compared with the total number of input values. The program generally predicts accurately the successive values in a periodic sequence

```
main () {

        const int size = 6;
        double* weight = new double[size];
        double* neurons = new double[size];
        for ( int loop = 0; loop < size; loop++ ) {
            neurons[loop] = 0;
            weight[loop] = 0;
        }

        char cInput = 'n';
        int rightPrediction = 0, count = 0;

        for ( ; ; ) {
            count++;
            cin >> cInput;
            if (cInput == 'e') break;
            int input = -1 + 2 * ( cInput == 'y' );
            int sum = 0;

            for ( int loop = 0; loop < size; loop++)
                sum += weight[loop] * neurons[loop];

            if ( sum * input > 0 )
                cout << endl << rightPrediction++
                     << " times correct of " << count << endl;
            else
                for ( int loop = 0; loop < size; loop++ )
                     weight[loop] += input *
                         neurons [loop]/ size;

            for ( int loop = size - 1; loop > 0; loop-- )
                neurons[loop] = neurons[loop - 1];

            neurons[0] = input;
        }
}
```

Your task is now to implement the above program in object-oriented form in the following format

```
#include <fstream.h>
#include <sstream.h>
#include <stdlib.h>

class Network {
        int iNumberOfNeurons;
        int iNumberOfTries;
        int iNumberOfCorrectPredictions;
        int *iWeight;
        int *iNeurons;
public:
// Use a string stream.
        string toString( )
// Implements the algorithm that predicts the subsequent
// value.
        void predict(char aInput);
// Constructor
        Network(int aNumberOfNeurons = 6);
// Destructor
        ~Network( );
// Assignment operator
        Network& operator = (const Network& aNetwork);
// Copy Constructor
        Network (const Network& aNetwork)
};

class OutputSystem {
        ostream *iOut;
        istream *iIn;
public:
        OutputSystem(char *aOutput) {
                if (!strcmp(aOutput, "terminal")) {
                        ...
                }
                else {
// NOTE: writing ostream iOut hides the class variable iOut!!!
                }
        }

// Writes aString to the ostream.
        void print(string aString)

        void read (char &c);
};

main () {
        Network N1(6);
        OutputSystem O1("terminal");
        char cInput = '\0';
        while (cInput != 'e') {
        O1.read(cInput);
        N1.predict(cInput);
        Network N2(N1);
        Network N3 = N2;
```

```
        O1.print(N3.toString( ));
        }
}
```

(A far more detailed discussion of the physics of this example can be found in Kinzel, W. and Reents, G, *Physics by Computer*, Berlin: Springer Verlag (1998), pp. 106–114.)

Chapter 20

The static keyword, multiple and virtual inheritance, templates, and the STL library

In this final programming chapter, we summarize several additional features of the C++ language. These include the **static** keyword, unions, bit fields, virtual and multiple inheritance, templates, and the STL library.

20.1 Static variables

Much like the **friend** construct, the **static** keyword, which can be applied to both variables and classes, is useful in practice, although it violates object-oriented programming principles. Adding the keyword **static** to a variable's definition statement extends its lifetime from the beginning to the end of the program, although it is not visible outside the block in which it is defined unless it can be accessed through the scope resolution operator. In other words, the program

```
main(){
      {
         static int i;
      }
      cout << i << endl;     // Error: i not defined in this scope
}
```

yields an error since **i** has the scope of the innermost block. Similarly, a variable defined within a function at global scope retains its value from one function call to the next, but it cannot be accessed outside the function body. Any **static** variable that is defined outside the body of a class is automatically initialized to zero if no specific initialization value is provided. The following program employs a **static** variable to determine the number of times a function is called

```
void f(){
        static int i = 1;
        cout << i++ << endl;
}
main(){
      f();
      f();
}                          // Output is 1 then 2
```

20.2 Static class members

If a member variable or function is declared **static** within a class definition, a single instance of this class member again exists over the entire lifetime of the program, irrespective of whether any actual objects of the class are ever instantiated. However, the static variable instance can now be accessed outside the class since it has an associated scope identifier. Consequently, if a class has a **static** member, its value is the same for all objects of the class. Since **static** members are independent of the state or even existence of the objects, they can be accessed through their name preceded by their class name as well as through the normal member-of or pointer-to-member operator. Class member variables, whether **public**, **protected**, or **private**, that are declared **static** *must* have a corresponding initialization statement at global scope (outside the class definition), since these variables exist even if no objects of the class are instantiated and, therefore, they could not be initialized by a constructor. The initialization statement does *not* include the **static** keyword. If no value is specified in the initialization statement, the variable is automatically initialized to zero as in the following example

```
class C {
        public:
        C(){cout << ++i << " ";}
        private:
        static int i;
};

int C::i;                                   // Zero by default

main (){
        C C1;
        C C2;
}                                           // Output: 1 2
```

A **static** member function like a **static** variable exists throughout program execution regardless of the number or existence of the class objects. Therefore, such a function can only change **static** class variables (of course, as illustrated above, non-**static** functions can change **static** data as well). Further, a **static** member function does not have a **this** pointer as it is not associated with any particular object. An example of **static** member variables and functions is given below. Different methods for changing the value of the **static** member variable **iStatic** are presented in **main( )**. Also, observe that the function **calculate( )** is actually independent of any particular **C** object and thus can be declared **static**.

```
class C {
        public:
                static int iStatic;
                int iVariable;
                C (int aVariable):iVariable(aVariable){}
                static void setStatic(int aStatic){iStatic =
                       aStatic;}
```

```
            static int calculate(C& aC){return iStatic *
                aC.iVariable;}
};

int C::iStatic = 2;                     // Zero by default; here 2

main(){
    C::setStatic(3);
    C::iStatic(4);
    C C1(2);
    C1.iStatic = 5;
    cout << C::calculate(C1) << endl;          // Output: 10
}
```

20.3 Pointer to class members

C++ provides a specialized *type*, **ClassName::***, that is associated with pointers to particular members of the class **ClassName.** That is, if a class **A** has a member **iA**, we can define a pointer, for example **pIA**, to this member using the syntax **A::*pIA = &A::iA.** To obtain the actual internal member variable of a particular *object*, **A1** of type **A**, the pointer is then dereferenced through the *pointer to member* operator, **.***, using the syntax **A1.*pIA**. Similarly, to obtain an internal member variable given a *pointer* to an object, the operator, **->***, is applied as in **A1->*pIA**. These concepts are illustrated below

```
class C{
public:
     int iC;
};

int C::*pC = &C::iC;

main (){
     C C1 = { 1 };
     C *C2 = &C1;
     cout << C1.*pC << " " << C2->*pC << endl;   // Output: 1 1
}
```

The definition statement for **pC** can as well be included inside the **main( )** function.

20.4 Multiple inheritance

A derived class may inherit the attributes (internal variables and functions) of any number of parent classes. A simple illustration of a class **C** that inherits the member variables and functions of two base classes, **A** and **B** is presented below

```
class A{
protected:
     int iA;
```

```
        A(int aA) : iA(aA){}
        void print(){ cout << iA << endl; }
        void printA(){ cout << iA << endl; }
};

class B{
protected:
        int iB;
        B(int aB) : iB(aB){}
        void print(){ cout << iB << endl; }
        void printB(){ cout << iB << endl; }
};

class C: public A,public B{
public:
        int iC;
        void print(){B::print();}
        void printC(){ cout << iC << endl; }
        C(int aA, int aB, int aC) : A(aA), B(aB){ iC=aC; }
};

main(){
        C C1(1, 2, 3);
        C1.print();              // Output: 2
        C1.printC();             // Output: 3
}
```

Since the constructor of class **C** passes arguments to the base class constructors of **A** and **B** through its initialization list, these are invoked before **C** is constructed. However, since **iA** and **iB** are inherited member variables of **C**, its constructor can also be written as

```
C(aA, aB, aC) : iA(aA), iB(aB), iC(aC) { }
```

By default, base class constructors are, as in simple inheritance, called before derived class constructors. Note that the **print( )** functions of the parent classes of **C** can be distinguished in class **C** through the scope resolution operator.

20.5 Virtual functions

Recall that a derived class object can be used anywhere in a C++ program that a base class object is expected. As an example, if a class, such as **Student**, is derived from a second class, for example **Person**, clearly a **Student** is simply a specialized form of a **Person**. Therefore, any attribute of a **Person** should also be applicable to a **Student**. However, if a derived class *overrides* one or more of the functions (methods) of the base class, the programmer may wish to specify whether the function in the derived class or the function in the base class should be used when a derived class object is employed in place of a base class object. This is possible through pointer or reference variables as opposed to standard variable types, since when, for example, a derived class pointer is used in place of a base class pointer, its memory address clearly remains intact, allowing features of the derived class to be addressed. On the other hand, if a standard derived class object variable is used in place of a base class object, the space occupied by the derived class cannot typically be accommodated within the memory allocation of the base class. Therefore, a conversion operation is applied, discarding the specialized features of the derived class.

Normally, however, the type of a pointer or reference is fixed at compile-time to the type specified in its definition statement – as is always the case with other variable types. That is, writing

```
class Person{
        public:
        void print(){cout << "Person" ;}
};

class Student : public Person{
        public:
        void print(){cout << "Student";}
};

main(){
        Student Student1;
        Person* PointerToStudent1 = &Student1;
        Person& RefToStudent1 = Student1;
        PointerToStudent1 -> print();
        RefToStudent1.print();
        Person Person2 = Student1;
        Person2.print();
}
```

yields the output **Person Person Person** since the pointer and reference variables are declared to be of type **Person*** and **Person&** respectively; therefore, even if they point at or refer to a **Student** object, the object is handed as a **Person**. However, this behavior can be altered for any function that appears in both the base and the derived class by declaring it **virtual** in the *base* class (additionally declaring the function **virtual** in the derived class, while unnecessary, is recommended as a reminder of its properties). In this case, pointers or references to derived class objects are associated at *runtime* with the type of the derived class. If a base class pointer is then assigned to different base or derived class objects because of, for example, varying user input, different object behaviors will result. Consequently, replacing the **Person** class above by the code

```
class Person{
        public:
        virtual void print(){cout << "Person";}
};
```

yields the output **Student Student Person** when the modified program is executed.

Once a function is declared virtual in a base class, it is virtual in all derived classes, including those that are derived through multiple layers of inheritance from the base class. If a virtual function is not defined in a particular derived class, the definition that is used is either the base class definition or the definition in the parent class that is closest in inheritance level to the derived class. *While constructors cannot be virtual, destructors should be declared virtual in the base class so that if a derived object is deallocated through a pointer to a base type at runtime, the destructor of the derived type is called.*

20.6 Heterogeneous object collections and runtime type identification

Virtual functions are particularly convenient for processing groups of objects that are derived from the same base class. Since every member of the derived class is also a member of the base class, these can be stored as a collection (for example, a data structure such as an array, list, or queue) of pointers or references to base class objects. If a function is declared **virtual** in the base class, its behavior when accessed through the collection will be that associated with its object type. Since the object type is only resolved at runtime, objects can be placed into the collection at runtime based on user selections or logical outcomes and the resulting collection queried by the program at a later time to determine its properties. For example, assuming that **print( )** is declared **virtual** in the **Person** base class a two-element heterogeneous object collection can be created and accessed in our previous example as follows

```
main(){
        Person** PArray = new Person*[2];
        Person P1;
        Student S1;
        int select;
        for (int loop = 0; loop < 2; loop++){
            cout << "Insert 0 to create a Person, 1 to create a\
              Student";
            cin >> select;                  // Sample input: 0 1
            cout << endl;
            if ( !select ) PArray[loop] = &P1;
            else PArray[loop] = &S1;
        }

        for ( int loop = 0; loop < 2; loop++ ) {
            PArray[loop] -> print( );       // Sample output: Person
                                            // Student
//           cout << endl << typeid(PArray[loop]).name( ) << endl;
//           Student *S2 = dynamic_cast<Student *>(PArray[loop]);
//           if (S2) cout << (typeid(S2) == typeid(Student *)) <<
//             endl;
        }
}
```

To convert explicitly a derived class pointer into a base class pointer or to downcast the base class pointer into a derived class pointer the **dynamic_cast<typename>**

operator can be employed. If the conversion fails, the operator returns the null pointer of the type specified by **typename.** Further, the type of a pointer can be established during program execution through the **typeid( )** function defined in the header file **typeinfo** or **typeinfo.h.** This function, which can be applied to both objects and type expressions, returns an object of the **type_info** class. One member of the **type_info** class is the **name( )** member function that returns the type of the original pointer as a string. The procedure can best be understood by examining the commented lines in the program above, which print out **Pointer** * (the type of the pointers stored in **PArray)** followed by **1** for each value of **loop** such that **PArray[loop]** is a pointer to a **Student**.

20.7 Abstract base classes and interfaces

Since a pointer or reference to a virtual function associates its implementation with its declared type, if base class objects are never constructed, the bodies of one or more virtual functions can be omitted in the base class. Such a class is then termed an abstract (base) class. A class that derives from this class must then supply bodies for all the missing functions in order for objects of the class type to be constructed. The base class constitutes an interface in the sense that all derived classes that will actually be used to define objects must conform to the interface by supplying the function definitions required by the base class.

More precisely, a base class may contain a function declaration (prototype) without supplying a corresponding function definition (body). The definition must then be specified in the derived classes and may in general be different for each derived class. Such a function is called a pure virtual function. A class that contains one or more pure virtual functions is further termed an abstract class.

In the **Student** example, to form an abstract base class the **print( )** function would be replaced in the **Person** class by

```
virtual void print() = 0;
```

after which the statements **Person Person2 = Student1; Person2.print( );** involving **Person** objects (but not the lines containing only pointers or references to **Person** objects) in **main( )** would be removed. The **print( )** function must then be defined in the **Student** class as well as all other (non-abstract) derived classes that inherit from **Person**.

Abstract classes are particularly useful for heterogeneous object collections as derived class elements can still be placed into arrays of *pointers* to base class objects as demonstrated below

```
#include <iostream.h>

class Person{
public:
        virtual void print() = 0;
};
```

```
class Student : public Person{
public:
        void print(){cout << "Student" << endl;}
};

class Worker : public Person{
public:
        void print(){cout << "Worker" << endl;}
};
main(){
        Student Student1;
        Worker Worker1;
        Person *PArray[2] = {&Student1, &Worker1};
        PArray[1] -> print( );                    // Output: Worker
}
```

Abstract base classes provide an *interface* that derived classes must conform to as opposed to standard inheritance, which instead provides an *implementation* that is adopted by the derived classes. In interface inheritance, the compiler verifies that the derived classes provide all required behaviors.

20.8 Virtual inheritance

An ambiguity in multiple inheritance occurs if, for example, two classes **Student** and **Worker** both inherit from a common base class **Person**, while a further class **StudentEmployee** inherits from both **Student** and **Worker**. In this case, if a data member or function such as **print( )** belongs to the **Person** class, it is inherited in **StudentEmployee** through both **Student** and **Worker**. That is, **print( )** is inherited twice (whether or not it is declared **virtual** in the base class or overridden in one or both of the **Student** or **Worker** classes), once through **Student** and once through **Worker**. This creates two related problems. First, **print( )** cannot be called in **StudentEmployee** unless it is preceded by a scope resolution operator, such as **Student::print( )**, as the compiler cannot resolve which of the two inherited **print( )** statements is intended (even if they are identical). Secondly, a pointer or reference to **StudentEmployee** cannot be employed where a pointer to a **Person** object is required. This is illustrated by the trivial program:

```
class Person{
public:
        void print(){cout << "person" << endl;}
};

class Student : public Person{
};

class Worker : public Person{
};
```

```
class StudentEmployee : public Student, public Worker{
};

main(){
        StudentEmployee SE1;
        SE1.print();
        Person *P1 = &SE1;
        P1 -> print();
}
```

which yields the error messages

```
Error E2014 virtuali.cpp 19: Member is ambiguous: 'Person::print'
and 'Person::print' in function main()
Error E2034 virtuali.cpp 20: Cannot convert 'StudentEmployee *' to
'Person *' in function main()
```

To insure that only one copy of **print( )** is present in **StudentEmployee**, the **Student** and **Worker** classes can be replaced by

```
class Student:virtual public Person{
};

class Worker:virtual public Person{
};
```

This facility, termed virtual inheritance, insures further that only one copy of the base class is inherited through the two derived classes. Both the function **print()** and the pointer to the base class in **StudentEmployee** are then unambiguous and the program can be successfully compiled and executed.

20.9 User-defined conversions

While many promotions and conversions among built-in types, such as those from **int** to **float** or **double**, are implicit in C++; conversions involving user-defined class types can be programmed either as conversion operators in class definitions or as single-argument constructors. In particular, suppose a class **A** is to be automatically converted to a class **B.** A conversion operator can be defined in the **A** class with the special signature **operator B( );** that implements the cast from **A** to **B**. While this is convenient for simple casts, especially if, for example, **B** is simply expressed in terms of one of the member variables of **A,** a more convenient method in most cases is to define a one-argument constructor with the signature **B(A)** in the **B** class. As an example of both procedures, to convert a **Fahrenheit** to a **Celsius** temperature using the formula $°F = 9/5\ °C + 32$ through a single-argument constructor and then from a **Celsius** temperature to an **double** we can write:

```
class Celsius;

class Fahrenheit{
public:
```

```
        double iDegrees;
        Fahrenheit(Celsius aCelsius);
        Fahrenheit(double aDegrees) : iDegrees(aDegrees){}
        operator double(){return iDegrees;}
};

class Celsius{
public:
        double iDegrees;
        Celsius (double aDegrees){iDegrees = aDegrees;}
        Celsius(Fahrenheit aFahrenheit){
                iDegrees = (aFahrenheit.iDegrees - 32)*
                        5 / 9;}
        operator double(){return iDegrees;}
};

Fahrenheit::Fahrenheit(Celsius aCelsius){
        iDegrees = aCelsius.iDegrees * 9 / 5 + 32;}

void printState (Celsius aC){aC <= 0. ?
        cout << "Below Freezing" << endl;
        cout << "Above Freezing" << endl;}

main (){
        Fahrenheit F(34.);
        Celsius C = F;
        cout << double(C) << endl;     // Output: 1.11111
        printState(F);                 // Output: above freezing
}
```

To prohibit single-argument constructors from automatically functioning as conversion operators, the keyword **explicit** can be included in their type definitions.

20.10 Function templates

Consider a class or function that has to be applied to variables or objects of several different types. Rather than write many versions of the same code with different class names appearing in, for example, declarations, definitions, and function arguments, the **template** keyword can be employed to transform the class names appearing in the code into generalized parameters, called metaparameters, that can be set to different types at compile-time. (The **typedef** construct can also partially accomplish this objective.) A separate copy of the template feature is then automatically generated during compilation for each distinct set of templated types. One application of templates is that of function templates (although this feature is deprecated and not supported by all current compilers). The procedure is illustrated by the following example, which copies arrays of any specified class type

```
template <class C1, class C2> void copy (C1 output[], C2 input [],
                int n){
        for (int loop = 0; loop < n; loop++) output[loop]
                = input[loop] + 32;
}
```

The keywords **class** and **typename** are interchangeable in the template argument list. The above function can be called without template metaparameters as in the first call to **copy( )** in the program below. The class identifiers **C1** and **C2** are then automatically determined from the types of the function parameters so that **C1** evaluates to **double** and **C2** evaluates to **int.** Alternatively, as in the next line of the program, the template metaparameters can be explicitly supplied in the metaparameter list. Here both arguments in the function call are accordingly interpreted as **double*** arguments. Finally, in the third call to **copy( )**, the template function arguments are both implemented as **char** arrays

```
main(){
	double aDouble[10], aDoubleNew[10];
	int aInt[10] = {1};
	copy (aDouble, aInt, 10);
	cout << aDouble[0] << endl ;        //Output : 33
	copy < double, double > (aDoubleNew, aDouble, 10);
	char c1[5], c2[5] = {'A'};
	copy (c1, c2, 5);
	cout << c1[0] << endl;              //Output: a
}
```

Default template metaparameter values cannot be specified in a function template, however a non-template specialization of a function can overload a template function with the same name. Hence, if a second non-template **copy( )** function were defined in the above program with specific parameter types such as

```
void copy(char output[], char input [], int n){
	for (int loop = 0; loop < n; loop++)
		output[loop] = input[loop] + 32;
}
```

this function would be called in place of the templated copy function whenever the **copy** is passed two character arrays as in the third function call in **main()** above.

20.11 Templates and classes

A **template** class is coded analogously to a function template. As a concrete example, which assumes that the classes **C1** and **C2** have appropriately overloaded stream insertion operators (<<)

```
#include <iostream.h>
#include <sstream.h>

template <class C1, class C2 = int, int aN = 1> class Logical{
	C1 iC1;
	C2 iC2;
	int iN;
public:
	Logical (C1 aC1, C2 aC2, int aN) : iN(aN)
		{iC1 = aC1, iC2 = aC2;}
```

```
        void and(){cout << (iC1 && iC2)
              << " " << iN << endl;}
        string asString();
};

template <class C1, class C2, int aN>
              string Logical<C1, C2, aN>::asString(){
        stringstream sout;
        sout << iC1 << ' ' << iC2 << endl;
        return sout.str();
}
```

Then the output of

```
main(){
        Logical<int, bool> Test(0, 0, 3);
        Test.and();
        cout << Test.asString();
}
```

is

```
0 3
0 0
```

(recall that **0** is automatically cast to **false** in an assignment to a **bool** variable). Note carefully the syntax of the scope label of the **asString( )** function when its body is supplied outside the class definition. The so-called non-type template parameter **aN** cannot be of floating point, class, pointer, or array type. In the code, **aN** can be assigned a compile-time constant of a compatible type (for example, if **aN** is of type **bool** it can be set to either of the **bool** constant values **true** or **false** or to a variable of type **const bool**). Default metaparameters (that is, metaparameters that are assigned default values) can be omitted from the template metaparameter list when declaring objects; in this case, the default values are employed. As for ordinary functions, default metaparameters must appear last in the template argument list.

A static member variable of a templated class has a different realization for each template instantiation, which occurs each time a distinct set of template metaparameters is passed to the template. That is, the program

```
template< class T > struct MyStatic{static int iI;};

template< class T > MyStatic < bool >::iI;
template< class T > MyStatic < char >::iI;

main(){
        MyStatic< bool > MB1, MB2;
        MyStatic< char > MC1, MC2;
        MB1.iI = 2;
        MyStatic< char >::iI = 'a';
        cout << MB2.iI << " << MyStatic <bool>:: iI
             << ' ' << MC2.iI << endl;        //Output: 2 2 97
}
```

yields the output 2 2 97. Referring to static template variables through the syntax **classname<typename>::staticvariablename** identifies the variable as static and is therefore recommended.

The choice of possible template arguments is quite broad as it can include further template metaparameters, any constant expressions except for **float** or **double** types, and the addresses of external objects, which include function names, references, and pointer variables. While a full discussion of templates is therefore necessarily lengthy, the simple example below provides a basic illustration of these features

```
template <class T = int> struct square{
        double operator () (T aX){return aX * aX; }
};

int cube(int x){return x * x * x;}

template < typename T, int (*aF) (int) > double test( int aI ){
        cout<<(*aF)(10) << ' ';
        return aF( aI );
}

main (){
        cout<<test<square<double>, cube>(6.0);   //Output: 1000 216
}
```

If a problem can be coded with either template classes or inheritance, template classes are generally preferable, since they lead to a simpler program structure and faster execution times. To design a template class, code should first be written and verified for a single specialized case, after which the program can be generalized through the introduction of an appropriate and minimal set of template metaparameters.

20.12 The complex class

Nearly all C++ compilers provide a template class implementation of complex numbers through the **<complex.h>** header file. For example, a representation of a complex number in which both the real and imaginary parts are **double** values is written

```
complex <double> c;
```

To initialize a **complex** object to a value such as $1+2i$, either of the following two statements can be employed

```
complex<double> c = complex<double>(1., 2.);
complex<double> c(1.,2.);
```

The real and imaginary parts of **c** can be accessed through the **real( )** and **imag( )** member functions, for example **c.real( )** and **c.imag( )**. All standard arithmetic operators, such as +, −, *, /, +=, . . . as well as the stream insertion and extraction operators << and >>, operate properly on **complex** objects. Additional functions in the **complex** class include **arg( ), conj( ), abs( ), polar(r, t)**, which yields re^{it} where r and t must be of a floating point type **cos( ), cosh( ), exp( ), log( ), log10( ), pow( ), sqrt( ), sinh( ), tan( ),** and **tanh( ).** The following program illustrates typical **complex** operations

```
#include <complex.h>

main(){
        complex<double> c1 = complex<double>(1., 2.), c2(-1., 1.);
        complex<double> c3 = (c1 + c2)/3.0;
        c1 = exp(M_PI * c3);
        c2 = polar(1., M_PI / 2.);
        cout << c1 << '\n' << c3 << '\n' << c2 << endl;
}
```

The output is

```
(-1,1.22461e-16)
(0,1)
(6.12303e-17,1)
```

20.13 The standard template library

While data structures such as lists, vectors, queues, and trees often appear in programming exercises, standard library implementations are far more reliable, comprehensive, and accessible. Accordingly, modern C++ compilers contain a common library of data structures named the STL (standard template library). The library possesses numerous advanced features that are beyond the scope of this text; however, a discussion of several basic techniques suffices to convey a working understanding of the relevant interfaces.

The basic data types implemented by the STL library are the **vector, dequeue, list, set, multiset, map, multimap, stack, queue,** and **priority_queue**. Each is defined through a corresponding **#include** statement, for example if a **stack** appears in a program, its definition must be preceded by the statement **#include <stack.h>** (generally the **.h** is optional). However, **queue** and **priority_queue, map** and **multimap**, and **set** and **multiset** share the include files **<queue.h>**, **<map.h>**, and **<set.h>**, respectively.

Although the member functions of all of the above classes differ, certain functions are common to almost all STL classes. These include the destructor, copy constructor and assignment operator, the **empty( )** function that returns **true** if the container is empty, the **max_size( )** function that can be used to set the maximum container size, the **size( )** function that returns the number of elements in the container, **erase( )** and **clear( )**, which erase a given number and all elements from the container respectively, and the comparison operators <, >, <=, >=, = =, and **!=** that compare the elements of two similar classes.

Another shared concept is the **iterator**, which is an object that can be used to step through the elements of an associated container (except for the **stack, queue, and priority_queue** containers). An **iterator** object, which is defined, for example for a vector, through the syntax

```
vector <object type>:: iterator it;
```

(the iterator name **it** can of course be arbitrarily chosen), possesses member functions **it.begin( )** and **it.end( )**. The former of these returns a pointer to the first member of the associated container (here **vector**) and the latter points to a fictitious container element one element beyond the end of the container. The iterator class includes an increment (++) and a decrement (−−) operator that displace the pointer by one container element, the comparison operators = =, !=, and the assignment operator, =. The iterators of the random access **vector** and **dequeue** classes further permit random access in the container through the index operator, for example **it[i]** or the corresponding pointer expression *(**it** + **i**).

We illustrate the above concepts through a simple example that stores and then prints four values {0, 1, 2, 3} in a four-component STL **vector** object. Note that the elements of an STL object are initialized to zero when defined

```
#include <iostream.h>
#include <vector.h>
#include <iterator.h>

main (){
        vector < int > aV(3);          // Elements initialized to zero
        vector < int > :: iterator itV;
        aV.push_back( 3 );             // Size expanded to 4 elements;
                                       // aV[3] = 3
        aV[1] = 1 ;
        aV.at(2) = 2;
        for (itV = aV.begin( ); itV < aV.end( ); itV++)
                cout << *itV << endl;
}
```

The output of this program is

```
0
1
2
3
```

The effect of the **push_back(3)** statement is to add an additional element **aV[3] = 3** to the end of the vector **aV**. The **at( )** function unlike the index operator **[ ]** implements bounds checking so that an attempt to write a value into a nonexistent vector element generates a runtime exception. Other important functions are the **sort( )** function of the **vector** and **dequeue** classes, and the **insert( ),**

remove ( ), and **resize( )** functions of the **vector** class that respectively insert or remove values at specified locations and resize the vector to a user-specified value. The **resize( )** function can be employed to avoid the automatic resizing performed by the **push_back( )** operation, which either increases the size of a full vector by one for each invocation or doubles its size.

The data types **set**, **multiset,** and **priority_queue** automatically sort values as they are inserted according to a comparator functor, which is a class that is employed in the same manner as a function. For simple built-in data types, the default comparator automatically sorts the values in increasing order as for the set below

```
#include <iostream.h>
#include <set>                                    //.h is optional here
#include <iterator>

main (){
        double a[3] = {1.3, 2.5, 0.3};
        set < double > S1(a, a + 3);
        set < double > :: iterator itS;
        S1.insert(0);
        S1.insert(1.3);
        for (itS = S1.begin( ); itS != S1.end(); itS++)
                cout << *itS << endl;
}
```

which yields the output

```
0
0.3
1.3
2.5
```

For user-defined data types, however, the user must specify the comparator functor. As an example using the set class, we have

```
#include <iostream.h>
#include <set>
#include <iterator>

class C{
public:
        int value1, value2;
};

class myGreater{
public:
        myGreater(){
        }
        bool operator() (const C &C1, const C &C2){
                return(C1.value1 < C2.value1);
        }
};
```

```
main (){
        C C1 = {1, 2};
        C C2 = {3, 4};
        set <C, myGreater> s;
        set < C, myGreater > :: iterator itV;
        s.insert(C1);
        s.insert(C2);
        for (itV = s.begin(); itV != s.end(); itV++)
                cout << (*itV).value1 << ' ';
}
```

which yields the output 1 3. The principle of operation of the functor is that since a constructor is absent in the class **myGreater**, the function **myGreater(C1, C2)** must be invoked without first instantiating a **myGreater** object. The member operator function **operator( )** of **myGreater** is then called directly.

Finally, STL contains a set of mathematical functions that can be used to operate on STL data structures as in

```
#include<iostream.h>
#include<algorithm>
#include<iterator>
#include<vector>
#include<numeric>                  // for accumulate

int tripleFunction( int aValue ){return 3 * aValue;}

int tripleAccumulateFunction( int aPartialSum, int aValue ){
       return aPartialSum + 3 * aValue;
}

main(){
ostream_iterator < int > myOut(cout, "loop element \n");
vector < int > myVector (20);
fill(myVector.begin(), myVector.end(), 5);
                                  // places 5 in all 20 positions
replace(myVector.begin(), myVector.begin() + 3, 5, 3);
                                  // changes first 3 values to 3
vector < int > tripleResult(3); // 3 element zero vector
transform(myVector.begin() + 2, myVector.begin() + 4,
       tripleResult.begin(), tripleFunction);
                                  // 3rd and 4th elements copied and
                                  // tripled
sort(tripleResult.begin(), tripleResult.end());
                                  // new vector sorted
copy(tripleResult.begin(), tripleResult.end(), myOut);
cout << endl;
cout << accumulate(tripleResult.begin( ), tripleResult.end( ), 0);
                                  // sums elements
cout << endl;
cout << accumulate(tripleResult.begin( ), tripleResult.end( ), 0,
       tripleAccumulateFunction);
                                  // triples and sums
}
```

which yields the output

```
0 loop element
9 loop element
15 loop element

24
72
```

In the program, an **ostream_iterator** object **myOut** is first created for **int** values. Iterators of this type are automatically associated with **cout** so that the **copy( )** function can later be employed to print out the elements of the container (here a **vector**) by sending these to the **ostream_iterator**. A twenty-element and a three-element **vector** object, which store **int** values, are defined that are by default initialized to zero. Subsequently, all elements of the twenty-element vector, **myVector,** are set to 5 and then 5 in the first three elements is replaced by 3. The third and fourth elements in **myVector** are tripled through the user-defined function **tripleFunction( )** that is passed as a parameter to the STL **transform ( )** function. The **transform( )** function further places the result into the first two positions of the three-element zero vector **tripleResult**. The three elements are sorted and the result printed. Finally, two forms of the **accumulate( )** function are used to process the elements of the **tripleResult** vector. The first simply adds the elements, while the second calls the user-supplied global function **tripleAccumulateFunction( )** to compute the sum of three times each element.

The **transform( )** function accepts certain predefined functors as arguments such as in

```
transform(inputVector1.begin(), inputVector1.end(),
        inputVector2.begin(),outputVector.begin(),
        minus<double>());
```

which subtracts the elements of **inputVector2** from the corresponding elements of **inputVector1** and places the result into **outputVector.** Many other such functors can be used in place of **minus<double>( )** such as **plus<double>( )** or **multiplies<double>( )**; further, the output vector can be the same as either of the input vectors. As well, the inner product of two vectors can be calculated through the statement

```
inner_product(inputVector1.begin(), inputVector1.end(),
        inputVector(2).begin(), 0., plus<double>(),
        multiplies<double>());
```

20.14 Structures and unions

A structure is identical to a class, except that the internal data members are public rather than private by default, and the keyword **class** is replaced by **struct**. That is to say, the code

```
class MyClass {
        int j;
public:
        double i:
};
```

is exactly equivalent to the code

```
struct MyClass {
        double i;
private:
        int j;
};
```

The **struct** keyword is often employed for classes without internal functions that only contain public variables. The **struct** is then effectively a generalized array formed from variables with more than one data type.

A **union** is identical to a **struct** except that all data members of the union share the same storage as can be seen from the program below

```
union U{
        int i;
        int j;
};

main () {
        U U1;
        U1.i = 3;
        U1.j = 0;
        cout << U1.i << endl;            // Output: 0
}
```

The **union** can be employed to conserve memory in programs for which the span of time during execution when each variable in the **union** is accessed does not overlap that of any other of the **union** variables. Unions can be also used to access different memory locations in a data type as illustrated by

```
#include <iostream.h>

union U{
        int i;
        char c[4];
};

main() {
        U U1;
        U1.i = 65;
        cout << U1.c[0] << endl;         // Output: A
}
```

which outputs A, corresponding to ASCII code 65. If a union contains two variables, such as an integer and a single character variable that occupy differing amounts of storage, the variables overlap at the least significant (rightmost) memory bits.

Structure (and class) definitions can be nested as in the example below. However, a nested class is *only* accessible to the members and friends of the class

```
struct S {
        int iS;
        struct C {
                int iS;
        };
        print( ) {C C1 = {2}; cout << iS << '\t' << C1.iS;}
};

main() {
        S S1 = {1} ;
        S1.print( );       // Output: 1 2
        cout << S1.C.iS; // Compile error: C not visible outside S
}
```

20.15 Bit fields and operators

Experimental applications, such as interfacing the computer with external devices, often require manipulation of single bits or sets of bits. This process is simplified through the introduction of a *bit field* type. The general form of a bit field is

```
struct myBitField {
        unsigned firstBit: 1;
        unsigned secondBit: 2;
        unsigned thirdBit: 1;
};
```

Then writing, for example, **myBitField MB;** generates a one-bit variable **MB.firstBit**, a two-bit variable **MB.secondBit**, and a second one-bit variable **MB.thirdBit**. These variables are ordered in either ascending or descending memory locations, depending on the computer hardware; however, the first bit appearing in the bit field is assigned to the least significant bit (bit zero) of the memory space reserved for the field.

C++ also provides operators that act on individual bits within a given variable. These are the bitwise logical operators and (**&**), or (|), exclusive or (ˆ), not (∼) and the right and left shift operators >> and << that shift the bit pattern of the variable right and left by a given number of bits respectively (and introduce 0 in place of bits that are dropped). The **&** and >> operators are illustrated by the two methods below for printing out the bit pattern associated with an arbitrary character variable. Note that the union overlaps the single-bit field with the rightmost, least-significant bit of the character variable

```
#include <iostream.h>

struct aBit {
        unsigned bit: 1;
};
```

```
union charBit {
        char c;
        aBit aB;
};

main() {
        charBit CB;
        CB.c = 'e';
        for (int loop = 0; loop < 7; loop ++) cout << " ";
        for ( int loop = 0; loop < 8; loop++ ) {
                (CB.aB.bit) ? cout << "1" : cout << "0";
                CB.c = CB.c >> 1;
// alternative constuction
//              (CB.c & 1) ? cout << "1" : cout << "0";
//              CB.c = CB.c >> 1;
        }

}
```

20.16 Assignments

Part I

In the problems below give the output of the code if the program will function properly. If there is an error or if a potential error condition, such as overflow, underflow, incorrect memory access, etc., arises so that the program result or runtime state is unpredictable, indicate instead where the error is and what type of error has been made. Some programs may have more than one error.

(1)
```
class C {
        int iA[3];
        public:
        void print ( ) {
                static int j;
                cout << iA[j++] << endl;
        }
        C(int aB[ ]) { for (int i = 0; i < 3; i++)
          iA[i] = aB[i]; }
};

main() {
        int a[3] = {1, 2, 3};
        C C1(a);
        C1.print( );
        C1.print( );
}
```

(2)
```
struct S {
        int iI, iJ;
};

union U {
        S iS;
        int iI[2];
};
```

```
main () {
        U U1;
        U1.iI[0] = 3;
        U1.iI[1] = 2;
        cout << U1.iS.iJ << endl;
}
```

(3)
```
class String {
public:
        String(char* ch){myString = ch;}
        String() {myString=NULL;}
        char* getString() {return myString;}
        char* myString;
};

template <class T,const int n> class MyClass {
public:
        void add(int m, T* T1) {a[m] = T1;}
        T* get(int m) {return a[m];}
        private:
        T* a[n];
};

main(){
        const int i = 10;
        MyClass<String, i> MyClass1;
        String String1("string1");
        String String2("string2");
        MyClass1.add(0,&String1);
        MyClass1.add(1,&String2);
        cout << MyClass1.get(1)->getString();
}
```

(4)
```
class B {
public:
        int x;
        virtual void print() {cout << x << endl;}
};

class D : public B {
        public:
        void print(int i) {cout << 3*i << endl;}
};

class E : public B {
        public:
        void print() {cout << 2*x << endl; }
};

main () {
        D D1;
        D* D2 = &D1;
        D2->x = 2;
        D2->print(2);
```

```
        E* E1 = (E*) D2;
        E1->print();
}
```

(5)
```
template <class T> T min (T a, T b){
        return a < b ? a : b;
}

class C {
public:
        double iC;
        operator int ( ) const {return iC;};
};

main () {
        C C1, C2;
        C1.iC = 3.0;
        C2.iC = 6.0;
        cout << min(C2, 3) << endl;
}
```

(6)
```
template <class T> T min (T a, T b){
        return a < b ? a : b;
}

class C {
public:
        double iC;
        operator int () const {return iC;};
};

main () {
        C C1,C2;
        C1.iC = 3.0;
        C2.iC = 6.0;
        cout << min(C2, C1) << endl;
}
```

(7)
```
class C {
public:
        virtual void print( ) = 0;
        int i;
};

class D : virtual public C {
public:
        virtual void print( ) = 0;
};

class E : virtual public C{
public:
        virtual void print( ) {cout << "In E," << i;}
} ;
```

```
class F : public D, public E {
public:
        void print( ) {cout << "In F," << i;}
};

void main ()
{
        F &F1 = *(new F);
        C &C1 = F1;
        C1.i = 10;
        F1.print();
}
```

(8)
```
class C {
public:
        virtual void print() = 0;
        int i;
};

class D : public C {
public:
        int i;
};

class E : public C {
public:
        virtual void print() {cout << "In E," << i;}
};

class F : public D, virtual public E {
public:
        void print() {cout << "In F," << D::i;}
};

main () {
        F &F1 = *(new F);
        D &D1 = F1;
        D1.i = 10;
        F1.print();
}
```

(9)
```
class C {
public:
        virtual void print( ) = 0;
        int i;
};

class D : public C {
public:
        int i;
        void print( ) {cout << "In D" << endl};
};

class E : public C {
public:
```

```
        void print() {cout << "In E" << endl;}
};

main () {
D D1;
E E1;
C C[3] = {D1, E1};
C[1].print( );
}
```

(10)
```
static int b = 3;

int myFun(int &i){
static int b;
b += i;
return(b);
}

main(){
int n = 2;
myFun(n);
cout << myFun(n);
}
```

(11)
```
class MyClass {
public:
        static int iP;
        MyClass(int aX) {iX = aX; iP++;}
        int iX;
};

int MyClass::iP = 3;

main() {
        MyClass::iP = 0;
        MyClass M1(2);
        MyClass M2(M1);
        cout << M1.iP << endl;
}
```

(12)
```
class Base {
public:
        void virtual print() {cout
              << "Base Class" << endl; }
};

class Derived : public Base {
public:
        void print() {cout
              << "Derived Class" << endl; }
};

main() {
        Derived D1;
        Base* B1 = &D1;
```

```
        B1->print();
}
```

(13)
```
class Base {
public:
        void print() {cout << "Base Class"
            << endl; }
};

class Derived : public Base {
public:
        void print() {cout << "Derived Class"
            << endl; }
};

main() {
        Derived D1;
        Base* B1 = &D1;
        B1->print();
}
```

Part II

(14) Generate a templated version of the linked list class in problem 11 of Chapter 19 such that the nodes can store an object of any given class type.

Use the same **main( )** program (with templated arguments) with nodes containing a class **C** with two internal variables **int iIntValue** and **double iDoubleValue.** Submit your program and the output. Additional functions, which should be introduced in the templated version of the program together with the **main( )** program and the signatures of a few representative templated functions, are given below

```
#include <sstream.h>

class C {
public:
        int iIntValue;
        double iDoubleValue;
        string asString( )
};

template <class T> class Node {
public:
        Node( T );
        Node( );
        string asString( )
};

template <class T> Node<T>::Node (T aData)

template <class T> Node<T>::Node ( )

template <class T>
        void List<T>::addNode(Node<T> *aNew, int aN)

template <class T> string List<T>::asString( )
```

```
typedef Node<C>* NCP;

main() {
        List<C> L;
        C C1 = {1, 2};
        C C2 = {2, 3};
        C C3 = {3, 4};
        NCP N1 = new Node<C>(C1);
        NCP N2 = new Node<C>(C2);
        NCP N3 = new Node<C>(C3);
        L.addNode(N1, 0);
        L.addNode(N2, 0);
        L.addNode(N3, 6);
        C C4 = {4, 5};
        NCP NewNode = new Node<C>(C4);
        L.addNode(NewNode, 2);
        cout << L.asString( );
        {
           List<C> L2;
           L2 = L;
           L2 = L2;
           cout << L2.asString( );
        }
        List<C> L3(L);
        L3 = L;
        cout << L3.asString( );
}
```

(15) Generate classes **Meter** and **Feet** with appropriate conversion operators such that writing **Feet (m)**, where **m** is a **Meter** variable and **Meters (f)** where **f** is a **Feet** variable, produce the correct values of the **Feet** and **Meter** internal variables, respectively. Use this program to convert 15 feet into meters and back into feet. Hand in the program with your output.

(16) Using static class variables, generate a simple class with one public **int** data member, a constructor, and a destructor that has a second **int** data member **iDataMember**, which indicates the number of objects of the class that exist at a given time (increment these data members by one when an object of the class is constructed and decrement them by one when an object is destructed). Put print statements into the constructor and destructor that display the number of active objects of the class every time this number changes. Hand in the program and your output for the following **main( )** program

```
main() {
        C C1(1);
        {
               C C2(0);
               {
                      C C3(3);
                      cout << C3.iDataMember;
               }
        }
}
```

(17) By modifying the program presented in Section 20.15, write a program using a bit field and a union that calculates and then displays the bit patterns for the integers 15 and −15. Hand in the program and your output.

(18) One procedure for finding the global extrema of a function (optimization) is the genetic algorithm. This method is patterned after the acceptance of favorable genetic mutations in natural selection. Consider a population with **numberOfGenomes** individuals, each containing one genome of a certain type. The genome in turn consists of **numberOfGenes** genes, each of which in our implementation can possess random values between the values 1 or −1. The benefit of a particular gene is determined by a **fitness( )** function; only the **numberOfGenomes** individuals with the highest value of this function survive to the next generation. Reproduction consists of selecting each pair of genomes out of the population and constructing from this pair a new genome that is a random mixture of the genes of the two parents. In other words, if one parent has a gene 0.8 in position 3 and the second parent has a gene −0.6 in this position, the position 3 gene of the offspring is −0.6 or 0.8 with equal probability. Of course, if both genomes have the same gene in position 3, the resulting genome conserves this property. After a number of generations, undesirable combinations of genes become gradually more infrequent in the population, while the opposite is true of gene combinations that maximize the **fitness( )** function.

Here we employ the genetics algorithm to illustrate the application of the STL classes. We employ the algorithm to determine the optimal Fourier series coefficients to a test function $f(x) = \text{sign}(x)$ over the interval from −1 to 1. While far more efficient procedures obviously exist for determining Fourier coefficients, the procedure below does not employ any specific numeric or analytic properties of the sinusoidal basis functions and can therefore be immediately adapted to arbitrary optimization problems. In our case, since the test function is antisymmetric, the Fourier series is restricted to antisymmetric sine functions, namely

$$s_n(x) = \sin(n\pi x)$$

that are orthogonal and are normalized to unity over the interval [−1, 1]. For each **Genome** object, the N_g values contained in the vector **iGene** correspond to the N_g Taylor coefficients denoted by g_i below

$$f_{approx}(x) = \sum_{i=0}^{N_g-1} g_i s_i(x)$$

At each generation, we generate $N_g(N_g - 1)/2$ new genomes by randomly mixing the genes of each pair of genomes in the population. The previous generation is stored in a **Genome** vector, then each pair of genomes in this vector is mixed and placed into a dynamically allocated new **Genome** and finally this **Genome** is placed into the **Genome** priority queue to contribute to the next generation of genomes. To improve the convergence, we assume that the N_g most-fit individuals survive, whether or not they belong to the previous generation or to the new generation. Therefore, the genomes are stored in a priority queue that accepts up to N_g genomes, which it automatically sorts according to the value of the **fitness( )** function, defined simply as

$$E(f_{approx}) = \sum_{j=0}^{N_f-1} \left(f_{approx}(x_j) - f(x_j)\right)^2$$

where N_f is denoted by **numberOfEvaluationPoints** in the program below. Thus if the fitness value of this genome is smaller than the smallest fitness value of the genomes in the set, the new genome replaces the old genome (note that, in this application, we search for genomes that *minimize* the fitness function).

After each generation, the program displays the fitness value of each surviving genome and graphs the difference between the approximated value of the test function and sign(x) over the specified interval.

The task is therefore to complete the following program, based on the description above. You must employ the STL classes where specified. This problem is also designed to illustrate the use of static class members.

```
#include <iostream>
#include <queue>
#include <vector>
#include <functional>
#include <math>
#include <string>
#include <strstream.h>
#include <algorithm>
#include <iomanip.h>
#include "dislin.h"
#include <numeric>

namespace std{

   class Genome {
   public:
      Genome (int aNumberOfGenes);
      Genome ( );
      double fitness( ) const;
      void setFitness(double aFitness);
      double numberOfGenes( );
      double & gene(int i);
      vector<double>& gene( )
      string toString( ) const;
   private:
      vector < double > iGene;
      vector < double >::iterator iGeneIterator;
      int iNumberOfGenes;
      double iFitness;
   };

// This class contains only static functions and variables
// and stores the required numerical routines.
   class Optimum {
   public:
```

```
      static double testFunction(double x) {
         if ( x > 0 ) return 1;
         else if ( x == 0 ) return .5;
         else return -1;
      }
      static void computeEvaluationPoints(double
         aLeftLimit, double aRightLimit)
// use the pushback member function of the vector class to
// place the grid point positions into the evaluationPoints
// vector.

      static void computeBasisFunctions( )
      sineFunctions = new vector<double>
                        [numberOfEvaluationPoints];
// Employ the push_back function to place the result for sin(
// x[loop] * pi * n ) into the appropriate position of the
// array sineFunctions[loop]

      static void computeTestSeries( ) {
// Use the transform function to set the values of the test
// function evaluated on the points in the evaluation point
// vector into the vector testFunction that will be used each
// time the fitness function is evaluated.

      static double computeFourierApproximation
            (Genome& aGenome, int aLocation)
// Use the inner_product function to rewrite
// the following code. You will need to use
// sineFunctions[aLocation]. begin( ).
// double sum = 0;
// for ( int loop = 0; loop < aGenome.numberOfGenes( );
//      loop++ )
// sum += aGenome.gene(loop) * sineFunctions[loop][aLocation];

      static void computeFitness(Genome& aGenome)
// Here first generate a vector of doubles and place the value
// of the Fourier approximation to the function at each
// evaluation point in this vector using the
// computeFourierApproximation( ) function. Next use the
// transform function to compute the difference between the
// elements of this vector and the vector, testSeries of exact
// values. Finally, compute the fitness function using the
// inner_product STL function and place it into aGenome using
// the setFitness method of the Genome class.

      static void draw(Genome aGenome)
// Dynamically allocate vectors of type float for the
// evaluation point positions and the Fourier series
// approximation. Plot the approximation and delete the
// dynamically allocated variables. (The x and y vectors can
// also be included as internal class variables to save
// computer time).
// Internal static data members:
   private:
```

```
        static int numberOfEvaluationPoints;
        static int numberOfGenes;
        static vector< double > testSeries;
        static vector< double > evaluationPoints;
        static vector< double >::iterator vectorIterator;
        static vector< double > * sineFunctions;
};

   Genome::Genome (int aNumberOfGenes) :
      iNumberOfGenes(aNumberOfGenes)
// Use the iGeneIterator to iterate through the genes of the
// Genome and set each using
// *iGeneIterator = 2 * (float(rand ( ))-.5) RAND_MAX;)/
// Then call the static computeFitness of the Optimum class
// using the current object as an argument.

// get and set member functions
   double Genome::fitness( ) const

   void Genome::setFitness(double aFitness)

   double Genome::numberOfGenes( )

   double& Genome::gene(int i)

   string Genome::toString( ) const
// Outputs the fitness of the Genome using a stringstream.

   class myGreater2
// This functor will compare two Genome objects and return a
// value of true if the first operator argument has a smaller
// fitness — in this case the Genomes are ordered in the
// priority queue with the genome with the largest value of
// the fitness function at the top of the queue.

   class GenomeQueue {
   public:
      void advanceGeneration( );
      void pushGenome (Genome aGenome)
// Place the genome argument into the priority queue. Pop the
// top (largest fitness value) element off the queue using the
// pop( ) member function of the PriorityQueue class if the
// number of elements in the queue after the new element is
// pushed onto it exceeds iNumberOfGenomes.

      void drawTop( )
// Use the static draw function of the Optimum class and the
// top() function of the Priority queue to graph the
// approximation corresponding to the genome with the largest
// fitness value in the queue.

      string toString( ) {
// Pop each element stored in iQueue, call its toString
// function, place this information into a stringstream and
// finally push the element onto a second storage queue and
```

Chapter 21
Program optimization in C++

Advanced features, such as templates, virtual function calls, pointers, and references, greatly enhance the flexibility of C++, but at the same time restrict the optimizations available to the compiler. Fortunately, proper structuring of code together with use of the extensive language resources of C++ can often be employed to circumvent these restrictions, as discussed below.

21.1 Compiling

Long compilation times constitute a major C++ performance issue. A header file that handles complex tasks, such as **fstream.h**, contains numerous declarations and definitions and often includes subsidiary header files. Further, each template invocation with a unique set of metaparameter types yields a new full implementation of the template code. Since multiple source files can generate the same template instance, some compilers employ or supply as an option a process called prelinking in which the required instances are determined at link-time followed by recompilation one or more times. Unfortunately, even longer compilation times then result if the template classes are numerous or possess a complicated structure. With these aspects of the language in mind, insuring that each header file is limited in size and dedicated to a single, specific, well-defined task will both shorten development times and enhance program transparency.

21.2 Critical code segments

Although advanced C++ constructs often carry substantial computational overhead, very often a large fraction of the computation time is spent on a few inner loops. These time-critical code segments can be located through a profiler, such as the **gprof** contained in Dev-C++. Subsequently, enhancing execution speed requires an understanding of the compiler optimizations. Many older procedures for enhancing program operation are ineffective or actually lead to increased compilation or execution times on current C++ compilers, since they prevent more efficient automatic compiler transformations. Some of these manipulations are:

- Replacing multiplication by addition, floating point artithmetic by integer arithmetic and **double** by **float (floats** are often represented as **doubles** and floating point arithmetic is extremely efficient).
- Insuring that the matrix indices furthest to the right are incremented in the innermost loops.
- Reversing loops so that the loop index runs backward from the largest value of the iterator to the smallest value.
- Unrolling loops so that a loop containing, for example, 20 iterations of one statement is replaced by a loop containing five iterations of four identical statements.
- Using the **register** keyword to instruct the compiler to place certain variables into internal CPU memory registers.
- Blocking or tiling loops to fit subcalculations into main or cache memory. While the performance improvements can be substantial, such techniques require precise knowledge of the computer hardware and are specific to a given computer system. The relative performance advantage will also decrease as the speed of main memory approaches that of cache memory.

Methods that will often improve performance are:

- Replacing division with multiplication and small integer powers by repeated products.
- Insuring that, for example, a product that is repeatedly evaluated in an inner loop and that always results in the same value is moved to an outer loop or outside all loops. Similarly, if the same calculation is repeatedly performed in an inner loop, its result should be stored in an appropriate array or matrix variable. As an example

```
for (int i = 0; i < 100; i++) {
        for (int j = 0; j < 100, j++)
                A[i][j] = sin(2.0 * M_PI * i / 100.0) * sin(2.0
                      * M_PI * j / 100.0);
}
```

can be replaced with

```
double sinConstant = 2.0 * M_PI * 0.01;
double sinArray[100];
for (int i = 0; i < 100; i++) sinArray[i] = sin(sinConstant * i);

for (int i = 0; i < 100; i++) {
      double sA = sinArray[i];
      for (int j = 0; j < 100, j++) A[i][j] = sA * sinArray[j];
}
```

- Improving the speed of numerical algorithms. For complex tasks, this normally involves employing a more sophisticated program library. However, for physical problems that are known not to generate exceptional cases, programs such as those found in this text will often perform more rapidly. The source code can further be tailored to the specific nature of the problem, for example if input variables do not change or all elements of an array are identical the code should be modified appropriately.

- Use of appropriate data structures. The efficiency of numerical methods that frequently access data can in many cases by improved by replacing arrays with appropriate data structures, such as representing a sparse array by a linked list.
- Selecting the highest correctly functioning optimization flag. Compilers generally offer optimization flags that implement certain assumptions about the program structure that are not always guaranteed to be valid. Consequently, a program should first be compiled and run without optimization for a few typical cases and the results then compared with the output at different optimization levels.
- Restricting the number of template metaparameters. As noted earlier, each time a different object or class type is passed to a template metaparameter, a new code instance is generated, leading to long compilation times.
- Moving functions out of templated classes. Even if a member function of a templated class does not employ template metaparameters, code for the function will still be regenerated for each instance of the template. Accordingly, such functions should be moved into a non-templated class, struct, or the global space.
- Inlining functions: Small functions that are called frequently during execution should be declared **inline** (recall, however, that functions appearing in the class definition are normally automatically inlined).

Numerous sophisticated C++ language features dramatically influence execution times. Several important cases are discussed individually below.

21.3 Virtual functions

Because a virtual function call is not resolved until runtime, the compiler is unable to determine the sequencing of instructions and processing of data around the function dispatch. Since this precludes normal code optimizations, virtual functions should be limited to infrequently called code sections or to large functions for which the time spent initializing the function is a small fraction of the computation time. Unfortunately, the greatest benefit of virtual functions is often realized in the opposite case. For example, classes that represent different types of matrices (for example sparse, diagonal, full) and that inherit from the same virtual base class may implement different overloaded versions of the parenthesis operator to address individual elements. This virtual operator is then called whenever a matrix element is accessed, potentially degrading inner loop performance.

Several procedures exist that simulate virtual function calls when these can be resolved at compile time. Such methods, however, preclude dynamic type resolution, and therefore structures such as the heterogeneous dynamic object collections of Section 20.6.

In particular, the polymorphism of virtual function calls can be replaced at compile-time by suitable template constructions. In one method, the type of the polymorphic object is specified at the point of declaration or definition through a template argument as in the **main( )** function below

```
template<class T> class Matrix;

class MatrixOne {
  public:
    void print( ) {cout << "in matrix one" << endl;}
};

class MatrixTwo {
  public:
    void print( ) {cout << "in matrix two" << endl;}
};

template <class T> class Matrix {
  public:
    T iT;
    void print( ) {iT.print( );}
};

template <class T> void matrixPrint( Matrix<T>& aMatrix ) {
  aMatrix.print( );
}

main() {
  Matrix<MatrixOne> M1;
  matrixPrint( M1 );
}
```

In this case, a **Matrix** object wrappers at the point of definition an object of the class specified by its template metaparameter. Each of its polymorphic functions then forward calls to this wrappered object. A drawback of this approach is that if a polymorphic function is not defined in all of the possible template metaparameter classes, unexpected compile-time errors will arise if a program attempts to access the omitted method. Similarly, any polymorphic method appearing in the subclasses, **MatrixOne** and **MatrixTwo**, must appear in the **Matrix** class; therefore, the base class will accumulate numerous member functions. Note that, since a matrix must be declared with a template metaparameter, dynamic type resolution is clearly precluded.

The above limitations (with the exception of dynamic type resolution) can be circumvented through a somewhat more involved procedure, illustrated by the following program (cf. Barton and Nackman, 1994).

```
#include <iostream.h>

const int dimension = 2;

template <class Leaf> class Matrix {
  public:
```

```
    double operator( )(int,int);
    void print( );
};

struct Methods {
  static void load(int aArray1[ ][dimension],
      int aArray2[ ][dimension]) {
    for (int loopOuter = 0; loopOuter < dimension; loopOuter++) {
      for (int loopInner = 0; loopInner < dimension; loopInner++)
        aArray1[loopOuter][loopInner] =
          aArray2[loopOuter][loopInner];
    }
  }

  template <class Leaf> static void printMatrix
        (Matrix<Leaf>& aMatrix) {
    for(int loopOuter = 0; loopOuter < dimension; loopOuter++) {
      for(int loopInner = 0; loopInner < dimension; loopInner++)
        cout << aMatrix(loopOuter, loopInner) << " ";
      cout << endl;
      }
    cout << endl;
  }
};

class DiagonalMatrix : public Matrix<DiagonalMatrix> {
  public:
    DiagonalMatrix(int aArray[ ][dimension])
      { Methods::load(iM, aArray); }
    double operator( )(int a1,int a2) {
      if(a1 == a2) return iM[a1][a2];
      else return 0;
    }
    void print( ) { cout << "Diagonal Matrix first element" <<
      iM[0][0] << endl; }
    int iM[dimension][dimension];
};

class FullMatrix : public Matrix<FullMatrix>{
  public:
    FullMatrix(int aArray[ ][dimension])
     { Methods::load(iM, aArray); }
    double operator( )(int a1, int a2) { return iM[a1][a2]; }
    void print( ) { cout << "Full Matrix first element" <<
     iM[0][0] << endl; }
    int iM[dimension][dimension];
};

template <class Leaf> double norm(Matrix<Leaf>& aMatrix){
  double sum = 0;
  for(int loopOuter = 0; loopOuter < dimension; loopOuter++) {
    for(int loopInner = 0; loopInner < dimension; loopInner++)
      sum += aMatrix(loopOuter, loopInner) *
        aMatrix(loopOuter, loopInner);
```

```
  }
  return sqrt(sum);
}

template <class Leaf>
  double Matrix<Leaf>::operator( )(int r, int c){
  return ((Leaf &)(*this))(r,c);
}

template <class Leaf> void Matrix<Leaf>::print( ){
  return static_cast<Leaf&>(*this).print( );
}

int main(){
  int matrix[dimension][dimension] = {1, 2, 3, 4};

  DiagonalMatrix D1(matrix);

  FullMatrix F1(matrix);
  D1.print( );
  F1.print( );
  cout << endl;

  Methods::printMatrix(D1);

  Methods::printMatrix(F1);

  cout << endl;

  cout << "The norm of D1 is " << norm(D1) << endl;
  cout << "The norm of F1 is " << norm(F1) << endl;
}
```

Here the **Matrix** class again accepts as a template metaparameter the name of the class to which it forwards its methods. However, each of these classes, for example **DiagonalMatrix**, is declared as a subclass of the **Matrix** class with itself as template metaparameter, for example **Matrix<DiagonalMatrix>**. Therefore, declaring a **DiagonalMatrix** object simultaneously generates an associated instance of the **Matrix** class with a **DiagonalMatrix** template argument. When a **DiagonalMatrix** object is passed to the function **printMatrix( )** that takes a templated **Matrix** argument, the function operates on the object as a **Matrix<DiagonalMatrix>** object (note here the use of a **struct** in the above program to encapsulate commonly used related numerical routines into a simulated namespace). In the function body, the **print( )** function of the **Matrix** class is called. This function is then delegated by the **Matrix** class to a function of an identical name in the **DiagonalMatrix** class, as the **Matrix** object is downcasted to a **DiagonalMatrix** object before the **print( )** function is executed. The downcasting operation is performed by converting the dereferenced **this** pointer of the **Matrix** object to a derived **Leaf** object at compile-time through either **((Leaf &)(*this))** or the equivalent expression **static_cast<Leaf&>(*this)**).

If the **Matrix** class possesses numerous methods, the above program can be simplified by providing a single internal method, **asLeaf( )**, in the **Matrix** class that performs the required downcasting

```
template <class Leaf> Leaf& Matrix<Leaf>::asLeaf( )
{
  return ((Leaf &)(*this));
}
```

Then each operator can be rewritten in a manner similar to

```
template <class Leaf> double Matrix<Leaf>::operator( )(int r, int c)
{
  return asLeaf( )(r,c);
}
```

A third closely related alternative stores a pointer to the **Leaf** class as an internal variable of the **Matrix** class, for example

```
template <class Leaf> class Matrix {
  public:
    double operator( )(int,int);
    void print( );
    Leaf *iLeaf;
    Matrix( ) {
          iLeaf = static_cast<Leaf*>(this);
    }
};
```

In this case the **print( )** and **operator( )** functions become simply

```
template <class Leaf> double Matrix<Leaf>::operator( )(int r, int c)
{
  return (*iLeaf)(r,c);
}

template <class Leaf> void Matrix<Leaf>::print( )
{
  return iLeaf->print( );
}
```

Since the base class can access features of a derived class, while the derived class in standard fashion has access to methods in the base class, the base class logically groups classes that require common functionality. That is, a method defined in the base class can be shared by all derived classes; however, methods that are specific to the derived classes that depend on this method can be delegated in the base class to the derived classes.

21.4 Function pointers and functors

The frequent use of non-inline functions, especially function callbacks, which occur when functions are used as arguments or are accessed through function pointers (especially when the function that is pointed to is often not resolved until runtime), can result in considerably increased computation times, even when the functions are not virtual. As mentioned briefly in the discussion of the STL library in the previous chapter, functions can be replaced by functors, which are constructorless objects that implement the function body through a redefinition of the parenthesis operator. By following the name of the function by the parenthesis operator, this operator will be invoked, inlining the body of the parenthesis operator into the code at the position of the function call. Function callbacks can then be replaced by templates which associate the name of the functor class when it appears as a function argument with an object of the functor class type as in the following program

```
class square {
   public:
       double operator ( ) (double aX) {
             return aX * aX;
       }
};

template < class T > double test( int aI, T aF ) {
       return aF( aI );
}

main () {
       cout << test (6, square( )) << endl;       // Output: 36
}
```

However, in this implementation, parameters that are passed to the functor must be passed as constructor arguments and stored in appropriate internal variables.

A second, far more straightforward, procedure is to employ the function name or pointer as a template metaparameter. Since the compiler automatically generates code for each unique template realization, the function body is automatically inlined at compile time. The above example becomes

```
double square (double aX) {
       return aX * aX;
}

template < double T(double) > double test( int aI ) {
       return T( aI );
}

main () {
       cout << test<square> ( 6 ) << endl;       // Output: 36
}
```

21.5 Aliasing

While access to arbitrary memory locations at runtime through pointers or reference variables greatly expands the flexibility of the C++ language, unusual program constructs arise that preclude program optimization. For example, suppose two vectors are repeatedly multiplied to construct the matrix outer product

```
const int n = 10;
void outerProduct( double aProduct[ ][n], double aX[ ]) {
        for (int outerLoop = 0; outerLoop < n; outerLoop++)
              for (int innerLoop = 0; innerLoop < n; innerLoop++)
                    aProduct[outerLoop][innerLoop] = aX[outerLoop] *
                          aX[innerLoop];
}

main () {
        double p[n][n];
        double *x = new double[n];
//      double *x = p[1];
        for (int loop = 0; loop < n; loop++) x[loop] = loop + 1;
        outerProduct(p, x);
        cout << p[1][2] << endl;                          // Output: 6
}
```

Clearly, the compiler can greatly improve the performance of the executable program by loading the elements of the vector **aX** into an internal memory register (or into low-level cache memory for a large-scale problem) and then performing the calculation.

Unfortunately, such an optimization is normally precluded in C++ since the compiler cannot exclude the possibility that the statement **double*x = p[1]**, that is commented out in the above program, is present in place of the uncommented definition on the prior line. In this case, once the first value, 1, is written to **p[1][0]**, the value of **x[0]** is automatically set to 1 as well; subsequently the remaining initial values of **x** are overwritten and the new output is 12. Since the pointer assignment can be located at an arbitrary code distance from the relevant computational algorithm, unless the compiler can analyze the entire program structure, it cannot optimize the code.

At present, most compilers perform basic analysis to determine if variables have unexpected aliases, possibly under the control of compilation options. However, such procedures are generally neither exhaustive nor error free. A suggested change to the future C++ standard would introduce a new keyword **restrict**, which the programmer can employ to indicate that a variable is free of aliases so that appropriate optimizations can be safely applied.

21.6 High-performance template libraries

Iterations involving small data arrays, such as algebraic products or functions of small vectors or matrices, can often be accelerated by replacing repeated operations over array elements with a single expression. While compilers will generally perform these "loop unrolling" procedures automatically in simple contexts, such as the inner product of two vectors, more complicated expressions require appropriate restructuring of source code.

Recursive template structures provide sophisticated methods for transforming source code. For example, the inner product of two four-element vectors, which would normally be evaluated through loop iteration as in

```
for (loop = 0; loop < 4; loop++) result += x[loop] * y[loop];
```

can be automatically replaced by the potentially faster algebraic expression

```
result = x[0] * y[0] + x[1] * y[1] + x[2] * y[2] + x[3] * y[3];
```

using templates. The template performs the following recursive sequence of steps

```
Vector<int> v1, v2;

f(v1, v2) → g<3> (v1, v2) → v1[3] * v2[3] + g<2>(v1, v2) → ...
```

which, together with the definition **g<0>(v1, v2) = v1[0]** * **v2[0]**, yields the desired expression at compile-time. The code for the above procedure requires template specialization; that is, calling a templated feature with specific argument values by eliminating the corresponding arguments from the template argument list and instead employing the specialized value directly as a parameter to the templated class, which is the origin of the syntax **template < > struct S<0>** in the program below

```
#include <iostream.h>

template <class T, int N> class Vector {
        public:
                T operator[ ] (int aI) {return iArray[aI];}
                T iArray[N];
};

template<class T, int N>
  inline float f (Vector<T, N> &aArg1, Vector<T, N> &aArg2) {
        return S<N-1>::g(aArg1, aArg2);
  }

template <int I> struct S {
        template <class T, int N>
        static T g(Vector<T, N> &aArg1, Vector<T, N> &aArg2) {
                return aArg1[I] * aArg2[I] +
                        S<I-1>::g(aArg1, aArg2);
        }
};
```

```
template < > struct S<0> {
     template <class T, int N>
     static T g(Vector<T, N> &aArg1, Vector<T, N> &aArg2) {
             return aArg1[0] * aArg2[0];
     }
};

main() {
      Vector <float, 4> x = {1, 2, 3, 4}, y = {2, 3, 4, 5};
      cout << f(x,y) << endl;
}
```

While, as previously noted, loop unrolling operations are generally performed by C++ compilers, recursive templates have been applied to far more complicated numerical algorithms. These and other rapid template procedures are collected in several numerical libraries, such as Blitz++, FFTW, PhiPAC, POOMA, A++/P++, that can in many cases handle matrix operations at roughly the same speed as languages such as FORTRAN that are designed for numerical computation. Since considerable time and sophistication are required to design such matrix classes, the reader is advised to employ preexisting class libraries wherever possible. In certain cases, however, user-defined template implementations can be very useful. For example, a scientific language such as FORTRAN will compute a small integer power of a quantity through iterated multiplications so that, for example, **pow(a, 8)** can be generated through three multiplication operations. C++ however applies a far less efficient numerical algorithm to perform this operation (therefore, such powers should be programmed explicitly in time-critical loops). While a template can instead be introduced for each individual power, such as

```
template <class T> inline T eighthPower( T aI ) {
      T aSquared = aI * aI;
      T aFourth = aSquared * aSquared;
      return aFourth * aFourth;
}
```

the techniques of this section can be employed to construct a general method for such cases.

21.7 Assignments

(1) Implement all three variants of the Barton and Nackman trick noted in the text (Section 21.3) and demonstrate that these yield the same result.

(2) Develop a template metaprogram along the lines of the program for the vector product presented in Section 21.6 that instead evaluates a product of three four-component vectors according to the formula

$$\vec{A} \cdot \vec{B} \cdot \vec{C} = \sum_{i=1}^{4} A_i B_i C_i$$

(3) Run and explain the output of the following program

```
#include <iostream.h>

template <int N> struct A {
   enum { c = N * A<N-1>::c };
};

template < > struct A<1> {
   enum { c = 1 };
};

main ( ) {
   cout << A<4>::c;
}
```

(4) Rewrite the **derivative** program of Section 12.1 in two ways according to the discussion of Section 21.4. In the first of these replace the argument of function type in the derivative( ) function by a suitable functor. In the second implementation, employ instead a suitable template metaparameter.

Part IV
Scientific programming examples

Chapter 22
Monte Carlo methods

In the final two chapters of this text, we present a more detailed discussion of two fundamental topics in modern scientific computation, namely Monte Carlo and partial differential equation solution methods. Our examination of Monte Carlo procedures ranges from simple integration applications to more complex variants that enable the rapid calculation of the probability distribution or density of states function for statistically unlikely events.

22.1 Monte Carlo integration

Monte Carlo methods are perhaps most frequently applied to multidimensional integration with complex integrands and boundaries. The procedure encloses the integration region inside a reference region, such as a rectangular prism of known area. Randomly placed points are generated inside the reference region and the ratio of the number of points that fall into the integration region to the total number of sample points is computed. Multiplying this ratio by the area of the reference region yields an estimate of the integral. Unfortunately, the flexibility of the procedure is compromised by its slow rate of convergence, as the accuracy is proportional to the number of sampling points.

We illustrate the Monte Carlo integration procedure by integrating a one-dimensional monotonic function between two user-supplied end limits. The routine generates random samples within a rectangular reference region delineated by upper and lower limits in the independent variable (x-limits) and between the values of the function at these two limits in the dependent variable (y-limits). (If the function is not monotonic, the y-limits should be placed below the minimum and above the maximum value of the function over the specified interval.) After each pair of random values (x_i, y_i) is generated, the variable **sum** is incremented by one if the associated point lies below $f(x_i)$. Finally, the ratio of the final value of **sum** to the total number of sample points is multiplied by the area of the reference region and the result is added to the difference in the x-limits multiplied by the value of the lower y-limit to arrive at an estimate of the integral.

```
double f(double aX) {return aX*aX;}

double monteIntegral(double aA, double aB, int
        aNumberOfRealizations){
    double yLower = f(aA); // Lower limit of rectangular region
    double yUpper = f(aB); // Upper limit of rectangular region
    double regionArea = (yUpper - yLower) * (aB - aA);
    int sum = 0;
    double randMax = float(RAND_MAX);
    // Generate points randomly within the rectangular region;
    // if they fall below the function to be integrated sum is
    // incremented by one.
    for (int loop = 0; loop < aNumberOfRealizations; loop++) {
        double xValue = rand( ) / randMax * (aB - aA) + aA;
        double yValue = rand( ) / randMax *
          (yUpper - yLower) + yLower;
        if (f(xValue) < yValue) sum++;
    }

    // The integral is the fraction falling below the
    // integrand added to the area of the rectangular region
    // between the horizontal axis limits and the lower y-limit
    return sum * regionArea / aNumberOfRealizations + yLower *
      (aB - aA);
}

main() {
     double a, b, res;
     int numberOfRealizations;
     cout << "Input a, b, numberOfRealizations:";
     cin >> a >> b >> numberOfRealizations;
     cout.precision(10);
     res = monteIntegral(a, b, numberOfRealizations);
     cout << "a = " << a << "\nb = " << b <<
         "\nnumberOfRealizations = " << numberOfRealizations
         << "\nIntegral is " << res;
}
```

22.2 Monte Carlo evaluation of distribution functions

Monte Carlo techniques can as well be applied to statistical quantities such as probability distribution functions. Consider a physical system composed of N subsystems, each of which is characterized by a randomly varying *local* parameter, s_i. This local parameter could be a spin that can only possess certain discrete values or a continuous variable, such as the length of a transmission line, a rotation angle, or a resistance. For each set of local parameters one or more *global* parameter values E^j can be measured or numerically evaluated. These might be the pulse propagation time, the net resistance or the magnetization. Since the local parameters are not known deterministic variables but instead fluctuate from one realization of the system to another, due for example to thermal effects, manufacturing uncertainties, or component aging, they are termed stochastic variables. The problem is then to predict the probability that the global variables possess certain values when averaged over all realizations. From these probability distribution functions, statistical properties, such as the average magnetization, mean time to failure, or bit-error rate, can be evaluated.

The most straightforward technique for modeling a stochastic system is to assign sets of random values to the local variables in a manner consistent with the known statistical distribution of each of these variables. For each set of values (a realization) the N global variables of interest are calculated. This result is then placed in a bin of an N-dimensional histogram that records the total number of realizations for which the global values fall within a certain limited region of the global variable space. The distribution of events in the histogram for a sufficiently large number of realizations then yields the desired probability distribution function. Mathematically, if the function $I_B(\vec{s}_i)$ is one within a histogram bin, B, and zero outside, then the pdf after N_R realizations is given at the location of this bin by

$$p_B(\vec{E}) = p(\{E^{(1)}, E^{(2)}, \ldots, E^{(M)}\} \in B) = \frac{1}{N_R} \sum_{i=1}^{N_R} I_B(\vec{s}_i) \tag{22.1}$$

The above considerations are best illustrated by a trivial example, namely that of a one-dimensional random walk. In this case, the underlying system variables s_i can take on the two discrete values ± 1 corresponding to displacement right or left by a unit distance on a line. The global quantity of interest is the total displacement from the origin after N steps. Averaging this quantity over many realizations of the system variables yields a discrete distribution function of the probability that the walk terminates at a given displacement from the origin.

To generate an object-oriented program for the random walk, we create a **DistributionFunction** class that contains internal variables corresponding to the number of steps in each random walk, the histogram array, the number of histogram bins (the number of possible termination positions), and the total number of realizations. For graphing purposes, we also include an array that contains the distance of each histogram bin from the origin normalized by the expected mean distance, which is $\sqrt{N}$ for unit length steps. The total number of bins is twice the number of steps in the random walk plus one to accommodate the initial point at the origin; however, the walk always terminates an odd number of points away from the origin for an odd number of steps and an even number of points from the origin for an even step number.

The **DistributionFunction** class below contains a constructor that dynamically allocates and initializes to zero the histogram array, **iHistogram**, of size **2 * iNumberOfSteps + 1** as well as an array **iNormalizedDistance** that stores the normalized distance from the origin of each grid point. In the **propagate( )** member function a value of $+1$ or -1 is generated randomly at each step and used to update the value of a **position** variable that is initialized to $x = 0$, corresponding to the array index value **iNumberOfSteps** at the beginning of each walk. After each random walk, the histogram element corresponding to the final position of the random walk is incremented by unity. Finally, a **plot( )** routine graphs the logarithm of the final histogram values (after adding one to the histogram elements to prevent logarithms of zero) with a set of markers

```
#include <iostream.h>
#include "dislin.h"

// iNumberOfSteps: Number of steps in the random walk
// iNumberOfRealizations: Number of random walks
// iNormalizedDistance: Distance from the center point normalized
// by the expectation value of the displacement after the random
// walk
// iNumberOfBins: Number of histogram bins
// iCenterPoint: Grid point corresponding to starting point of
// random walk
class DistributionFunction {
        public:
        DistributionFunction(int aNumberOfSteps, int
          aNumberOfRealizations) : iNumberOfSteps(aNumberOfSteps),
          iNumberOfRealizations(aNumberOfRealizations),
          iNumberOfBins(2 * aNumberOfSteps + 1) {
               iCenterPoint = aNumberOfSteps + 1;
               iHistogram = new float[iNumberOfBins];
               iNormalizedDistance = new float[iNumberOfBins];
               for ( int loop = 0; loop < iNumberOfBins; loop++ ) {
                       iHistogram[loop] = 0;
                       iNormalizedDistance[loop] = (loop -
                         iCenterPoint) / sqrt(iNumberOfSteps);
               }
        }

// At each step in the random walk, the walker moves one
// position to the right or left
        DistributionFunction& propagate ( ) {
               for ( int outerLoop = 0; outerLoop <
                         iNumberOfRealizations; outerLoop++ ) {
                       int position = iCenterPoint;
                       srand(outerLoop);
                       for ( int innerLoop = 0; innerLoop <
                         iNumberOfSteps; innerLoop++ )
                              position += 2. * (rand( ) % 2) - 1;
                       iHistogram[position] += 1;
               }
               return *this;
        }

// The logarithm of the histogram of final positions is plotted
// after adding one to each element and normalizing so that the
```

```
// largest element is plotted as 1.
        DistributionFunction& plot( ) {
                metafl("XWIN");
                float maxValue = FLT_MIN;
                for ( int loop = 0; loop < iNumberOfBins; loop++ )
                        if (iHistogram[loop] > maxValue) maxValue =
                            iHistogram[loop];
                for ( int loop = 0; loop < iNumberOfBins; loop++ )
                        iHistogram[loop] = log10((iHistogram[loop] + 1)
                            / (maxValue + 1));
                            // Plot points with square markers
                qplsca(iNormalizedDistance, iHistogram,
                            iNumberOfBins);
                return *this;
        }

        private:
        int iNumberOfSteps;
        int iNumberOfBins;
        int iCenterPoint;
        int iNumberOfRealizations;
        float *iHistogram;
        float *iNormalizedDistance;
};

main() {
        int numberOfSteps = 40;
        int numberOfRealizations = 1000000;
        DistributionFunction D1(numberOfSteps,
                        numberOfRealizations);
        D1.propagate( ).plot( );
}
```

The function **srand( )** in the above program generates a new value of an internal random number *seed* parameter each time it is called with a different argument. As a consequence, the random number sequence is started at a different position each time the **rand( )** function is called, decreasing correlations between the random values. Since the probability distribution function is approximately Gaussian (of the form $ae^{-x^2/2a^2}$), graphing the logarithm of the distribution function yields an inverted parabola with every odd-numbered histogram element equal to zero.

Running the above program, however, results in a somewhat incorrect result for the distribution function, as shown below. In particular, observe the large difference between the first non-zero histogram value and the zero level, which indicates that unlikely events are not adequately sampled by the random number generator. This feature persists when the number of realizations is increased (Fig. 22.1).

More reliable random number generators are readily available; one of these is the freeware random number generator **randomc.zip** contained in the \random subdirectory of the CD-ROM that you should unzip into your current program

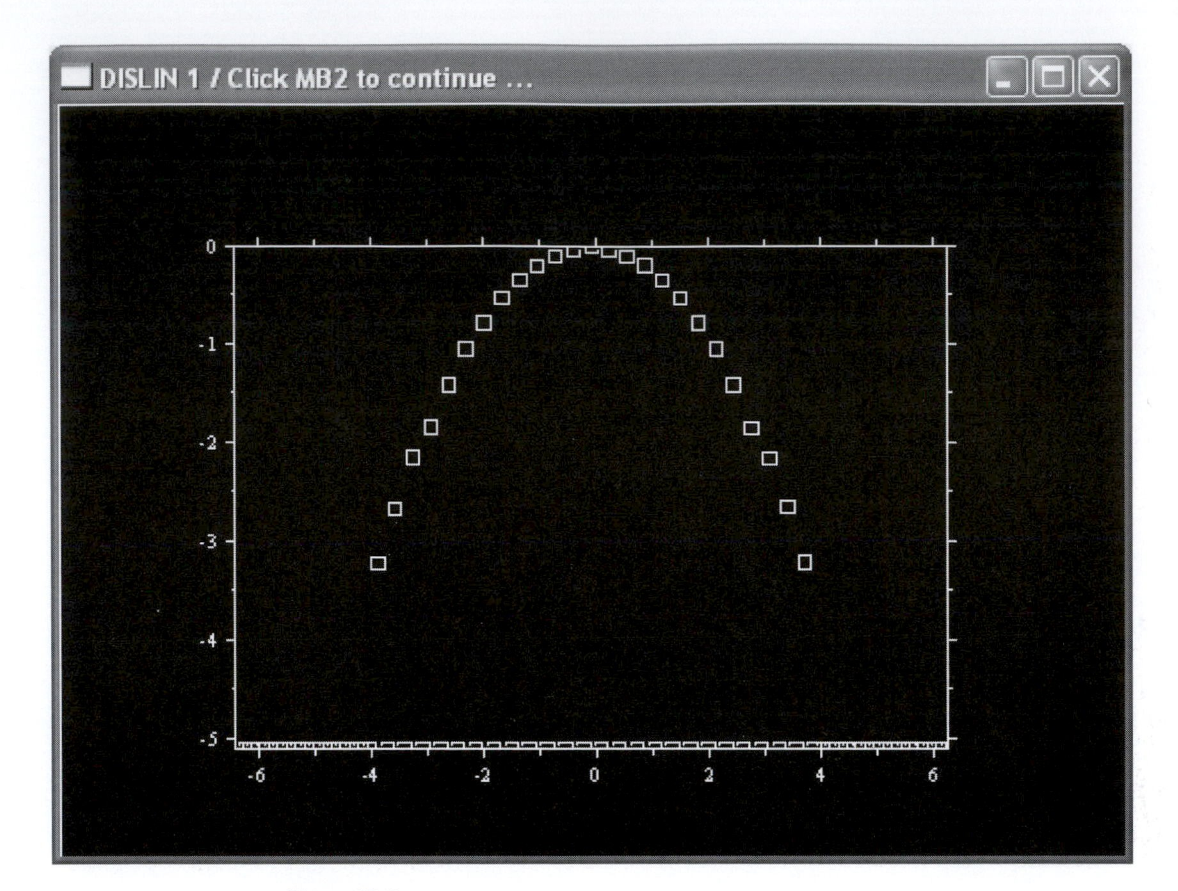

Figure 22.1

directory. With this set of random number routines, writing

```
double r = rg.Random( );          // Random floating number in [0 1]
```

generates a random floating point number lying in the interval [0, 1] while

```
int k = rg.IRandom(0, 10);
```

yields a random integer from 0 to 10. This and other features of the random number generators are described in the files **randomc.htm** and **ex-ran.cpp**. Accordingly, insert the lines

```
#include "randomc.h"
#include "mersenne.cpp"           // members of class TRandomMersenne
```

at the end of the include file section at the beginning of your program and replace the **propagate( )** function with the alternative code

```
DistributionFunction& propagate ( ) {
   long int seed = time(0);                        // random seed
```

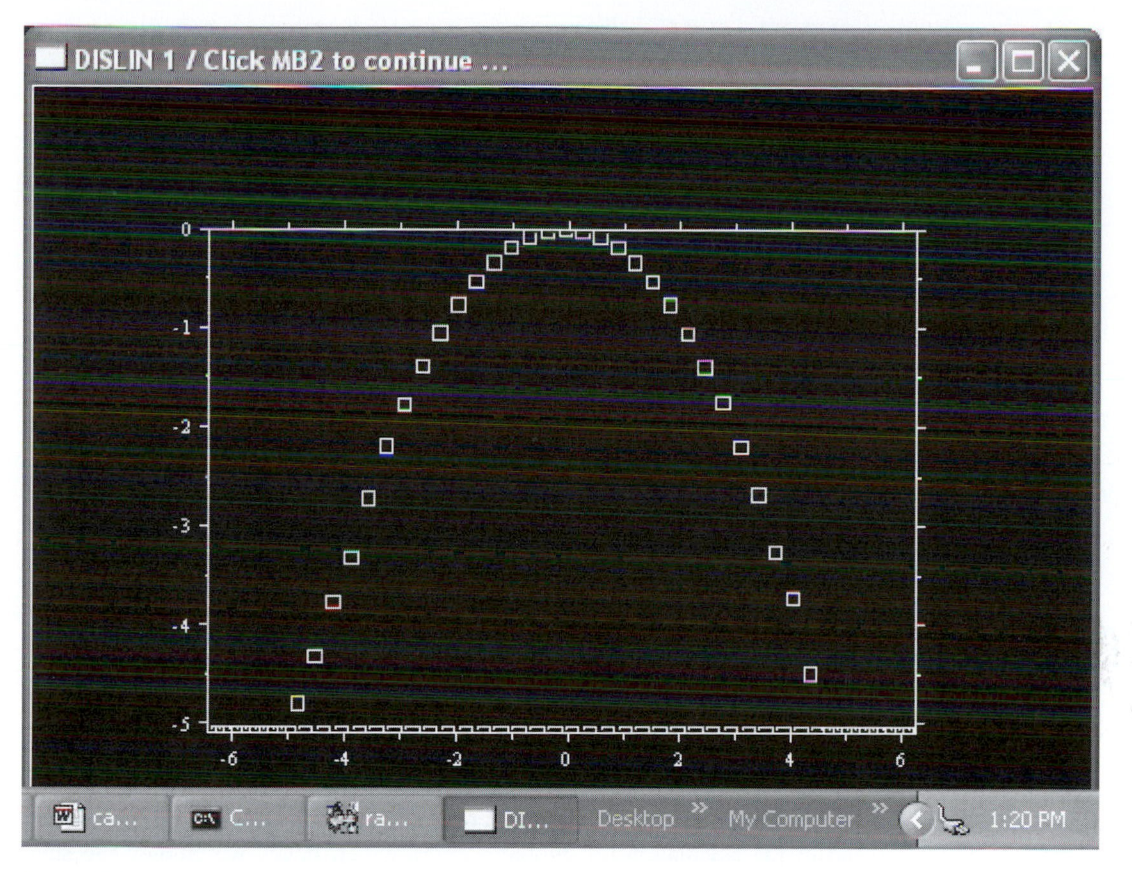

Figure 22.2

```
   TRandomMersenne rg(seed);
   for ( int outerLoop = 0; outerLoop
           < iNumberOfRealizations; outerLoop++ ) {
      int position = iCenterPoint;
      for ( int innerLoop = 0; innerLoop < iNumberOfSteps;
            innerLoop++ )
         position += 2 * rg.IRandom(0,1) - 1;
      iHistogram[position] += 1;
   }
   return *this;
}
```

The resulting probability distribution function is shown in Fig. 22.2.

22.3 Importance sampling

The standard Monte Carlo procedure described above is simple to program and closely models the physical random walk process. However, estimating the probability of infrequent events in the tail regions of the probability distribution functions requires many realizations. Fortunately, the region of local parameter space that leads to the low-probability events of interest can often be estimated from physical or mathematical analyses. The sample space can then be weighted such that these events occur more frequently than in randomly generated samples. Since the resulting biased histogram, $I_B(\vec{s})$, overestimates the likelihood of these low-probability events it must be multiplied by a likelihood ratio

$$L(\vec{s}) = p(\vec{s})/p_B(\vec{s}) \tag{22.2}$$

that is the quotient of the unbiased, $p(\vec{s})$, and the biased pdf. This yields the physical probability density distribution

$$p(\vec{E}) = \frac{1}{N_R}\sum_{i=1}^{N_R} I_B(\vec{s}_i)L(\vec{s}_i) \tag{22.3}$$

The likelihood ratio is typically determined separately for each realization but can also be a known or predetermined function of the histogram bin parameters. In the first case, the ratio corresponds to the probability that the biased realization would occur in a normal unbiased simulation.

In the random walk illustration, we can bias the probability of moving away from the origin to a value, **importanceValue** = 0.6. The likelihood ratio can here be obtained for any final position using probability theory in which case the histogram values are multiplied only once by the values of the ratio at the end of the calculation (cf. problem 2). However, below we employ the standard procedure that generates the likelihood ratio for each realization of the random walk by initializing its inverse, designated by **product** in the program below, to unity and then multiplying this quantity by the probability of each step in the random walk. That is, **product** is multiplied by 0.6 for every step away from the origin and by 0.4 for every step towards the origin. (If the random walker is at the origin, the corresponding factor is 0.5.) At the end of the random walk, the histogram element corresponding to the final position is incremented by the likelihood ratio, **1/product,** eliminating the bias associated with the path. Note that the bias can now be changed as the program proceeds and additional information regarding the sample state distribution becomes available.

In the following code, the variable **sign** corresponds to the direction of the step relative to the origin. Further, since the likelihood ratios are large, at the end of the calculation the smallest non-zero histogram element is added to all elements of the histogram to facilitate plotting

```
#include <iostream.h>
#include <math.h>
```

```
#include "dislin.h"
// iNumberOfSteps: Number of steps in the random walk
// iNumberOfRealizations: Number of random walks
// iNormalizedDistance: Distance from the center point normalized
// by the expectation value of the displacement after the random
// walk
// iNumberOfBins: Number of histogram bins
// iCenterPoint: Grid point corresponding to starting point of
// random walk
class DistributionFunction {
public:

   DistributionFunction(int aNumberOfSteps, int
   aNumberOfRealizations, double aLimit) :
   iNumberOfSteps(aNumberOfSteps),
   iNumberOfRealizations(aNumberOfRealizations),
   iLimit(aLimit), iNumberOfBins(2 * aNumberOfSteps + 1) {
      iCenterPoint = aNumberOfSteps + 1;
      iHistogram = new float[iNumberOfBins];
      iNormalizedDistance = new float [iNumberOfBins];
      for ( int loop = 0; loop < iNumberOfBins; loop++ ) {
         iHistogram[loop] = 0;
         iNormalizedDistance[loop] = (loop -
            iCenterPoint)/sqrt(iNumberOfSteps);
      }
   }

// At each step in the random walk, the walker moves one
// position to the right or left. The probability of moving away
// from the origin is iLimit unless the walker is at the center in
// which case it is 0.5.
   DistributionFunction& propagate ( ) {
      for ( int outerLoop = 0; outerLoop <
            iNumberOfRealizations; outerLoop++ ) {
         srand(outerLoop);
         int position = iCenterPoint;
         float product = 1;
         float limit;
         int sign;
         for ( int innerLoop = 0; innerLoop < iNumberOfSteps;
               innerLoop++ ) {
            if ( position == iCenterPoint ) {
               limit = 0.5;
               sign = 1;
            } else {
               sign = (position -
                iCenterPoint)/abs(position - iCenterPoint);
               limit = iLimit;
            }
            int j = (rand( ) / float(RAND_MAX) < iLimit);
            product *= iLimit * j + (1 - iLimit) * (1 - j);
            position += (2 * j - 1) * sign;
```

```
         }
         iHistogram[position] += 1 / product;
      }
      return *this;
   }

// The logarithm of the histogram of final positions is plotted
// after adding the minimum non-zero value recorded in the
// histogram to each element and normalizing so that the largest
// histogram value is one.
   DistributionFunction& plot( ) {
      metafl("XWIN");
      float maxValue = FLT_MIN;
      float minValue = FLT_MAX;
      for ( int loop = 0; loop < iNumberOfBins; loop++ ) {
            if (iHistogram[loop] > maxValue) maxValue =
               iHistogram[loop];
            if (iHistogram[loop] < minValue &&
               iHistogram[loop]!= iHistogram[0])
                  minValue = iHistogram[loop];
      }
      for ( int loop = 0; loop < iNumberOfBins; loop++ )
        iHistogram[loop] = log10((iHistogram[loop] + minValue) /
          (maxValue + minValue));
                                    // Plot points with square markers
      qplsca(iNormalizedDistance, iHistogram, iNumberOfBins);
      return *this;
   }

private:

   int iNumberOfSteps;
   int iNumberOfBins;
   int iNumberOfRealizations;
   int iCenterPoint;
   double iLimit;
   float *iHistogram;
   float *iNormalizedDistance;
};

main( ) {
   int numberOfSteps = 40;
   int numberOfRealizations = 1000000;
   double importanceLimit = 0.6;
   DistributionFunction D1(numberOfSteps,
      numberOfRealizations, importanceLimit);
   D1.propagate( ).plot( );
}
```

The result for the probability distribution function is displayed in Fig. 22.3.

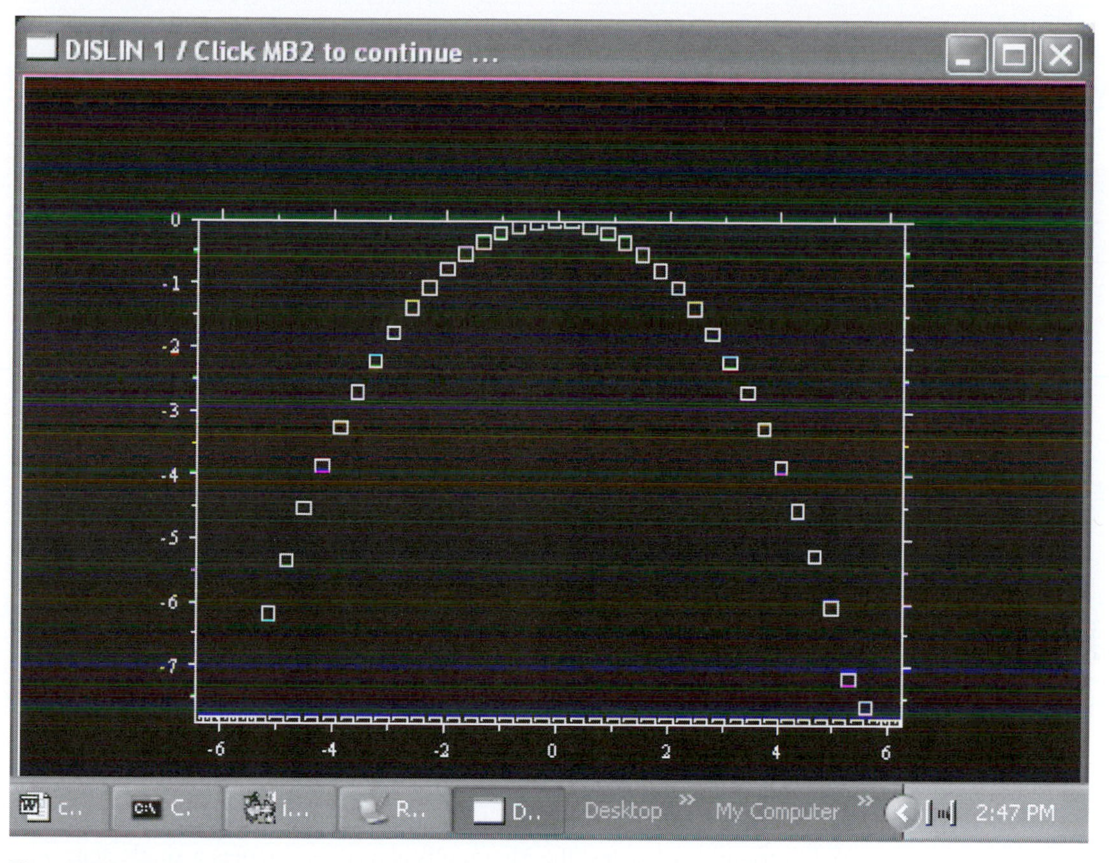

Figure 22.3

22.4 The Metropolis algorithm

The applicability of importance sampling is limited for complex problems, especially those involving several global variables, by the difficulty of determining the regions of optimal local parameter values towards which the system should be biased. Further, if M equivalent minima exist in widely separated regions of the local parameter space, a method that preferentially biases the sample states toward one of these minima can underestimate the probability distribution by the factor $1/M$. In such cases, *statistical* methods that maximize the entropy of the calculation (minimize the amount of additional information introduced or required by the numerical method) are preferable.

An effective method for biasing the local sample space towards large or small values of a single global variable or set of global variables (in the case of the one-dimensional random walk, the variable is of course the final distance from the origin) is provided by the technique of Markov chains. Here each new system realization (local variable set) is slightly modified from the previous realization. A rule is supplied that governs the acceptance or rejection of each such transition. If the transition is rejected, the previous result for the global variable is accepted an additional time.

The Metropolis algorithm is a Markov chain procedure that we employ in our example to increase the probability with respect to a single global system variable, E, of generating a sample state by a factor $e^{\beta E}$ (the method is used in other applications with obvious modifications to find the probability of events with low E). The resulting histogram is accordingly biased toward large values of E, the degree of bias increasing with β, corresponding to small temperatures in statistical mechanics. The required rule is to accept all transitions that raise the value of E, while permitting transitions that lower E by an amount $\Delta E < 0$ with a probability $e^{\beta \Delta E}$. Since a finite probability always exists for a transition to smaller E, successive states in the Markov chain can escape from local maxima, although after an average number of transitions that grows exponentially with the height of the maximum.

To understand the origin of the transition rule, consider a system consisting of a single pair of states. The probability of transition from the state with smaller E to the state with $E+\Delta E$ is unity for $\Delta E > 0$, while the probability of a transition in the reverse direction is $e^{-\beta \Delta E}$. However, as the calculation proceeds, the number of transitions in both directions must be equal. Accordingly, the number of times the upper state is visited must exceed the corresponding number for the lower state by the factor $e^{-\beta \Delta E}$. As this argument can be extended to any number and distribution of states, the Metropolis algorithm samples a single state with global variable E with a probability proportional to $e^{\beta E}$. Therefore, the physical probability distribution function, or in statistical mechanics the density of states function, for sampling a state around E is obtained by multiplying the calculated Metropolis distribution by the likelihood function $L(E) = e^{-\beta E}$.

More generally, consider an ensemble of realizations and designate the observed number of samples with system parameters $\vec{s}$ divided by the total number of samples after a certain number of simulation steps by $n(\vec{s})$. To obtain samples with a probability density $w(\vec{s})$, note that if the probability of a transition from a state $\vec{s}$ to a second state $\vec{s}\,'$ in the following simulation step is $P_{s \to s'}$, the change in $n(\vec{s})$ associated with transitions to and from $\vec{s}$ is

$$\begin{aligned} \Delta n(\vec{s}) &= n(\vec{s}\,')P_{s' \to s} - n(\vec{s})P_{s \to s'} \\ &= n(\vec{s}\,')P_{s \to s'} \left(\frac{P_{s' \to s}}{P_{s \to s'}} - \frac{n(\vec{s})}{n(\vec{s}\,')} \right) \end{aligned} \tag{22.4}$$

Clearly, if e.g. $n(\vec{s})/n(\vec{s}\,') < P_{s' \to s}/P_{s \to s'}$, $\Delta n(\vec{s}) > 0$, implying that states are on average arriving at $\vec{s}$ from $\vec{s}\,'$, and increasing the left-hand side of the inequality.

Asymptotically, therefore, the two ratios are identical. Further, the ratio of the transition probabilities is proportional to the probabilities of acceptance, A, of the transitions. The acceptance rule that for some specified function w, $w(\vec{s}) > w(\vec{s}\,')$, $A_{s' \to s} = 1$ and $A_{s \to s'} = w(\vec{s}\,')/w(\vec{s})$, while for $w(\vec{s}) < w(\vec{s}\,')$, $A_{s \to s'} = 1$ and $A_{s' \to s} = w(\vec{s})/w(\vec{s}\,')$, yields in both cases, $n_{\text{equilibrium}}(\vec{s})/n_{\text{equilibrium}}(\vec{s}\,') = w(\vec{s})/w(\vec{s}\,')$. In our example, $w(\vec{s}) \propto \exp(\beta E(\vec{s}))$ and the corresponding ratio of densities is $\exp(\beta(E(\vec{s}) - E(\vec{s}\,')))$.

In the following Metropolis algorithm program for the random walk probability distribution function, the Markov chain is implemented simply by changing the direction of a single, randomly chosen step in the random walk for each new realization. The entire random walk is executed and the modification is accepted according to the Metropolis transition rule, where ΔE is taken as the difference in the absolute value of the final positions of the new and old random walks. If a transition is rejected the histogram bin associated with the previous transition is incremented by unity, consistent with a self-transition. The final histogram is multiplied with the likelihood function in the **plot( )** routine below to generate the unbiased probability distribution function

```
#include <iostream.h>
#include "dislin.h"
#include "randomc.h"
#include "mersenne.cpp"        // members of class TRandomMersenne

// iNumberOfSteps: Number of steps in the random walk
// iNumberOfRealizations: Number of random walks
// iNumberOfBins: Number of histogram bins
// iNormalizedDistance: Distance from the center point normalized
// by the expectation value of the displacement after the random
// walk iCenterPoint: Grid point corresponding to starting point
// of the random walk
class DistributionFunction {
public:
  DistributionFunction(int aNumberOfSteps,
    int aNumberOfRealizations, double aKT) :
    iNumberOfSteps(aNumberOfSteps),
    iNumberOfRealizations(aNumberOfRealizations),
    iNumberOfBins(2 * aNumberOfSteps + 1), iKT(aKT) {
      iCenterPoint = iNumberOfSteps + 1;
      iHistogram = new float[iNumberOfBins];
      iNormalizedDistance = new float[iNumberOfBins];
      for ( int loop = 0; loop < iNumberOfBins; loop++ ) {
      iHistogram[loop] = 0;
      iNormalizedDistance[loop] = (loop -
      iCenterPoint) / sqrt(iNumberOfSteps);
      }
    }

  DistributionFunction& propagate ( ) {
    long int seed = time(0);   // random seed
```

```
 TRandomMersenne rg(seed);          // make instance of random
                                    // number generator
 int* randOld = new int[iNumberOfSteps];
 int* randNew = new int[iNumberOfSteps];
 // Initialize first set of steps in the Markov chain
 for ( int loop = 0; loop < iNumberOfSteps; loop++ )
   randOld[loop] = 2 * rg.IRandom(0,1) - 1;
 // Determine the final position of the walker for
 // this initial set of steps
 int positionOld = iCenterPoint;
 for ( int loop = 0; loop < iNumberOfSteps; loop++ )
   positionOld = positionOld + randOld[loop];
 // Loop over the number of random walks
 for ( int outerLoop = 0; outerLoop < iNumberOfRealizations;
   outerLoop++ ) {
 // In the new random walk, one step at a random location,
 // flipPosition, is reversed
   for ( int innerLoop = 0; innerLoop < iNumberOfSteps;
             innerLoop++ )
      randNew[innerLoop] = randOld[innerLoop];
   int flipPosition = rg.IRandom(0, iNumberOfSteps - 1);
   randNew[flipPosition] *= -1;
   // The new random walk location is computed
   int positionNew = iCenterPoint;
   for ( int innerLoop = 0; innerLoop < iNumberOfSteps;
         innerLoop++ )
     positionNew = positionNew + randNew[innerLoop];
   // The transition is accepted according to the
   // Metropolis condition; otherwise the old configuration
   // is retained
   if ( rg.Random( ) < exp((abs(positionNew - iCenterPoint) -
     abs(positionOld - iCenterPoint)) / iKT) ) {
      positionOld = positionNew;
      randOld[flipPosition] *= -1;
   }
 // The histogram element for the final state is finally
 // augmented by one.
 iHistogram[positionOld] += 1;
 }
 delete [ ] randNew;
 delete [ ] randOld;
 return *this;
}

DistributionFunction& plot( ) {
 metafl("XWIN");
 float maxValue = FLT_MIN;
 float minValue = FLT_MAX;
 // The biased histogram values are corrected by
 // multiplication with the likelihood ratio before plotting
 for ( int loop = 0; loop < iNumberOfBins; loop++ )
    iHistogram[loop] = iHistogram[loop] * exp(- abs(loop -
       iCenterPoint) / iKT );
 // The logarithm of the histogram of final positions is
```

```
      // plotted after adding the minimum non-zero value recorded
      // in the histogram to each element and normalizing so that
      // the largest histogram value is one.
    for ( int loop = 0; loop < iNumberOfBins; loop++ ) {
      if (iHistogram[loop] > maxValue) maxValue = iHistogram[loop];
      if (iHistogram[loop] < minValue &&
          iHistogram[loop]!= iHistogram[0])
          minValue = iHistogram[loop];
    }
    for ( int loop = 0; loop < iNumberOfBins; loop++ )
      iHistogram[loop] = log10((iHistogram[loop] + minValue) /
        (minValue + maxValue));
// Plot points with square markers
    qplsca(iNormalizedDistance, iHistogram, iNumberOfBins);
    return *this;
  }
private:
  int iNumberOfSteps;
  int iNumberOfBins;
  int iNumberOfRealizations;
  int iCenterPoint;
  double iKT;
  float *iHistogram;
  float *iNormalizedDistance;
};

main() {
  int numberOfSteps = 40;
  int numberOfRealizations = 1000000;
  double kT = 5;
  DistributionFunction D1(numberOfSteps, numberOfRealizations, kT);
  D1.propagate( ).plot( );
}
```

While the Metropolis algorithm works well in our probability density function calculation, as can be seen from Fig. 22.4 below, for small **kT** the sample space can become too strongly biased toward events with large E, so that the small-E region of the pdf is undersampled. Such calculations, like importance sampling, can require that the pdf be calculated for different biasing parameters (**kT** values) after which the resulting functions are joined in the regions where their slopes agree.

22.5 Multicanonical methods

A drawback of the Metropolis and importance sampling methods is that, while the biasing function enhances the number of events at large E, statistical accuracy for small E is considerably lowered. An optimal choice for the function should, however, sample all regions of the distribution function with equivalent statistical accuracy, while enabling rapid escape of successive realizations even from pronounced extrema.

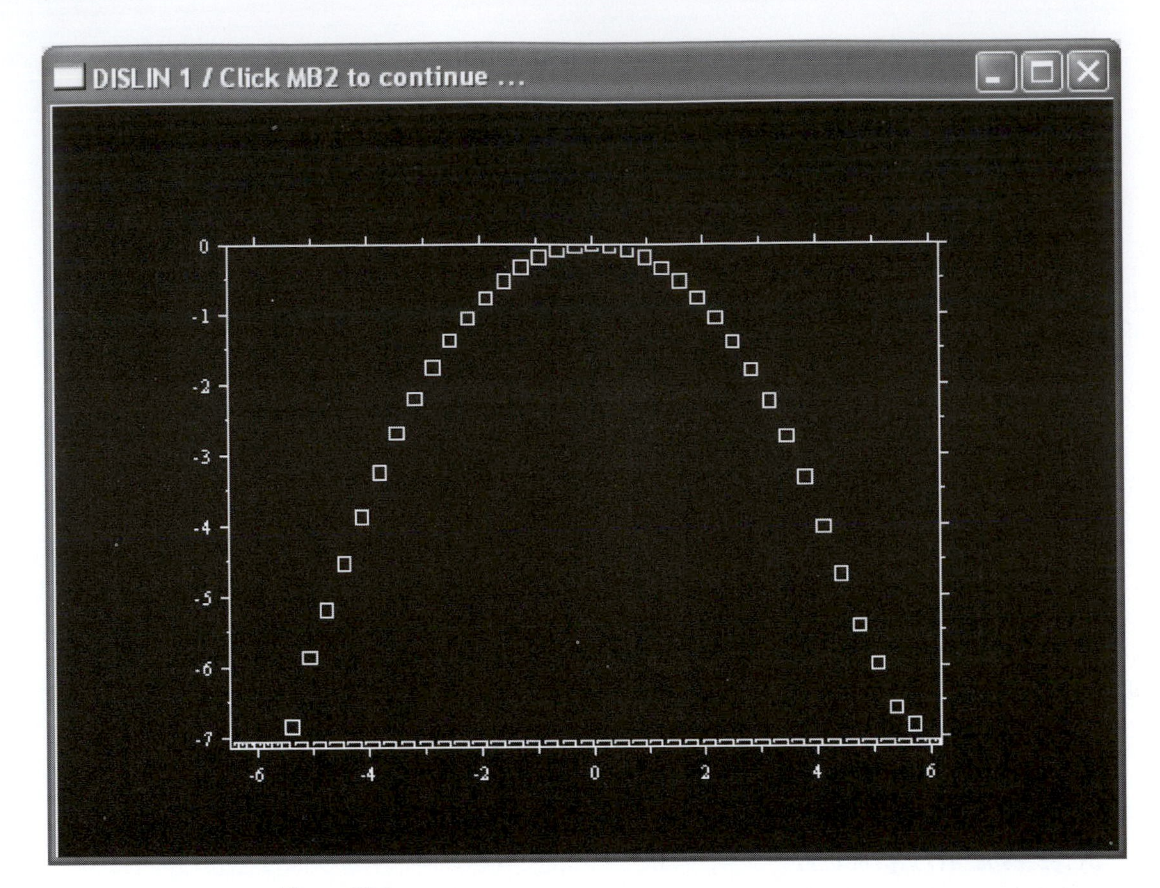

Figure 22.4

For the system realizations to sample the global parameter space uniformly in E and simultaneously avoid capture by extrema, the values of the E should execute a random walk as the number of realizations increase. The likelihood ratio, Eq. (21.2), is then proportional to the probability distribution function, since the biased pdf is constant.

An obvious problem with the above strategy is that the probability distribution function is not initially known. However, it can be determined iteratively through the following procedure. First, introduce an array, h, that will become proportional to the pdf after a sufficient number of iterations and a second histogram array, H, that will store the results of each individual iteration. Initialize all elements of each array to unity, $H = h_0 = 1$, and perform a Markov chain simulation with the rule that transitions that increase h are accepted with a probability $h_0(\vec{E}_n)/h_0(\vec{E}_{n+1})$ and those that decrease h with unit probability; in the initial step this is the standard Monte Carlo procedure. Label the resulting intermediate histogram H_0. The relationship

$$h_{i+1} = h_i H_i \tag{22.5}$$

yields the first approximation, h_1, to the true probability distribution function. The procedure is iterated with the new histogram, h_1, in place of h_0 to generate a new histogram H_1. The probability distribution function is approximated after N_I iterations by h_{N_I}.

The conceptual foundation of Eq. (22.5) can again be established through an analysis of a system of two states with equal occupation probabilities but different values of E. Suppose that the estimate, h_i, of the probability density function for one of the states is a factor of 2 too small. In this case, the probability of a transition out of the underestimated state decreases by a factor of 2, while the incoming probability remains constant. As a result, the subsequent histogram value, H_i, for the underestimated state will be enhanced by approximately a factor of 2, so that the product of this intermediate histogram and the previous estimate of the probability distribution function will far better approximate the true constant probability density. As a byproduct, successive iterations increasingly sample low-probability regions of the distribution function.

The multicanonical random walk program is easily obtained from the Metropolis code. With the improved random walk generator, we have

```
#include <iostream.h>
#include "dislin.h"
#include <math.h>
#include "randomc.h"
#include "mersenne.cpp"        // members of class TRandomMersenne

// iNumberOfSteps: Number of steps in the random walk
// iNormalizedDistance: Distance from the center point normalized
// by the expectation value of the displacement after the random
// walk
// iNumberOfRealizations: Number of random walks per multicanoni-
// cal iteration iNumberOfIterations: Number of multicanonical
// iterations
// iNumberOfBins: Number of histogram bins
// iCenterPoint: Grid point corresponding to starting point of the
// random walk
class DistributionFunction {
public:
  DistributionFunction(int aNumberOfSteps, int
         aNumberOfRealizations, int aNumberOfIterations) :
         iNumberOfSteps(aNumberOfSteps),
         iNumberOfRealizations(aNumberOfRealizations),
         iNumberOfBins(aNumberOfSteps * 2 + 1),
         iNumberOfIterations(aNumberOfIterations),
         iCenterPoint(aNumberOfSteps + 1) {
      // iHistogram is the current estimate of the pdf (initially
      // unity)
      iHistogram = new float[iNumberOfBins];
      iNormalizedDistance = new float[iNumberOfBins];
      for ( int loop = 0; loop < iNumberOfBins; loop++ ) {
         iHistogram[loop] = 1;
```

```
        iNormalizedDistance[loop] = (loop - iCenterPoint) /
                sqrt(iNumberOfSteps);
    }
}

DistributionFunction& propagate ( ) {
    long int seed = time(0);   // random seed
    TRandomMersenne rg(seed);  // make instance of random number
                               // generator

    int* randOld = new int[iNumberOfSteps];
    int* randNew = new int[iNumberOfSteps];
    // Initialize first set of steps in the Markov chain
    for ( int outerLoop = 0; outerLoop < iNumberOfSteps;
         outerLoop++ )
         randOld[outerLoop] = 2 * rg.IRandom(0,1) - 1;
    int positionOld = iCenterPoint;
    // Determine the final position of the walker for this initial
    // set of steps
    for ( int innerLoop = 0; innerLoop < iNumberOfSteps;
         innerLoop++ )
         positionOld = positionOld + randOld[innerLoop];
    float* histogramNew = new float[iNumberOfBins];
    // Initialize the histogram for the iteration to unity.
    for ( int loop = 0; loop < iNumberOfBins; loop++ )
         histogramNew[loop] = 1;
    // Loop over multicanonical iterations
    for ( int iterationLoop = 0;
         iterationLoop < iNumberOfIterations;
         iterationLoop++ ) {
    // Loop over the number of random walks
         for ( int outerLoop = 0; outerLoop <
           iNumberOfRealizations; outerLoop++ ){
             srand(outerLoop);
    // In the new random walk, one step at a random location,
    // flipPosition, is reversed
           for ( int innerLoop = 0; innerLoop < iNumberOfSteps;
             innerLoop++ )
                  randNew[innerLoop] = randOld[innerLoop];
           int flipPosition = rg.IRandom(0, iNumberOfSteps - 1);
           randNew[flipPosition] *= -1;
           // The new random walk location is computed
           int positionNew = iCenterPoint;
           for ( int innerLoop = 0; innerLoop < iNumberOfSteps;
             innerLoop++ )
                  positionNew = positionNew + randNew[innerLoop];
    // The transition is accepted according to the multicanonical
    // condition; i.e. probability of acceptance is the ratio of
    // the probability at the old state to that at the new state
    // otherwise the old configuration is retained
          if ( rg.Random( ) < iHistogram[positionOld]/iHistogram
             [positionNew] ){
           positionOld = positionNew;
```

```
          randOld[flipPosition] *= -1;
        }
        histogramNew[abs(positionOld)] += 1;
    }
        // The estimate of the pdf for the next iteration is the
        // product of the old pdf estimate and the histogram
        // associated with the current iteration.
        for ( int loop = 0; loop < iNumberOfBins; loop++ ) {
          iHistogram[loop] = iHistogram[loop] *
          histogramNew[loop]; }
        }
    }
    delete [ ] histogramNew;
    delete [ ] randNew;
    return *this;
 }
        // The logarithm of the estimated probability density
        // scaled to its maximum value is graphed
  DistributionFunction& plot( ) {
    metafl("XWIN");
    float maxValue = FLT_MIN;
    for ( int loop = 0; loop < iNumberOfBins; loop++ )
        if (iHistogram[loop] > maxValue) maxValue =
          iHistogram[loop];
    for ( int loop = 0; loop < iNumberOfBins; loop++ )
        iHistogram[loop] = log10(iHistogram[loop] / maxValue);
    qplsca(iNormalizedDistance, iHistogram, iNumberOfBins);
    return *this;
  }
private:
  int iNumberOfSteps;
  int iNumberOfBins;
  int iNumberOfRealizations;
  int iNumberOfIterations;
  int iCenterPoint;
  float *iHistogram;
  float *iNormalizedDistance;
};

main() {
  int numberOfSteps = 40;
  int numberOfRealizations = 1000000;
  int numberOfIterations = 2;
  DistributionFunction D1(numberOfSteps, numberOfRealizations,
    numberOfIterations);
  D1.propagate( ).plot( );
}
```

The multicanonical method yields adequate statistics throughout the pdf, as is evident from Fig. 22.5. However, the method can be sensitive to computational parameters and boundary conditions. More exhaustive treatments of the multicanonical and related methods can be found in the references.

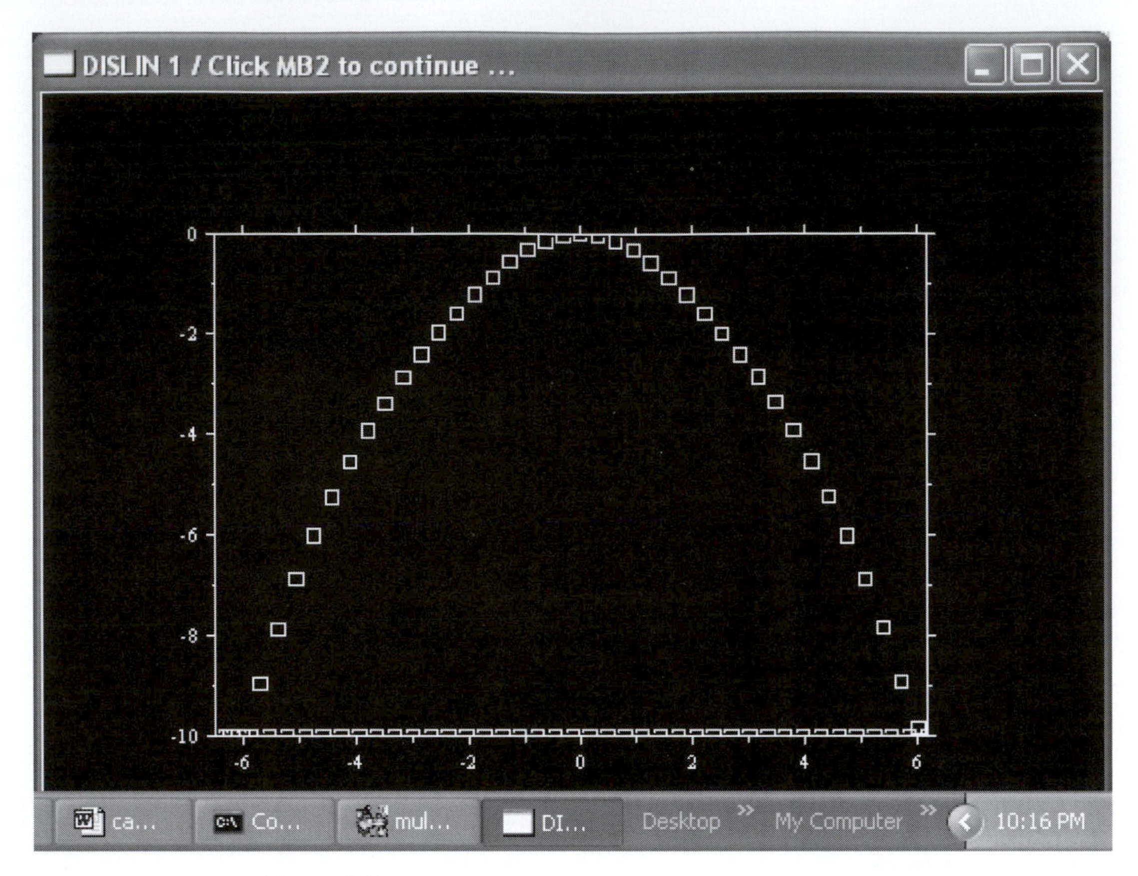

Figure 22.5

22.6 Assignments

(1) Modify the Monte Carlo integration program to find the area of a unit circle. You may use symmetry to restrict your calculation to positive x- and y-values. Graph your results for the logarithm of the error as a function of the logarithm of the number of evaluation points. Use 2000, 4000, 8000, ... , $2^{10}\cdot 1000$ steps.

(2) Write a program that calculates the probability of N heads after M coin flips. Bias the distribution such that the probability of a head is reduced to 0.35. Noting that the probability of N heads in this case is ${}_MC_N\ (0.35)^N(0.65)^{M-N}$, find the unbiased distribution by modifying the importance sampling program so that the importance function is calculated only once analytically or numerically and incorporated at the end of the calculation. Use 20 flips and 100,000 trials.

(3) By steadily increasing the value of β as the calculation proceeds (e.g. as the realizations are looped through), show that the Metropolis algorithm can be modified to generate the global maximum length configuration for the random walk problem. As an example, let **kT** in the program vary linearly from 5 to 0.5 as the calculation progresses

and print out the final configuration of steps together with the final endpoint of the walk.

(4) Consider a two-dimensional random walk with unit length steps in which the angle that the walker describes with respect to any fixed axis is a uniformly distributed random variable on $[0, 2\pi]$. Modify the multicanonical program above to generate the probability distribution function for the distance traveled from the origin after 40 steps. Your histogram bins should have unit width.

Chapter 23
Parabolic partial differential equation solvers

Partial differential equations contain derivatives in two or more variables and describe the evolution of *continuous* physical systems for which one or more degrees of freedom exist at every point in the system. Discretizing a continuous partial differential equation, however, yields a set of coupled one-dimensional ordinary difference equations. Hence, as will be demonstrated in this chapter, many of the numerical methods developed for ordinary differential equations can be generalized with little modification to partial differential equations.

23.1 Partial differential equations in scientific applications

We will consider two partial differential equations, namely the diffusion or heat equation and the Schrödinger equation. To conceptualize the meaning of the equations, we examine a one-dimensional example, namely a non-uniformly heated metal bar with a temperature distribution $T(x,t)$.

If the temperature distribution is discretized on a grid x_i, the temperature at x_i at a later time increases by an amount proportional to $T(x_{i+1},t) - T(x_i,t)$ as a result of a differing temperature at x_{i+1} and by an additional amount proportional to $T(x_{i-1},t) - T(x_i,t)$ from the temperature at x_{i-1}. Therefore

$$T(x_i, t_{j+1}) - T(x_i, t_j) = \bar{c}\left(T(x_{i+1}, t_j) - 2T(x_i, t_j) + T(x_{i-1}, t_j)\right) \tag{23.1}$$

where the constant $\bar{c}$ is clearly dimensionless. Assuming that the term on the right-hand side of Eq. (23.1) differs from zero, for small time increments $\Delta t = t_{j+1} - t_j$ and temperature distributions that are continuous and differentiable with time, $T(x_i, t_{j+1}) - T(x_i, t_j)$ is proportional to Δt. However, if the temperature distribution is linear in the spatial coordinate, the positive temperature change at x_i resulting from the neighboring material at x_{i+1} exactly balances the negative temperature change induced by the material at x_{i-1}. Thus, the change in temperature with time is proportional to the degree to which the temperature at a given point *exceeds* the *average* of the temperature at the two surrounding grid points. Since the dominant contribution then comes from the parabolic curvature of the function and is, therefore, described by a quadratic equation in x, if the difference in spatial increments is Δx, the expression on the right-hand side of the equation is proportional to $(\Delta x)^2$. This result can be derived directly from the Taylor series expansion of T, assuming that the temperature distribution is twice differentiable and that its second derivative is non-zero at x_i.

Accordingly, we write the change in temperature as

$$\frac{T(x_i, t_{j+1}) - T(x_i, t_j)}{\Delta t} = \bar{c}\frac{(\Delta x)^2}{\Delta t}\frac{T(x_{i+1}, t_j) - 2T(x_i, t_j) + T(x_{i-1}, t_j)}{(\Delta x)^2} \tag{23.2}$$

which in continuous form yields the parabolic diffusion equation

$$\frac{\partial T(x, t)}{\partial t} = D\frac{\partial^2 T(x, t)}{\partial x^2} \tag{23.3}$$

The thermal diffusion constant D now has units of squared distance over time and is independent to leading order of the point spacing in both the time and the space variable.

Clearly, if the average of the temperatures to the right and left neighbors of a grid point exceeds the temperature at the grid point so that the field curvature is positive (directed upward), there will be a net transfer of heat into the material at the grid point and its temperature will rise with time. On the other hand, a temperature distribution such as $T(x, t = 0) = T_0 + A\sin(\pi x/L)$ for $0 < x < L$ with uniform negative curvature will decay smoothly to a value of T_0 as time progresses, assuming that the temperatures at the boundaries are held fixed at T_0 for all time. In fact, Eq. (23.3) indicates that the *velocity* with which the temperature distribution changes is proportional to its curvature and is therefore most negative for the sine function at $L/2$, where the magnitude of the curvature is largest. As time passes, the temperature distribution accordingly decreases most rapidly in the central region as the entire distribution relaxes to a constant function.

Now consider a two-dimensional surface with a non-uniform temperature distribution $T(\vec{x}, t)$ (the three-dimensional generalization is trivial). If the temperature is discretized on a two-dimensional grid, for small time differences the temperature change at a given grid point will again be determined by the difference between its temperature and the average temperature at its neighboring grid points. Assuming equal grid point spacing in the x and y directions, Eq. (23.2) then generalizes to

$$\frac{T(x_{i,k}, t_{j+1}) - T(x_{i,k}, t_j)}{\Delta t} = \bar{c}\frac{(\Delta x)^2}{\Delta t}\frac{T(x_{i,k+1}, t_j) + T(x_{i,k-1}, t_j) + T(x_{i+1,k}, t_j) + T(x_{i-1,k}, t_j) - 4T(x_{i,k}, t_j)}{(\Delta x)^2} \tag{23.4}$$

In the continuous limit, we obtain the partial differential equation

$$\frac{\partial T(x, y, t)}{\partial t} = D\left(\frac{\partial^2 T(x, y, t)}{\partial x^2} + \frac{\partial^2 T(x, y, t)}{\partial y^2}\right) \tag{23.5}$$

Replacing t by it modifies the diffusion equation so that it describes wave motion instead of relaxation. This yields the parabolic wave equation, one example of which is Schrödinger's equation

$$iE\frac{\partial \phi}{\partial t} = -\frac{\hbar^2}{2m}\frac{\partial^2 \phi}{\partial x^2} \tag{23.6}$$

Writing Eq. (23.6) in the generic form

$$\frac{\phi(x_j, t_{i+1}) - \phi(x_j, t_i)}{\Delta t} = iD\frac{\phi(x_{j+1}, t) - 2\phi(x_j, t) + \phi(x_{j-1}, t)}{(\Delta x)^2} \tag{23.7}$$

we observe that if $\phi(x, 0)$ is initially of the form $A\sin(\pi x/L)$ for $0 < x < L$, since for all points, x_j, twice the value of the wavefunction at the point exceeds the average of the wavefunction at adjacent points, the change in T is initially directed in the $-i$ direction. However, once ϕ evolves to a negative imaginary sin function, the finite difference operator becomes positive and purely imaginary. The wavefunction then experiences a negative and real evolution with time, that is along the negative horizontal direction in the complex plane. Accordingly, the field rotates in the complex plane, while preserving its form.

A different type of wave equation is the so-called hyperbolic equation of the form

$$\frac{\partial^2 S(x,t)}{\partial t^2} = D\frac{\partial^2 S(x,t)}{\partial x^2} \tag{23.8}$$

In this case, the *acceleration* rather than the velocity of the field is proportional to its curvature. As a result, the acceleration at each point on the wavefunction at $t = 0$ in our $A\sin(\pi x/L)$ example possesses its maximum negative value, while a quarter of a period later the field distribution is constant but the acceleration rather than the velocity is zero. Therefore at this time the distribution reaches a maximum velocity directed toward negative field values. The acceleration then reverses sign and becomes increasingly positive until the field distribution reaches the maximum negative displacement $S_0 - A\sin(\pi x/L)$, leading to conventional wave motion.

We restrict our subsequent discussion to the parabolic diffusion and wave equations, as the hyperbolic equation exhibits numerous additional subtleties because of the presence of both forward and backward traveling solutions.

23.2 Direct solution methods

As noted in the introduction, every discretized partial differential equation can be viewed as a set of coupled ordinary differential equations, as is clearly evident from, for example, Eq. (23.2) or Eq. (23.4). Accordingly, the simplest procedure to evolve an initial field in time is the direct solution method that applies a forward difference approximation to the time derivative, as in the Euler method. For parabolic equations we obtain

$$T(x_i, t_{j+1}) = \bar{c}\left(T(x_{i+1}, t_j) - 2T(x_i, t_j) + T(x_{i-1}, t_j)\right) + T(x_i, t_j) \tag{23.9}$$

A simple program that applies Eq. (23.9) to the initial function $T_0 + A\sin(\pi x/L)$ with boundaries held at a constant temperature $T(x_0, t) = T(x_{N-1}, t) = T_0$, so that the field is advanced over the points $i = 1, 2, \ldots, N-2$, is presented below. This program employs template classes in order to provide the user with the option to model either diffusion or parabolic wave propagation by setting the value of the **bool** flag **isDiffusion** in **main( )** to **true** or **false**, respectively. Note that to be able to use either a **double** or a **complex <double>** type in the **plot( )** routine, we first cast the variable to a **complex <double>** (a variable that is already a **complex <double>** will be unaffected by this cast). The **abs( )** function of the **complex** class is then employed to generate a **double** value for the field modulus.

```
#include <iostream.h>

#include <math.h>
#include "dislin.h"
#include <complex.h>

// The initial sin(x) temperature profile
double fieldProfile(double aPosition, double aBoundaryTemperature,
       double aAmplitude) {
  return(aBoundaryTemperature + aAmplitude * sin(M_PI * aPosition));
}

// The temperature profile (iField) and the associated grid point
// positions (iPosition).
template <class T> class Field {
  T *iField;
  double *iPosition;
  int iNumberOfPoints;              // total number of grid points
  double iPropagationConstant;      // c bar in textbook.
  T iType;                          // Equals 1 for diffusion, i for
                                    // propagation

public:
  Field(int aNumberOfPoints, double aBoundaryTemperature,
       double aAmplitude, double aPropagationConstant, T aType) :
       iNumberOfPoints(aNumberOfPoints),
       iPropagationConstant(aPropagationConstant), iType(aType) {
    double deltaPosition = 1.0 / (iNumberOfPoints - 1);
    iPosition = new double[iNumberOfPoints];
    iField = new T[iNumberOfPoints];
    for ( int loop = 0; loop < iNumberOfPoints; loop++ ) {
       iPosition[loop] = loop * deltaPosition;
       iField[loop] = fieldProfile(iPosition[loop],
              aBoundaryTemperature, aAmplitude);
    }
    iField[0] = iField[aNumberOfPoints - 1] =
              aBoundaryTemperature;
  }

  ~Field() {
    delete [ ] iField;
    delete [ ] iPosition;
  }
```

```
  Field& propagate(int aNumberOfPropagationSteps) {
    T *fieldSave = new T[iNumberOfPoints];
    for ( int loopSteps = 0; loopSteps < aNumberOfPropagationSteps;
                loopSteps++ ) {
      for ( int loopField = 0; loopField < iNumberOfPoints ;
                loopField++ )
        fieldSave[loopField] = iField[loopField];
      for ( int loopField = 1; loopField < iNumberOfPoints - 1;
                loopField++ ) {
        iField[loopField] = fieldSave[loopField] +
                iType * iPropagationConstant *
                (fieldSave[loopField - 1] - 2.0 *
                fieldSave[loopField] + fieldSave[loopField + 1]);
      }
    }
    delete [ ] fieldSave;
    return *this;
  }

  Field& plot() {
    metafl("XWIN");
    float* newPosition = new float[iNumberOfPoints];
    float* newField = new float[iNumberOfPoints];
    for ( int loop = 0; loop < iNumberOfPoints; loop++ ) {
      newPosition[loop] = abs(complex<double>(iPosition[loop]));
      newField[loop] = iField[loop];
    }
    qplot(newPosition, newField, iNumberOfPoints);
    delete[ ] newPosition;
    delete[ ] newField;
    return *this;
  }
};

main() {
  int numberOfPoints = 200;
  double boundaryTemperature = 0;
  double amplitude = 1;
  double timeStep = 0.001;
  double diffusionConstant = 0.01;
  double gridPointSpacing = 1.0 / (numberOfPoints - 1);
  double propagationConstant = diffusionConstant * timeStep /
    (gridPointSpacing * gridPointSpacing); // c bar in textbook
  cout << propagationConstant;
  int numberOfLoops = 10;
  int numberOfPropagationSteps = 10000;

  // Propagation between one graph and the next with the
  // diffusion equation
  bool isDiffusion = true;
  if ( isDiffusion ) {
    double ci = 1.;
    Field <double> F(numberOfPoints, boundaryTemperature,
          amplitude, propagationConstant, ci);
```

```
    for ( int loop = 0; loop < numberOfLoops; loop++ ) {
      F.propagate(numberOfPropagationSteps).plot( );
    }

  // Propagation between one graph and the next with the parabolic
  // wave equation
  } else {
      complex <double> ci(0.,1.);
      Field <complex<double>> F(numberOfPoints,
        boundaryTemperature, amplitude, propagationConstant, ci);
      for ( int loop = 0; loop < numberOfLoops; loop++ ) {
        F.propagate(numberOfPropagationSteps).plot( );
      }
  }
}
```

23.3 The Crank–Nicholson method

While the direct solution procedure is simple to program and conceptually transparent, it retains the same divergence difficulties of the Euler method for ordinary differential equations, as can be immediately seen by increasing the variable **timeStep** to 0.021 in the diffusion implementation corresponding to $\bar{c} =$ **propagationConstant** ≈ 0.504. (The method diverges when **propagationConstant** > 0.5 for which **timeStep** is slightly less than 0.021 for the input parameters supplied in the program.) The procedure is far less stable for wave propagation, for which the critical step size is reduced by a factor of approximately 100 for the given input values.

The simplest stable propagation procedure for parabolic equations is the Crank–Nicholson algorithm that we outline below for the condition of zero field at the boundaries. This algorithm employs the propagation equation, Eq. (23.2), in the more accurate form

$$T(x_i, t_{j+1}) - T(x_i, t_j) = i\bar{c}\left(T(x_{i+1}, t_{j+1/2}) - 2T(x_i, t_{j+1/2}) + T(x_{i-1}, t_{j+1/2})\right) \quad (23.10)$$

together with the approximation

$$T(x, t_{j+1/2}) \approx \frac{T(x, t_j) + T(x, t_{j+1})}{2} \quad (23.11)$$

Introducing the tridiagonal matrix

$$\mathbf{H} = \begin{pmatrix} -2 & 1 & 0 & 0 & \dots & 0 \\ 1 & -2 & 1 & 0 & \dots & 0 \\ 0 & 1 & -2 & 1 & \dots & 0 \\ 0 & 0 & 1 & -2 & \dots & 0 \\ \vdots & \vdots & \vdots & \vdots & \ddots & 1 \\ 0 & 0 & 0 & 0 & 1 & -2 \end{pmatrix} \quad (23.12)$$

Eq. (23.10) yields the matrix equation

$$\left(1-\frac{i\bar{c}\mathbf{H}}{2}\right)\mathbf{T}(\vec{x},t_{i+1})=\left(1+\frac{i\bar{c}\mathbf{H}}{2}\right)\mathbf{T}(\vec{x},t_i) \tag{23.13}$$

in which $\mathbf{T}(\vec{x},t)$ is the column vector

$$\mathbf{T}(\vec{x},t)=\begin{pmatrix} T(x_0,t) \\ T(x_1,t) \\ \vdots \\ T(x_{N-1},t) \end{pmatrix} \tag{23.14}$$

Implicit in the above formalism is the zero boundary condition $T(x_{-1}=0,t)=T(x_N=L,t)=0$ for all values of t since at, for example the left boundary, the second derivative operator is approximated by the forward expression

$$\left.\frac{\partial^2 T(x,t)}{\partial x^2}\right|_{x=x_0} \rightarrow \frac{T(x_1,t)-2T(x_0,t)}{(\Delta x)^2} \tag{23.15}$$

Observe that if $\mathbf{H}$ were a constant rather than a matrix, the norm of $\mathbf{T}(\vec{x},t_{i+1})$ would equal that of $\mathbf{T}(\vec{x},t_i)$, since $|(1+iA)/(1-iA)|=1$ for any real number A. In fact, since $\mathbf{H}$ is real and symmetric and therefore Hermitian, it only possesses real eigenvalues. For this reason, if $\mathbf{T}$ is proportional to any eigenvector, $\mathbf{T}_\mathrm{i}$, of $\mathbf{H}$ with $\mathbf{HT}_\mathrm{i}=\lambda_\mathrm{i}\mathbf{T}_\mathrm{i}$, its norm is conserved. However, any $\mathbf{T}$ can be expressed as a linear combination of the eigenvectors of $\mathbf{H}$, which form a complete set. Consquently, the algorithm is norm-conserving and is termed unconditionally stable.

A consequence of the above argument is that to model radiation losses of the field out of the computational window, the boundary conditions must introduce a non-Hermitian component into $\mathbf{H}$. In fact, radiation losses into a semi-infinite space can be modeled exactly in one spatial dimension, as described in the references to this chapter.

Eq. (23.13) can be replaced by a more compact numerical algorithm. Adding and subtracting two within the parenthesis on the right-hand side of this equation, we obtain

$$\begin{aligned}\mathbf{T}(\vec{x},t_{i+1}) &= \left(1-\frac{i\bar{c}\mathbf{H}}{2}\right)^{-1}\left(2-\left(1-\frac{i\bar{c}\mathbf{H}}{2}\right)\right)\mathbf{T}(\vec{x},t_i) \\ &= \left(2\left(1-\frac{i\bar{c}\mathbf{H}}{2}\right)^{-1}-1\right)\mathbf{T}(\vec{x},t_i)\end{aligned} \tag{23.16}$$

Therefore, to advance the field we first solve the following tridiagonal system of equations for χ

$$\left(1-\frac{i\bar{c}\mathbf{H}}{2}\right)\chi=\mathbf{T}(\vec{x},t_i) \tag{23.17}$$

after which the propagated field is constructed from

$$\mathbf{T}(\vec{x},t_{i+1})=2\chi-\mathbf{T}(\vec{x},t_i) \tag{23.18}$$

This procedure is implemented for both the parabolic diffusion and the wave propagation equation in the program below. In order to increase the generality of the code several of the functions are purposely not optimized; for example, memory for the intermediate variable χ is allocated and then deallocated every time the **propagate( )** function is called. Running the code with various values of the time step demonstrates the unconditional stability of the procedure

```
#include <iostream.h>
#include <math.h>
#include <complex.h>
#include "dislin.h"
// The initial sin(x) temperature profile
double fieldProfile(double aPosition, double aBoundaryTemperature,
       double aAmplitude) {
  return(aBoundaryTemperature + aAmplitude *sin(M_PI *aPosition));
}

// Gaussian elimination method for tridiagonal matrices.
// a = lower co-diagonal, b = diagonal, c = upper co-diagonal.
// These and the input vector are not overwritten.
struct MatrixSolver {
  template <class T> static void tridiagonalSolver(
       T *a,T *b,T *c,T *inputVector,T *outputVector,
        int numberOfPoints) {
    T* psave = new T[numberOfPoints];
    T bsave = b[0];
    outputVector[0] = inputVector[0] / bsave;
    for ( int loop = 1; loop < numberOfPoints; loop++ ) {
       psave[loop] = c[loop - 1] / bsave;
       bsave = b[loop] - psave[loop] * a[loop - 1];
       outputVector[loop] = (inputVector[loop] - a[loop - 1] *
             outputVector[loop - 1]) / bsave;
    }
    for ( int loop = numberOfPoints - 2; loop > -1; loop-- )
      outputVector[loop] -= psave[loop + 1] *
             outputVector[loop + 1];
    delete [ ] psave;
  }
};

// The temperature profile (iField) and the associated grid point
// positions (iPosition).
template <class T> class Field {
  T *iField;
  double *iPosition;
  int iNumberOfPoints;           // total number of grid points
  double iPropagationConstant;   // c bar in textbook.
  T iType;                       // Equals 1 for diffusion, i for
                                 // propagation

public:
  Field(int aNumberOfPoints, double aBoundaryTemperature,
      double aAmplitude, double aPropagationConstant, T aType) :
      iNumberOfPoints(aNumberOfPoints),
      iPropagationConstant(aPropagationConstant), iType(aType) {
```

```
// The grid point spacing (for iNumberOfGrid points in the grid
// and two points,-1 and N, at the boundaries at 0 and 1 that are
// not within the computational grid
      double deltaPosition = 1.0 / (iNumberOfPoints + 1);
      iPosition = new double[iNumberOfPoints];
      iField = new T[iNumberOfPoints];
// The point positions and the initial field distribution
      for ( int loop = 0; loop < iNumberOfPoints; loop++ ) {
        iPosition[loop] = (loop + 1) * deltaPosition;
        iField[loop] = fieldProfile(iPosition[loop],
               aBoundaryTemperature, aAmplitude);
      }
   }

   ~Field() {
     delete [ ] iField;
     delete [ ] iPosition;
   }

   Field& propagate(int aNumberOfPropagationSteps) {
     T *fieldNew = new T[iNumberOfPoints];
     T *codiagonal = new T[iNumberOfPoints];
     T *diagonal = new T[iNumberOfPoints];
// The diagonal and co-diagonal elements of H are initialized
// here.
     for ( int loopField = 0; loopField < iNumberOfPoints ;
          loopField++ ) {
       codiagonal[loopField] = iType * (iPropagationConstant /
          2.0);
       diagonal[loopField] = 1.0 - 2.0 * iType *
          (iPropagationConstant / 2.0);
     }
     for ( int loopSteps = 0; loopSteps <
          aNumberOfPropagationSteps; loopSteps++ ) {
       MatrixSolver::tridiagonalSolver (codiagonal, diagonal,
          codiagonal,iField, fieldNew, iNumberOfPoints);
       for ( int loopField = 0; loopField < iNumberOfPoints;
          loopField++ )
         iField[loopField] = 2.0 * fieldNew[loopField] -
                         iField[loopField];
     }
     delete [ ] fieldNew;
     delete [ ] diagonal;
     delete [ ] codiagonal;
     return *this;
   }

   Field& plot() {
     metafl("XWIN");
     float newPosition[1000];
     float newField[1000];
     for ( int loop = 0; loop < iNumberOfPoints; loop++ ) {
       newPosition[loop] = iPosition[loop];
       newField[loop] = abs(complex<double>(iField[loop]));
     }
```

```
      qplot(newPosition, newField, iNumberOfPoints);
      return *this;
   }
};

main() {
      int numberOfPoints = 200;
      double boundaryTemperature = 0;
      double amplitude = 1;
      double timeStep = 0.001;
      double diffusionConstant = 0.01;
      double gridPointSpacing = 1.0 / (numberOfPoints - 1);
      double propagationConstant = diffusionConstant *
         timeStep / (gridPointSpacing * gridPointSpacing);
      int numberOfLoops = 10;
      int numberOfPropagationSteps = 10000;
      bool isDiffusion = 0;
// Propagation between one graph and the next with the diffusion
// equation
      if ( isDiffusion ) {
         double ci = -1.;
         Field <double> F(numberOfPoints, boundaryTemperature,
            amplitude, propagationConstant, ci);
         for ( int loop = 0; loop < numberOfLoops; loop++ ) {
            F.propagate(numberOfPropagationSteps).plot( );
         }
// Propagation between one graph and the next with the parabolic
   wave equation
      } else {
         complex <double> ci(0.,1.);
         Field <complex<double> > F(numberOfPoints,
            boundaryTemperature, amplitude, propagationConstant, ci);
         for ( int loop = 0; loop < numberOfLoops; loop++ ) {
            F.propagate(numberOfPropagationSteps).plot( );
         }
      }
 }
```

23.4 Assignments

(1) Modify the Crank–Nicholson program above to implement the Neumann boundary condition that the derivative of the temperature distribution is zero at the left and right boundaries and run the program with the following parameters. Hand in a copy of the last graph produced by the program.

```
int numberOfPoints = 200;
double boundaryTemperature = 0;
double amplitude = 1;
double timeStep = 0.001;
double diffusionConstant = 0.01;
int numberOfLoops = 10;
int numberOfPropagationSteps = 10;
```

(2) Solve the diffusion problem using the three-level Richardson procedure, namely

$$T(x_i, t_{j+1}) = \bar{c}\left(T(x_{i+1}, t_j) - 2T(x_i, t_j) + T(x_{i-1}, t_j)\right) + T(x_i, t_{j-1}) \qquad (23.19)$$

and demonstrate that your procedure is always unstable by running the diffusion problem given in the text for several small values of the step length. Hand in a graph of the final field for the parameter values

```
int numberOfPoints = 200;
double boundaryTemperature = 0;
double amplitude = 1;
double timeStep = 0.001;
double diffusionConstant = 0.01;
int numberOfLoops = 10;
int numberOfPropagationSteps = 10;
```

Repeat this calculation for the standard diffusion equation formalism given in the text and show graphically that the result is convergent.

(3) Now change your procedure to the Du-Fort Frankel method defined by

$$\begin{aligned} T(x_i, t_{j+1}) = {} & \bar{c}\left(T(x_{i+1}, t_j) - \left[T(x_i, t_{j+1}) + T(x_i, t_{j-1})\right] + T(x_{i-1}, t_j)\right) \\ & + T(x_i, t_{j-1}) \end{aligned}$$

and demonstrate that the method is now stable for any value of **timeStep**. Hand in a final field graph for the parameters given above and **timeStep** = 1.0 as an example of this behavior.

(4) A simple procedure for implementing a boundary condition that approximates the behavior of the field on an infinite region by the field on the finite grid is to assume that the field is dominated by a single outgoing wave component at the grid boundaries (see, in particular, G.R. Hadley, *Optics Letters*, Vol. 16, p. 624). In this case, for the left boundary we can make the approximation that

$$\frac{\partial T(x,t)}{\partial x} \approx \frac{T(x_1,t) - T(x_0,t)}{\Delta x} \approx -ikT(x,t) \approx -\frac{ik}{2}\left(T(x_1,t) + T(x_0,t)\right)$$

Using similar arguments at the right boundary, show that this method can be used to suppress the outward-traveling field components. Use the following input parameters and hand in the changed code sections and the last graph that is output by the graphics routines

```
int numberOfPoints = 100;
double boundaryTemperature = 0;
double amplitude = 1;
double timeStep = 0.0001;
double diffusionConstant = 0.01;
int numberOfLoops = 10;
int numberOfPropagationSteps = 10000;
```

Appendix A
Overview of MATLAB

For small programs or rapid prototyping of ideas or methods, the commercial MATLAB® language, where available, often constitutes a practical alternative to C++ or FORTRAN. Although the MATLAB product is accompanied by a large assortment of manuals, a summary of the features and difficulties of the language from a scientific programming perspective is presented below. Once aware of the basic commands and language structure, the reader should be able to program efficiently by accessing the on-line help provided within a MATLAB session for information on specialized commands.

1. *Running MATLAB:* To start a MATLAB session, either click on the MATLAB icon or, for command line installations, type the command **MATLAB**. Statements can then be entered interactively. A MATLAB session is terminated by typing **quit**. MATLAB lines end at a carriage return, comma or semicolon; however, only a semicolon suppresses the output of the statement from being written to the terminal. To continue a statement on a new line, the line should be terminated by a continuation character given by three periods, **...**. Any text to the right of a comment character **%** is ignored by the MATLAB interpreter. The command **diary on** writes a log of the subsequent commands entered from the keyboard into a disk file named **diary** until **diary off** is issued. A list of active (defined) variables is generated by depressing the workspace browser menu button or typing **who**.
2. *Using help:* The command **lookfor subject** searches for all commands involving the operation "**subject**". Typing **help** lists the MATLAB components for which help exists, while **help commandName** types out help on the command **commandName**.
3. *MATLAB programs*: MATLAB program and function files have a **.m** extension; that is, to run a MATLAB program, create a file such as **test.m** with a program or notepad editor, for example

   ```
   s = 2;
   s * ...% this illustrates the comment and continuation symbols
   s
   ```

   At a MATLAB prompt, type **cd** followed by the directory (including if necessary the partition name, for exmple **cd e:\testDirectory**) where **test.m** resides, press the carriage return and then type **test**. The program **test** can be also called from within another **.m** file.

4. *Input and output*: The variable **G** can be entered into a *program* (a **.m** file) from the keyboard by **G = input('user prompt')** To switch between program segments using a GUI interface a switch variable can be entered by using the command

```
switch = menu('Select the method', 'Method A', 'Method B',
         'Method C');
```

which assigns the value 1, 2, or 3 to **switch**, depending on the selected push button. Numbers can be retrieved from a graphical input field using

```
a = inputdlg('fieldlabel'), b = str2num(a{1})
```

The **more** command pages subsequent output; that is, the output is sent to the terminal one screen at a time. To write out subsequent floating point output with 16 digits of precision, type in **format long e**, to revert to the default 5 digits, type **format short e**.

5. *Loading vectors and matrices:* A symbol **A** can represent a scalar, vector, or matrix of any dimension. A row vector is entered as

```
V = [1 2 3 4];
```

while a column vector can be written as

```
V = [1
     2
     3
     4];
```

or

```
V = [1; 2; 3; 4];
```

or, using the transpose operator `.'`

```
V = [1 2 3 4].';
```

A matrix can be loaded in any of the following ways:

```
M = [1 2; 3 4];

M = [1 2
3 4];

M_1 = [1 3 2 4]; M = reshape(M_1,2,2);

M_1 = [1
3
2
4]; M = reshape(M_1,2,2);
```

The component $(M)_{12}$ is accessed through the notation **M(1,2).** The command **size(M)** returns the row and column dimensions of a matrix.

6. *Matrix operations:* To illustrate addition and subtraction in expressions involving matrices, first let **S = 2** and **M = [1 2; 3 4],** that is

$$\mathtt{M} = \begin{pmatrix} 1 & 2 \\ 3 & 4 \end{pmatrix}$$

Then

$$\mathtt{S} + \mathtt{M} = \mathtt{S}^{*}\ \mathtt{ones(2)} + \mathtt{M} = \begin{pmatrix} 3 & 2 \\ 3 & 6 \end{pmatrix}$$

while

$$\mathtt{S}^{*}\ \mathtt{eye(2)} + \mathtt{M} = \begin{pmatrix} 3 & 2 \\ 5 & 4 \end{pmatrix}$$

Failing to differentiate between these is a very common mistake. Multiplication similarly has different meanings depending on variable type. Multiplying or dividing a scalar **S** with a matrix **M** multiplies or divides each element in **M** by **S.** Further, **M*M** symbolizes normal matrix multiplication, while

$$M.^{*}\ M = \begin{pmatrix} 1 & 4 \\ 9 & 16 \end{pmatrix}$$

represents component-by-component multiplication. Similarly **M^2** is **M*M** while **M. ^2** instead squares the individual elements of **M**. The dot operator functions analogously for all arithmetic operations, such as **M . / N**, which yields a matrix whose (i, j)th element is simply $(M)_{ij}$ / $(N)_{ij}$. Most functions such as **cos(M)** operate on the individual elements of **M** (yielding here the matrix formed after taking the cosine of each element). A vector or matrix can be constructed from component vectors or from submatrices through the same notation that applies to scalar quantities, so that V14 = [[1 2] [1 2]] = [1 2 1 2] while **M44 = [M M; M M]**; is a matrix of twice the dimension of **M**.

Vectors are implemented as row or column matrices. Consequently, to perform the operation **V*M** for a 2 × 2 matrix **M, V** must be a 1 × 2 vector. Tracking matrix and vector dimensions can be simplified by appending the dimension information to the variable name; as for example **V12**.

If **V12** is the row vector **[1 i]**, where **i** (and **j**) is by definition $\sqrt{-1}$, the complex conjugate of the transpose of **V12** is the column vector **[1; –i]** obtained from **Vcc21 = V12′;.** This is often confused with the non-conjugated transpose **V12.′** that equals **[1; +i]**. The two vector–vector products are then the inner product **V12*Vcc21′** or **2** and the outer product **Vcc21 *V12,** which is **[1 i; -i 1]**.

7. *Matrix functions*: A few matrix functions ending in the letter **m** such as the matrix exponential **expm** act on matrix arguments and are defined (although not implemented) through power series expansions such as

```
expm(A) = A + A^2/2! + A^3/3! + ...
```

The determinant, trace, inverse, logarithm, and square root of **M** are similarly given by the functions **det(M)**, **trace(M)**, **inv(M)**, **logm(M)**, and **sqrtm(M)**, respectively.

The LU decomposition of M is given by **[L,U] = lu(M);**. The eigenvalues, arranged in ascending order, and the corresponding eigenvectors of a matrix **M** are placed in the columns of the matrix **Mvec** and the diagonal elements of **Mval** in **[Mvec, Mval] = eig(M);** (many analogous linear algebra routines exist).

8. *Solving linear equation systems*: The quotient of two matrices in MATLAB written as **M / N** denotes **M * inv(N)**, while **M \ N** is **inv(M) * N**. Accordingly, the solution of the linear equation system **X12 * M22 = Y12** is **X12 = Y12 / M22**, while that of **M22 * X21 = Y21** is **X21 = M22 \ Y21**.
9. *Saving variables*: A variable **M** is stored in the binary MATLAB file '**filename.mat**' through **save filename M** or in an ASCII file '**filename.dat**' with **save filename.dat M −ascii.** The variable is later recovered from the **.mat** file using **load filename** or from **filename.dat** by **load filename.dat; M = filename;**. The entire workspace can be saved and loaded through the commands **save filename** and **load filename**.
10. *String manipulation:* Strings such as test are stored as vectors of ASCII values. Thus, for example, if **V = 'string', ['a' V]** yields the 8-element vector **a string**. Integers and floating point numbers are translated into strings through the functions **int2str( )** and **num2str( )**, respectively.
11. *Vectorized iterators*: A *vectorized* loop is written as, for example, **V = sin(pi: -pi/10: -1.e-4)**; which constructs the vector **V = [sin(π) sin(9π / 10) ... sin(π/10) 0]**. Particularly useful is the **linspace(S1, S2, N)** (and **logspace)** command, which generates **N** equally (logarithmically) spaced points between **S1** and **S2**, including the endpoints. Another vectorized operation is constructed from the colon alone, **:**, which loops through all the rows or columns of a matrix, for example **.M(:,1)=V(:);** places all the elements of **V** into the first column of **M**. A matrix can also be mapped into a vector using the colon operator; for the matrix **M** defined above, writing **V = M(:);** yields the column vector **V = [1; 2; 3; 4]**.
12. *Control logic and iteration:* The logical operators in MATLAB are ==, <, >, ==, >=, <=, ~= (not equal) and the and, or, and not operators **&**, | and ~, respectively. The principal control constructs are illustrated below (note that **elseif** is a single word)

```
V1 = 0: 0.1: 20;
loop = 0;
m = 0;
for l = .9: - .1: .3
loop = loop + 1;
        if V1(loop) == 2
                break
        elseif rem(l*5, 3) == 0 & V1(loop) > 0.2
                while m < V1(loop)
                        m = m + 1;
                end
        else
                m = m + 2;
        end
end
```

A very common error is to write **for l = .9, -.1, .3** in the iterator above. This assigns the vector **[.9,-.1,.3]** to **l** without generating an error message. Also using **i** or **j** as loop variables (or ordinary variables) eliminates their use as complex numbers – *one must generally be very careful not to use* **i** *and* **j** *as variable names.*

13. *Files and function files:* A function, which is called through the syntax **[MO, VO, . . .] = functionname(MI,VI, . . .)**, is stored in the file **functionname.m**. To pass internal variables to other program units, a **global** statement, which includes the names of the variables to be transferred separated by spaces (not commas), must be present in each of these units as in

```
%File functionName.m:

function [VO, SO] = functionName(VI, SI)
global GS iConv
GS = length(VI);
VO = [VI(1), VI(2)^2];
iConv = 1
if SI > 30
          SO = 1;
          return
end
iConv = 0
SO = 0;

%File test.m:

global GS iConv
Vin = [1 1];
[V, S] = functionName(Vin, 2);
GS
iConv
```

14. *Built-in constants and functions*: Important predefined scalar quantities are, for example, **pi**, the unit complex number **i** or **j**, and **eps**, the smallest number recognized by the machine (machine epsilon). Important scalar functions: **rem(n, m)**, the remainder of n/m, which is positive or zero for $n > 0$ and negative or zero for $n < 0$ as in C++, **ceil( )**, **floor( )**, which round up and down, and **fix**, which rounds toward zero. Important vector functions are: the discrete forward and inverse Fourier transforms, **fft( )** and **ifft( )**, and the functions **mean( )**, **sum( )**, **min( )**, **max( )**, and **sort( )**. Interpolation of data is performed by **y1 = interp1(x, y, x1, 'method')**, where 'method' is **'linear'** (default), **'spline'** or **'cubic'**, **x** and **y** are the input x- and y-coordinate vectors and the scalar or vector **x1** contains the x-coordinate(s) of the point(s) at which interpolated value(s) is desired.

 Several functions accept other functions in single quotation marks as their arguments, for example **fmin('functionname', a, b)**, where **a** and **b** are the lower and upper endpoints for a minimum search. The function **fmins('functionname', x)** searches for a root in two dimensions closest to the vector **x**, **fzero('functionname', a)** finds the zero closest to **a**, **quad('functionname', a, b, tol)** integrates a function from **a** to

b to within an error **tol**, and **ode23( )** or **ode45( )** solve systems of coupled first-order ODEs. The function **roots([1 3 5])** returns the roots of the polynomial $x^2 + 3x + 5$.

15. *Graphic operations*: Writing **plot(X1, Y1, X2, Y2, ...)** with column or row vectors as arguments plots multiple (X,Y) lines. Consequently, **plot(V, 'g.')**, where **V** is complex, graphs the real against the imaginary part of **V** in green with point marker style. Logarithmic graphs are plotted in the same manner with **semilogy( )**, **semilogx ( )**, or **loglog( )**. Three-dimensional grid and contour plots are created with **mesh(M)** or **mesh(VX, VY, M)** and **contour(M)** or **contour(VX, VY, M)**, where **VX** and **VY** are vectors that contain the x and y positions of the grid points along the axes. A plot generated in a program (**.m**) file is not necessarily displayed until a **drawnow**, **pause(seconds)**, **end** or subsequent **plot** statement is reached. The commands **hold on** and **hold off** disable erasing so that additional curves can be overlaid.

    The figure window may also be broken into an $m \times n$ matrix of smaller plots of which the pth is selected for the next plot (proceeding first along columns) by the command **subplot(m, n, p).** The command **clf** clears the figure window, while **figure(n)**, for integer n, opens a new graphics window called **Figure n**. Functions are most easily plotted using the command **fplot('functionname', [a b])**, where **a** and **b** are the beginning and final x values. Axis defaults can be overridden with **axis([xmin xmax ymin ymax])**. Additional optional arguments in the axis command allow simple manipulation of the most common graphical functions. Labels are set by **xlabel('xtext')** and **ylabel('ytext').** Commands such as **print –deps** or **print –dtiff** yield encapsulated postscript and TIFF graphics files of the current plot window, respectively (**help print** displays all options), while **orient landscape** changes the printer or graphics output to landscape format.

16. *Memory management:* User-defined variable names hide built-in variable and function names, since MATLAB does not permit overloading. Therefore, if the program defines a variable such as **length,** the built-in function **length(x)** ceases to function. The variable **M** is destroyed through the command **clear M**. Each program should begin with **clear all** to remove all preexisting variables. This avoids puzzling errors generated by resizing variables defined during a previous run. Previously allocated memory is, however, not released until garbage collection is activated through the **pack** command.

17. *Structures:* In MATLAB a structure can be defined by appending the member-of operator **(.)** to a variable name as in

```
Spring.position = 0;
Spring.velocity = 1;
Spring.position = Spring.position + deltaTime * k / m *
                  Spring.velocity
```

Appendix B

The Borland C++ Compiler

The Dev-C++ program was chosen for this book because of its ease of use and functionality. However, the Borland C++ command line compiler yields faster compilation times and avoids perplexing errors generated when flags in the integrated development environment are incorrectly set. Accordingly, the Borland compiler is also included on the CD-ROM. Current updates are available at

http://www.borland.com/bcppbuilder/freecompiler/

The following text should be substituted for Chapter 2 for users of Borland C++. With small modifications, these instructions can also be applied to other command line compilers running on a Windows system.

If, however, you are installing both Borland C++ and Dev-C++ on the same computer, be sure to avoid installing the Borland C++ version of DISLIN into the same directory as the Mingw DISLIN version. Assuming, for example, that DISLIN for Borland C++ is accordingly installed in a directory X:\disbcc, replace all occurrences of X:\dislin by X:\disbcc in the discussion below.

B.1 Borland C++ installation

To install the C++ software package, follow the following sequence of steps:

1. Insert the accompanying CD-ROM into your computer.
2. Double click on the My Computer icon on your desktop and then on the CD-ROM icon and finally on the folder entitled BorlandC++.
3. Double click on the package icon labeled FreecommandLinetools. The installation program will start. When prompted to specify a directory, you may change the default directory C:\borland\bcc55 to any desired directory. To install to a different partition such as D: or E: (also known as the drive letter), that we will represent here by X:, replace C: in the default directory by X: (however, to be consistent, with the notation of this book, do not change the \borland\bcc55 part of the directory location). Repeat this step for the package icon labeled TurboDebugger, being sure to use X: in place of C: if you changed the default directory location for the command line tools.

4. Next identify the two files in your present directory on the CD-ROM entitled ilink32.cfg and bcc32.cfg. Place your arrow over the first of these and click the right mouse button (a right-click). From the resulting pop-up menu, select copy. Then return to the desktop and double-click on the My Computer icon a second time. Find the icon for the X: partition (this icon for this folder resembles a hard disk) and double click on it. Double click on the folder entitled Borland and then double click in turn on the bcc55 and then the Bin folder. Now right click on an empty space in the resulting window and select paste. Return to the original window where ilink32.cfg was located, right click on bcc32.cfg and select copy. Then go back to the window for X:\Borland\bcc55\Bin and again right-click and select paste. You should now have copied both the ilink32.cfg and the bcc32.cfg file into this window (directory).
5. If you have installed the software to some other partition than C:, you will still need to edit the ilink32.cfg and bcc32.cfg files. *This and the subsequent steps must be repeated if the software is later moved from one partition to another!* From the Start button either select Run and then type notepad in the text entry box to the right of Open: or select Programs and then Accessories and finally Notepad to start the Notepad editor. (Such a chain of selections will be represented subsequently in this text by Start->Programs->Accessories->Notepad.) *Note: the exact location of programs such as Notepad on the start menu may differ slightly among different versions of Windows*. From the File entry on the menu bar select Open (or click inside the Notepad editor window and, while holding down the Ctrl key on the keyboard, type O (abbreviated Ctrl-O). From the Open file menu navigate to X:\borland\bcc55\bin. In the text box labeled File Name: type **bcc32.cfg**. This should bring up the contents of the bcc32.cfg file in the editor window, namely

```
-I"c:\Borland\Bcc55\include"
-L"c:\Borland\Bcc55\lib"
```

   Change the C: to X:, where X: again represents the partition (drive letter) that you installed Borland on. Next save the file by either selecting File->Save from the menu bar or typing Ctrl-S. *Be sure that you have saved the file – failure to save files before exiting is the source of many unexpected errors during programming*. Select File->Open and edit and save ilink32.cfg in a similar fashion.
6. You have arrived at the last but potentially most troublesome step in the installation, namely setting the PATH environment variable. (For later reference, this variable is set to the locations of all the directories where commands entered into the command window will search for programs. That is if a program entitled, **foo.exe** is resident in a directory X:\MyDirectory, typing **foo** into the command window will produce an error message unless either the command window is in the directory X:\MyDirectory or this directory is added to the path.)

   One procedure for changing the PATH variable and setting the DISLIN environment variable that functions in all contexts is to open a command (MS-DOS) window and type at the command prompt

```
PATH=X:\Borland\Bcc55\bin;X:\dislin\win;%PATH%
SET DISLIN=X:\DISLIN
```

However this procedure must be followed each time a new command window is opened. To change the PATH statement permanently requires a procedure that is dependent on the Windows version and is described below (these instructions also add the required modifications for the DISLIN installation in the following section).

(a) Windows 95 and 98: Enter Notepad and navigate to the root directory for your operating system (generally C:\). Type **autoexec.bat** in the File Name: text box and add the statements

```
PATH=%PATH%;X:\Borland\Bcc55\bin;X:\dislin\win
SET DISLIN=X:\DISLIN
```

to the bottom of this file. Save the file and exit.

(b) Windows NT and 2000: Login as administrator and navigate to the Control Panel. Double click on the Control Panel and then double click on the System icon. Click on the Advanced tab and then depress the Environment Variables pushbutton. In the System Variables listbox find the entry marked Path. Click on this entry to highlight it and select Edit. Right click in the text entry field labeled Variable Value and use the right-arrow on your keyboard to advance to the end of the text string. Without introducing a space, add

;X:\Borland\Bcc55\bin;X:\dislin\win

to the end of the string (be *sure* that you have a *single* semi colon character between each directory entry). Now press OK and then press New below the System Variables listbox. Write in the Variable Name field DISLIN and in the Variable Value field X:\DISLIN and press OK again. Press OK one more time to exit the Environment Variables menu page.

(c) Windows ME: Select Run from the Start menu and then select Startup Config and then click on the Environment tab, and then follow the above instructions.

(d) Windows XP: Select Start then Control Panel. In the upper left-hand corner there should be an icon labeled Switch to Classic View (if the icon is instead labeled Switch to Category View, ignore this step). Click on this icon and then follow the instructions above for Windows NT and 2000.

B.2 Compiling and running a first program

1. Entering and working inside a command (MS-DOS) window: You are now ready to compile, run, and debug C++ programs using the Borland C++ compiler and debugger. This is simply accomplished from the Command window, which is also known as the MS-DOS window (in early Windows versions), shell, or command line interpreter. Accordingly select Start->Programs->Accessories->Command Prompt from the Start button. In early versions of Windows the Command prompt is instead called MS-DOS prompt and is accessed by selecting "Programs" from the Start Menu. In either case, a command window should open. If the first line inside the command window does not read C:\ (or X:\) if you have installed the windows system on partition X:), type

```
cd C:\
```

(or **cd X:**) to enter the root directory. If you enter

```
help | more
```

into the command window at the command (C:>) prompt (you may have to click first inside the window to make it active) you can see a list of all the available commands (pressing the space bar will display the next screen). Again at the C:> prompt enter

```
mkdir MyPrograms
```

to generate a subdirectory (folder) called MyPrograms and then type **cd MyPrograms** or equivalently **cd\myprograms** to enter the subdirectory. The command

```
dir
```

will list the files in this subdirectory (currently none) while

```
del filename
```

deletes the file named **filename,** which must include the proper 3-letter extension such as program.cpp

```
rename oldfilename newfilename
```

will rename a file

```
copy filename newfilename
```

copies a file. Typing

```
rmdir
```

from the directory above an empty subdirectory removes it. In specifying file names, the asterisk * can be used to match any sequence of numbers or letters except for the period that separates the file name from the file extension, while a question mark matches any single valid character. Thus **del** * will delete all files without a 3-letter extensions, **del *.cpp** will delete all files that have the extension **.cpp, del h?.*** will delete all files that start with **h** and have two letters in the file name, and **del** *.* will delete all files in the directory.

2. Creating and editing a program: To write a computer program, a program editor, which saves text as a plain unformatted file, must be employed as opposed to a word processor that adds formatting instructions and often converts the text to a binary representation when saving. Type

```
notepad test.cpp
```

at the command prompt and click on "Yes" in the "Do you want to create this file" dialog box. *Important: Always create a new file in this fashion from the command line. If you create a new file by starting notepad and then saving it within notepad under a name such as test, the file will automatically be called test.txt and will not compile until it is subsequently renamed using the command rename test.txt test.cpp.* Note that the file directory and file name now appear on the title bar of the notepad window. In the editor, type (remember to memorize the shaded lines)

```
// Hello world v. 1.0
// Aug. 11 2000
// (your name)
// This program tests the C++ environment

#include <iostream.h>

main () {
cout << "Hello World" << endl;
}
```

Note that any text on a line to the right of two forward slashes is treated as a comment and is not read as part of the C++ program. Three common yet surprisingly difficult to detect errors should be memorized before entering a program such as the above. First, *variable names are frequently misspelled*, leading in many cases to unexpected consequences, especially when characters that look nearly alike are substituted for one another. In particular 0 and O and 1 and l are very difficult to distinguish at low screen or printer resolution and can lead to subtle errors such as when 1 < 10 (1.0 < 10) is used in place of 1 < 10 (the letter *l* < 10). Secondly, *the semicolon at the end of a statement is often omitted* and two successive lines are then read as a single statement. This often produces strange error messages with distorted line numbering that confuse the debugging process. Finally, *neglect of fine details of the program syntax*, such as the use of double quotation marks rather than single quotation marks, braces instead of parentheses above, and the lack of a semicolon after the **#include** statement, lead to extreme difficulties for the beginning programmer.

3. Saving the file and running the program: Select File -> Save from the menu bar at the top of the editor. Click on the command window to make it active (do not close the editor in case you need it later to correct your program). Note that your program can be simply printed by instead selecting File->Print. Next in the command window type **bcc32 test.cpp** or **bcc32 test**. (A more rapid executable file, as discussed in more detail in Chapter 21, can be obtained by writing instead **bcc32 –O1 test.cpp** or for yet faster code **bcc32 –O2 test.cpp**.) If you obtain error messages, return to the notepad, correct your program and save it a second time. Otherwise, type **dir** to see the new **.obj** and **.exe** files that have been created. The **.exe** file is the actual executable file while the **.obj** file is an intermediate file that will be described in more detail in the following chapter. Type **test** in the command window to run the program (**.exe** file). The line "Hello World" should appear on the screen as output.

B.3 Installing the optional program editor

You may wish at this point to substitute a dedicated programming editor that provides features such as keyword highlighting and automatic language-dependent indentation for the Notepad. To install the freeware Crimson Editor included on the CD, navigate again to the CD folder in My Computer, open this folder, navigate to the Crimson Editor subdirectory, and double-click on the icon labeled cedt360r. When prompted, type in an appropriate installation directory (usually the default directory is appropriate since the installed program only requires 2MB of disk space). After installation, the editor is accessible through the Programs option in the Start menu. *Again as in the case of the Notepad editor, when a new file is saved for the first time it must be given a name with a .cpp extension, for example MyProgram.cpp.*

The Crimson editor can be automatically configured to run Borland C++ when Ctrl-F7 is pressed using the following instructions, which are copied from the Crimson Editor home page:

The following example assumes that C++ Builder is installed in the 'C:\Borland\BCC55' directory.

Open the Tools->*Configure User Tools* page

Write the following arguments into the indicated text boxes:

- Menu Text: *Compile C/C++ Source Code*
- Command: *C:\Borland\BCC55\bin\bcc32.exe*
- Argument: *$(FileName)*
- Initial dir: *$(FileDir)*
- Hot key: *Ctrl+F7 – to enter this press Ctrl-F7 while the cursor is in this text box (other choices can be selected the same way)*
- Capture output: *Yes – i.e. place a check in this check box*
- Save before execute: *Yes*

You can now press Ctrl+F7 while editing 'test.c' to compile the source code. (Pressing F10 then opens up a new command window from which the executable file can be run).

In the same manner, you may be able to configure the editor to run the .exe file as follows (however output is not captured in the window at the bottom of the editor)

Again open the Tools->*Configure User Tools* page. In the top list box click on the first slot entitled Empty directly under the slot entitled Compile C++ Source Code. Now enter the following arguments:

- Menu Text: *Run Binary Executable*
- Command: *command.com*
- Argument: */C $(FileTitle)*
- Initial dir: *$(FileDir)*
- Hot key: *Ctrl+F8*
- Capture output: *Yes*
- Save before execute: *No*

Pressing Ctrl+F8 while editing 'test.cpp' now executes 'test.exe'.

Finally it is possible to enter a procedure that automatically generates graphs in conjunction with the DISLIN software discussed below, namely:

Again open the Tools->*Configure User Tools* page.

In the top list box click on the first slot entitled Empty directly under the slot entitled Compile C++ Source Code. Now enter the following arguments:

- Menu Text: *Graph*
- Command: *bcclink.bat*
- Argument: -a*$(FileTitle)*
- Initial dir: *$(FileDir)*
- Hot key: *Ctrl+F10*
- Capture output: *No*
- Save before execute: *No*

Pressing Ctrl+F10 after editing 'graphtest.cpp' now automatically includes the graphics libraries and produces a screen with the graphical result.

B.4 Using the Borland turbo debugger

A critically important technique for locating errors in code is to inspect the values of internal variables within the program during execution (as the program is run). The simplest method is to introduce additional code lines that write these values to the terminal or to a file. Alternately, variables can be inspected at runtime through use of a debugger. In this section, we introduce an error into our Hello World program in order to illustrate both procedures.

Return to the program you created in the section entitled Creating and running a first program by clicking inside the appropriate notepad window. Incorporate both a programming error and debugging lines into the program as follows

```
#include <iostream.h>

main () {
       int i = 0;
       cout << "The value of i is" << i << endl;
       i = 5;
       cout << "The value of i is" << i << endl;
       int j = 0;
       cout << "The value of j is" << j << endl;
       int k = i / j;
       cout << k << endl;
}
```

Save and then click again inside the command window and type **bcc32 test** and then **test**. The values of the various variables should appear on the screen before the program terminates abnormally due to the invalid division by zero. From this information, the error is trivially located.

To inspect the intermediate variable values using a debugger, return to the command window and type **bcc32 –v test** (the –v flag adds debugging information to the executable program). Now type **td32 test**. Click on various menu items on the menu bar to see the commands available in the debugger. Click on the menu item "View" and the subitem "Variables" to see the value of each variable in the program. If you are unable to select menu items with the mouse, use the following alternate procedure. Depress the Alt key on the keyboard and, while holding it down, press the v key to select the menu item "View" from the menu bar – this action is generally referred to as Alt-v. Then type v to select the option "Variables."

Note: You may have to employ the scroll bars to see the bottom of the window where there is a listing of commands and their shortcut keys. These scroll bars can be eliminated by resetting the properties – number of lines and buffer size – of all MS-DOS (command) windows by right-clicking on the command window (or MS-DOS) icon anywhere it appears and selecting Properties from the drop-down menu and changing the window height. You may later be prompted to indicate whether this change should apply to the present window only or to all windows associated with the icon, in this case choose the latter selection. However, if your command window prompt appears in several places on your desktop and menu system, this sequence of steps may have to be repeated for each icon.

Now using the same procedure select Run -> Step Over multiple times from the menu bar to execute the program line by line. Note when the variable **i** is set to 0 and then to 5. Finally, select File -> Quit to return to the command prompt.

B.5 Installing DISLIN

DISLIN is a high-performance professional scientific graphics package that is available for numerous C, C++ and FORTRAN programming environments. The installation steps for the product are:

(a) Return to the CD-ROM and double click on the folder labeled dislinbcc. Find the icon labeled setup and double click on this icon to install the product. Be sure to set when prompted for a directory name the partition letter to the same partition (drive letter) X: that you used for Borland C++, while retaining the rest of the default directory name.
(b) At this point, all the required environment variables should have been correctly set when you performed step 6 in the Borland C++ installation section above. To check this, open up a command (MS-DOS) window and type **X:** and press the enter key, where X represents the letter of the drive on which you installed Borland C++ and DISLIN (if you are not already in this partition). Now type **cd \dislin\examples** and then **clink –a exa_c**. This should compile, link, and run the sample program.
(c) If you are installing a new version of DISLIN from the website you will still need to copy the file **bcclink.bat** from the **dislinbcc** directory of the CD-ROM into your **X:\dislin\win** directory. Alternatively, for installations from the website you can edit

the file **bcclink.bat** in your **X:\dislin\win** directory directly. Using, for example, the notepad editor, open this file and find the lines

```
:COMP
@ set_ext=c
@ set_int=% _dislin%
@ if %_opt1%==-cpp set_ext=cpp
```

Replace **=c** by **=cpp** in the second line and **cpp** by **c** (twice) in the last line and save the file.

In either case, you are now ready to use the graphics package.

B.6 A first graphics program

You will now write a sample graphics program in order to become familiar with the code lines required by all DISLIN programs. From any command window, navigate to your programs directory using **X:** followed by **cd \MyPrograms**. Type the command **notepad graphicstest.cpp** and enter the following lines into the editor window

```
#include <iostream.h>    // Required by DISLIN!
#include "dislin.h"      // Required by DISLIN! Includes the
                         // plotting package

main () {
       int numberOfPointsToBePlotted = 2;
       float x[2] = {0, 1};
       float y[2] = {0, 2};
       qplot(x, y, numberOfPointsToBePlotted);
}
```

It is extremely important to observe that the command ***#include <iostream.h >*** *must be present and must be situated before* **#include "dislin.h"** *for DISLIN to function properly on a C++ file.* (The line **#include <dislin.h>** can be substituted for **#include "dislin.h"**.) Save your file and type **bcclink –a graphicstest** from the command line. *Do not forget to use the –a flag*! A graph of the two points should appear.

If needed, you can now generate a TIFF, Adobe PDF or postscript file in place of the screen plot by placing one of the lines

```
metafl("TIFF");
metafl("PDF");
```

or

```
metafl("POST");
```

into the **main( )** program before the line containing the call to **qplot( )**. A file named **dislin**.xxx, where xxx is respectively **tif**, **pdf**, or **eps**, is then produced in your directory when the program is executed (other graphics formats are also possible). If an .eps file called, for example, **dislin.eps** already exists, a new .eps file will instead be called **dislin_1.eps.**

A procedure for printing the contents of any window, which can also be applied to a graph window, is to depress the left mouse button inside the graph window and subsequently press the Print Screen key while holding down the ALT key. (Using the CTRL key instead of the ALT key instead captures the contents of the entire screen.) You can then open any application program that accepts graphics, such as Paint (Start->Programs->Accessories->Paint), or an appropriate word processor and select Edit->Paste from its menu bar to insert a bitmap of the captured window. The bitmap can then be printed through the application program's print functions.

B.7 The help system

The help subsystems of the Borland and DISLIN products are not integrated with the standard Windows help system. To display an electronic version of the DISLIN manual, instead open a command window and type **disman** or **dishlp.** The second of these displays a list of DISLIN functions; clicking on one of the names in the list then brings up a description of the selected function. Icons for **disman** and **dishlp** also exist under the MyPrograms -> DISBCC submenu that is accessible from the Windows Start menu.

To obtain help on the Borland C++ product, double click on the My Computer icon on your desktop and then click in turn on X:, Borland, bcc55, help and finally one of the help icons such as bcb5tool. Additional help files can be downloaded from: www.borland.com/techpubs/bcppbuilder/v5/updates/std.html

Appendix C

The Linux/Windows Command-Line C++ Compiler and Profiler

The freeware Linux C++ compiler, named g++, that forms the basis of the Dev-C++ development environment, also exists in a pure command line form for Windows systems. We have included an easily installed Windows version of g++ in the CD-ROM that accompanies this book.

To install the package simply double click on the MinGW-3.1.0–1 install icon contained in the mingw folder on the CD-ROM. This will by default create a directory (folder) called mingw containing a further subdirectory, bin, that contains the C++ compiler and numerous additional utilities. Uninstallation proceeds though the 'Add or Remove Programs' icon within the Windows Control Panel. As well, a slightly older, mingw package, mingw.zip, is included in the same CD-ROM directory. This implementation contains help files for the mingw utilities in the subdirectory info-html directly under the mingw directory. This version can be installed by unzipping the file to, e.g. the C: drive (while specifying if requested the option of retaining the original subdirectories) and uninstalled simply by deleting the resulting C:\mingw folder together with all subdirectories.

Assuming that the software was installed to the C: drive, to run the compiler or the additional utilities the directory C:\mingw\bin must be added to the system path. Either this can be done permanently following the instructions in the previous appendix for DISLIN, or it can be done at the start of every command line session by typing the command

```
PATH=C:\mingw\bin;%PATH%
```

after a MS-DOS or Command window has been opened (the g++ Dev-Cpp compiler can similarly be run from a command line after typing **set PATH=X:\dev-cpp\bin;%PATH%,** where X: is the drive letter of the directory (folder) in which the program was installed, inside a command window). Then, to compile a file named **myprogram.cpp** using the g++ compiler type either

```
c++ -o myprogram.exe myprogram.cpp
```

or

```
mingw32 -g++ -o myprogram.exe myprogram.cpp
```

at the command prompt. (The speed of the executable file can be increased by appending the –O1 or –O2 flag, c.f. Chapter 21.) If **–o myprogram.exe** is not included in the above statement, the executable file will be called **a.exe** by default and can then be run by typing simply **a**.

The g++ compiler contains a profiler **gprof.exe** that displays the amount of time spent in each function for programs with sufficiently long execution times. The programmer can then attempt to optimize the functions that require the most processing time.

As an example of the **gprof** utility, consider the following program, **gproftest.cpp**

```
#include <iostream.h>

int firstRandom() {
        return ( (2 * (rand() % 2)) ? 1 : 0 );
}

int secondRandom() {
        return ( (3 * (rand() % 3)) ? 1 : 0 );
}

main() {
        int numberOfRealizations = 100000000;
        int result, first = 0, second = 0;
        for (int loop = 0; loop < numberOfRealizations; loop++) {
             if (result = firstRandom()) second += secondRandom();
             first += result;
        }
        cout << double(first) / numberOfRealizations << "\t"
             << double (second) / numberOfRealizations << endl;
}
```

Typical output values are

```
0.500316     0.333547
```

To profile the amount of time spent in each function, enter

```
g++ -g -o gproftest.exe gproftest.cpp -pg
```

Next run the executable file **gproftest.exe** by typing

```
gproftest
```

A new file **gmon.out** is created that is read by the profiler. To run the profiler, type

```
gprof -a -b gproftest.exe > output.dat
```

This creates a file **output.dat** containing the lines

```
Flat profile:
Each sample counts as 0.01 seconds.
  %    cumulative     self               self     total
time       seconds  seconds  calls     ns/call  ns/call        name
40.49         4.49     4.49                                    mcount
24.17         7.17     2.68 50031593   53.57    53.57    secondRandom()
23.90         9.82     2.65 100000000  26.50    26.50     firstRandom()
10.37        10.97     1.15                                     main
 0.99        11.08     0.11                                     rand
```

Evidently the two functions consume nearly equal amounts of processing time even though **secondRandom( )** is called half as many times as **firstRandom( )**. However, calculating the remainder after division by 3 requires more computational time than the remainder after division by 2 as can be verified by replacing 3 by 2 in **secondRandom( )**. The degree of profiler overhead is evidenced by the 40% of the execution time taken by **mcount( )**, a profiling function.

To generate meaningful profiler results, a program must execute for longer than a few tenths of a second. Additionally, if the program is changed, *all three steps in the above instructions must be repeated in order*. Finally, the program must run and terminate in a normal fashion. An exceptional condition that causes the program to cease prematurely or termination through, for example, an **exit( )** statement yields misleading or no program statistics, respectively.

Appendix D
Calling FORTRAN programs from C++

Since numerous scientific programs are only available in FORTRAN, various techniques have been developed for integrating FORTRAN and C++ code. One procedure is to translate FORTRAN source code directly into C through a program such as the Linux-based f2c. However, the resulting code is complex, while the translator does not handle all FORTRAN features.

Below we instead discuss calling FORTRAN routines from C++. This requires compatable FORTRAN and C++ compilers, typically from the same software suite or manufacturer. We will here apply the g++/g77 Linux-based compiler pair to a simple example that can be adapted to numerous practical problems. Unfortunately, inter-language calls are compiler-dependent and the code, therefore, cannot be directly transported to other programming environments. Recent FORTRAN compilers additionally contain numerous new language features that are difficult to interface with C++. In the latter case, the FORTRAN routine should be called from within a second, simplified FORTRAN function.

The FORTRAN language has several conventions that affect a C++ calling program:

1. FORTRAN parameters are passed by reference to subroutines and functions. Consequently, a C++ program must employ reference arguments when calling a FORTRAN function.
2. Arrays are stored in memory in FORTRAN programs in row-major instead of column-major order. In other words, in a C++ program, the elements of the matrix M(2,3) are stored in memory sequentially as {M(0,0), M(0,1), M(0,2), M(1,0), M(1,1), M(1,2)}, while in a FORTRAN program the same matrix is stored as {M(1,1), M(2,1), M(1,2), M(2,2), M(1,3), M(2,3)}. Thus a 2×3 matrix in FORTRAN is read as a transposed 3×2 matrix in C++.
3. A FORTRAN subroutine that has no return arguments, maps to a C++ void function.
4. FORTRAN character strings are not automatically null-terminated and can only be accessed in the context of the g77 environment if the libg2c.a library is included at link time.
5. Since FORTRAN is not case-sensitive, C++ names that map to corresponding FORTRAN identifiers can be any mixture of capital and small letters.

6. Generally, an underscore character is appended by the compiler to the end of the names of visible FORTRAN symbols, such as subroutines and functions, and common block names (sometimes flags must be used during FORTRAN compilation to insure this). Therefore, in the C++ calling program, the corresponding names must be similarly terminated by an underscore.
7. Open files cannot be easily shared between C++ and FORTRAN programs.

Further, in a C++ program any non-C++ program structures must be prefaced by the **extern "C"** directive, which indicates that the module definition will be introduced by the linker from a program written in a different language. To illustrate, we will call the following FORTRAN routine, which we assume is placed in a file **fortranprogram.f**, from a C++ program

```
      double precision function example( input )

      double precision input

      integer matrix(2, 3)
      character*10 characterarray
      common /fortrancommon/ matrix, characterarray

      do 20 loopouter = 1, 3
         do 10 loopinner = 1, 2
            matrix(loopinner, loopouter) = loopouter * loopinner
10       continue
20    continue

      characterarray(1:4) = "test"
      example = input * input
      write(6, *) characterarray
      return
      end
```

The required C++ code, placed in a separate program **cplusprogram.cpp**, is given below

```
#include <iostream.h>
typedef struct {
   int iMatrix[3][2];
   char iCharacterArray[10];
}  commonStruct;

extern "C" {
   double example_(double&);
}
   int main() {
   extern commonStruct fortrancommon_;
   double input = 10.0;
   double result = example_(input);
   cout << result << "\t" << fortrancommon_.iCharacterArray
```

```
        << "\t" << fortrancommon_.iMatrix[2][1] << endl;
    return 0;
}
```

Observe the following features:

1. A **double precision** variable in FORTRAN corresponds to a C++ **double**, while FORTRAN **integer** and **character** variables map to C++ **int** and **char [ ]** types, respectively.
2. The 2 × 3 two-dimensional integer array, **matrix**, in the FORTRAN common block is defined as a 3 × 2 **int** array in the C++ **struct**.
3. The names of both the FORTRAN common block **fortrancommon** and the FORTRAN function **example( )** acquire a trailing underscore in the C++ routine.
4. The common block, **fortrancommon**, in the FORTRAN program is declared through the line **extern commonStruct fortrancommon_;** in the C++ program, where **commonStruct** is a **struct** defined to store the same variable types as in the FORTRAN common block.
5. The **extern "C"** linkage directive is employed to indicate that the function **example** and **struct fortrancommon_** are associated with the corresponding FORTRAN routine and common block.
6. The variable **input** is passed by reference to the FORTRAN routine **example( ).**
7. The character array **characterarray** in the C++ program is correctly zero-terminated when declared in the **extern commonStruct fortrancommon_;** statement. Consequently, it remains zero-terminated when returned from the FORTRAN program.

To compile and run the programs from a mingw installation (g77 is included in the Dev-C++ environment so that the above programs can also be compiled after typing **set PATH=X:\dev-cpp\bin;%PATH%** from within a command window), set your path environment variable as described in Appendix B and enter at the command prompt:

```
g++ -c cplusprogram.cpp
g77 -c fortranprogram.f
g++ -o cplusprogram cplusprogram.o fortranprogram.o -lg2c
cplusprogram
```

This will compile the two programs separately, link them together with the required libraries and finally run the resulting executable file to produce the desired screen output.

Appendix E
C++ coding standard

While a condensed description of C++ programming conventions has been introduced in Section 4.10, a more complete set of conventions is presented below. This material is partially adapted from the somewhat different but more complete discussion in the document "*C++ Language Coding Standard*", Naval Command, Control and Ocean Surveillance Center RDT&E Division (NRaD), now Space and Naval Warfare Systems Center, San Diego and AHNTECH, Inc., San Diego CA (1995).

E.1 Program design language

In employing C++ as a program design language for sequential program development, a description of each program unit should be supplied in the form of a *prolog*. These prologs introduce the various header and source files, classes, and functions that will appear in the program. Before the classes and functions bodies are inserted, the last line of the corresponding prologs should read

```
// Implementation not yet introduced
```

Conventions for prologs are:

- A source or header file prolog should give the name of the file, the author, and creation date and a description of the contents

```
//
// FILE        : systemmodel.h
//
// REVISION    : 1.0
//
// DATE        : Sept 20, 1981
//
// AUTHOR      : Author's Name
//
// DESCRIPTION : Model of OC-192 communication system
//
```

- The class prolog should name and describe the class

```
//
// CLASS        : Oscilloscope
//
// DESCRIPTION : Simulates key functions of a 20MGHz oscilloscope
//
```

- A function prolog should provide the name of the function and, where relevant, its enclosing class, together with a description of the function and a list of input parameters, return values, and any output values that are returned by reference. Preconditions and postconditions should also be supplied.

```
//
// FUNCTION        : Oscilliscope::setScale
//
// DESCRIPTION     : Sets the scale for the simulated
//                   oscilloscope plot
//
// INPUTS          : timeScale:double - horizontal scale
//                 : voltage:double - vertical scale
//                 : numberOfPoints:int - number of points to be
//                   plotted
//
// OUTPUTS         : isInRange - true if the signal voltage is
//                   within scale
//
// PRECONDITIONS   : 2 < numberOfPoints < 1000
//
// POSTCONDITIONS : Return value from plotting routine must be 1
//                   (success).
//
```

E.2 Comments

- Blank lines should be used to group blocks or lines of related code. They should, therefore, be placed before and after declarations or definitions, comments, logically related statements, and function or class bodies.
- Comments should be placed immediately above or to the right of the statements being clarified. C style comments (/* . . . */) should be avoided. The comments should state the purpose of the following statements in more detail than can be directly inferred from the code lines.
- Comments should ideally accompany each definition, block or loop structure and **return** statement.

E.3 Layout

The source and header files should follow a preset order of statement groups. These are:

- The prolog
- **#include** system and library files
- Application specific #include files
- **#include** files contained in the local directory
- Class header files.
- **#define** statements.
- Global variables that are referred to in several files
- Global variables that are only defined in the given file
- Global function definitions
- Class definitions
- Member function definitions
- **main( )** program

E.4 Continuation lines

Continuation lines should be one tab stop past the indentation of the line they precede. However, lists of variables can be aligned on successive lines after an equality sign or directly underneath the variable in the initial line of the statement; in this case one variable should be placed on each line as in (these formatting details are usually managed by a dedicated program editor):

```
int myFunction(int aArgument1,
               double aArgument 2,
               double aArgument3);
```

E.5 Constants and literals

- Constants should not be defined using the **#define** macro but rather as **const** variables:

```
const double pi = 3.14159;
```

- String literals that are used more than once should be introduced at global scope and placed in a pointer variable:

```
char *outOfBoundsErrorMessage = "The index is out of bounds";
```

E.6 Variables and definitions

- Variable names should be descriptive or, if unacceptably long, should be self-evident abbreviations of multiple word phrases. The first letter of a variable name should be uncapitalized, while successive words in the name should be capitalized.
- Global variables are often written in capital letters.

- Variables should be initialized when defined and should be defined consistently, either at the beginning of the block in which they appear or as close as possible to the point at which they are used (the convention of this text).
- Only one variable should be defined in each definition statement.
- **int** should always be present in variable and function definition statements in cases where the type would otherwise default to **int** as in:

```
const i;            // should be replaced by const int i;
sum (int* aArray); // should be replaced by int sum
                    // (int* aArray);
```

- An alternate convention is to indent within all definitions so that the variable and function names are aligned after a tab stop as in

```
int         mass;
double      *trajectory;
void        print();
```

E.7 Functions

- Function names should describe the action taken by the function and should follow the same convention as variable names. Parameter names should similarly be descriptive.
- One convention is to place a space after opening parenthesis and before closing parentheses, while a second convention is not to places spaces before or after parentheses. A program should follow one of these conventions consistently.
- A function argument should start with the letter **a.**
- The name of a function that returns a **bool** should start with **is** or **enable**.
- A function body generally should not exceed 50 statements or two code pages.

E.8 Operators

- There should be spaces to the right and left of a binary operator but no additional spaces should be placed around a unary operator.

E.9 Classes

- Class, **struct**, and object names should be capitalized.
- Internal class member variables should start with the letter **i**, such as **iRadius**.
- Set member functions should have the name of the internal variable without the leading **i** after the word set, for example **Circle& setRadius(double aRadius)**;
- Get member functions should have the name of the internal variable without the leading **i**, for example **double radius( );.** Another convention is to begin the function name with the word **get**.

- The keyword **private:** should not be omitted at the beginning of the class definition before private class members.
- Code after the access specifiers **private:**, **public:**, or **protected:** (these keywords are preferably not identified) should appear on a subsequent line.
- The **public** interface of the class should appear first in the class definition, followed by the **protected** and finally the **private** section.
- Class definitions should be packaged into header files, generally with one header file for each class and surrounded by preprocessor directives so that definitions are only read once by the compiler. The associated member function implementations are placed into separate **.cpp** source code files. An individual **include** source file should generally be limited to a few pages, excluding prolog, and should only contain features that are associated with a single task.
- Each constructor should initialize all internal class variables. Internal class variables that are set equal to one of the constructor arguments should be initialized in an initialization list before the opening brace of the constructor function.

E.10 Typedefs

- The name associated with a new type identifier introduced through a **typedef** should be capitalized.

E.11 Macros

- Inline functions and **const** variables should be used in place of macros (**#define**).

E.12 Templates

- Template declarations should, if space permits, be placed on a single line

```
template <class T> int print( T aErrorMessage );
```

E.13 Control structures

- If a single statement follows a control structure, an enclosing brace should still be employed to make the block structure explicit and to simplify the addition of further statements later. Alternatively, the statement should be written on the same line as its associated control structure.
- Loop variables should be named, **loop, loopOuter, loopInner**, etc., and not single letters such as **i, j** that may overwrite other similarly named variables.
- Complex logical statements should be split across lines with each condition documented:

```
if ( aFlag == true    &&    // Insures input data correct
     radius < 0       ||    // Geometrical constraints satisfied
```

```
    fileStatus == 1)       // File is open
{ .. statements }
```

- Case statements in **switch** blocks should be indented after the colon:

```
switch (...) {
       case 'A' :
              ...statements ...
              break;
       default:
              ...statements ...
              break;
}
```

References and further reading

The following is far from complete, but lists some helpful texts arranged either by subject or by chapter.

Numerous texts on general C++ *programming are available at every level. Some well-known books are:*

Adams, J., Leestma, S., and Nyhoff, L. (1995). *C++: An Introduction to Computing*, Englewood Cliffs, NJ: Prentice Hall.

Bronson, G.J. (1997). *Program Development and Design using C++*, Boston, MA: PWS Publishing.

Cline, M.P. and Lomow, G.A. (1995). *C++ FAQs: Frequently Asked Questions*, Reading, MA: Addison Wesley Longman.

Cohoon, J.P. and Davidson, J.W. (1999). *C++ Program Design: An Introduction to Programming and Object-Oriented Design*, 2nd edn, Boston, MA: WCB McGraw-Hill.

Cooper, J.W. and Lam, R.B. (1994). *A Jump Start Course in C++ Programming*, New York, NY: John Wiley & Sons.

Deitel, H.M. and Deitel, P.J. (1998). *C++: How to Program*, 2nd edn, Upper Saddle River, NJ: Prentice Hall.

Eckel, B. (2000). *Thinking in C++, Volume 1: Introduction to Standard C++*, 2nd edn, Upper Saddle River, NJ: Prentice Hall.

Horstman, C.S. (1997). *Practical Object-Oriented Development in C++ and Java*, New York, NY: John Wiley & Sons.

Horstman, C.S. (1999). *Computing Concepts with C++ Essentials*, 2nd edn, New York, NY: John Wiley & Sons.

Johnsonbaugh, R. and Kalin, M. (1995). *Object-Oriented Programming in C++*, Englewood Cliffs, NJ: Prentice Hall.

Kafura, D. (1998). *Object-Oriented Software Design and Construction with C++*, Upper Saddle River, NJ: Prentice Hall.

Lee, P.A. and Phillips, C. (1996). *The Apprentice C++ Programmer: A Touch of Class*, London: PWS Publishing Company.

Lippman, S.B. and Lajoie, J. (1998). *C++ Primer*, 3rd edn, Reading, MA: Addison Wesley Longman.

Meyers, S. (1998). *Effective C++*, 2nd edn, Reading, MA: Addison Wesley Longman.

Nagler, E. (1997). *Learning C++: A Hands-On Approach*, Minneapolis/St.Paul: West Publishing.

Oualline, S. (2003). *How Not to Program in C++: 111 Broken Programs and 3 Working Ones, or Why does 2+2 = 5986*, San Francisco, CA: No Starch Press.

Perry, J.E. and Levin, H.D. (1996). *An Introduction to Object-Oriented Design in C++*, Reading, MA: Addison Wesley Longman.

Pohl, I. (1993). *Object-Oriented Programming Using C++*, Redwood City, CA: The Benjamin/Cummings Publishing Company.

Pohl, I. (1997). *C++ Distilled: A Concise ANSI/ISO Reference and Style Guide*, Reading, MA: Addison Wesley Longman.

Reiss, S.P. (1999). *A Practical Introduction to Software Design with C++*, New York, NY: John Wiley & Sons.

Savitch, W. (1995). *Problem Solving with C++: The Object of Programming*, Menlo Park, CA: Addison Wesley Longman.

Schildt, H. (1994). *C++ from the Ground Up*, Berkeley, CA: Osborne McGraw-Hill.

Skansholm, J. (1997). *C++ from the Beginning*, Harlow: Addison Wesley Longman.

Staugaard, A.C. (1997). *Structured and Object-Oriented Techniques: An Introduction using C++*, Upper Saddle River, NJ: Prentice Hall.

Strostrup, B. (1997). *The C++ Programming Language*, 3rd edn, Reading, MA: Addison Wesley Longman.

Winston, P.H. (1994). *On to C++*, Reading, MA: Addison Wesley Longman.

Books that introduce both C++ *and scientific programming are:*

Barton, J.J. and Nackman, L.R. (1994). *Scientific and Engineering C++: An Introduction with Advanced Techniques and Examples*, Reading, MA: Addison Wesley Longman.

Buzzi-Ferraris, G. (1993). *Scientific C++: Building Numerical Libraries the Object-Oriented Way*, Harlow: Addison Wesley Longman.

Hanley, J.R. (2002). *Essential C++ for Engineers and Scientists*, Boston, MA: Pearson Education.

Texts on C++ *and numerical methods include:*

Flowers, B.H. (2000). *An Introduction to Numerical Methods in C++*, 2nd edn, New York, NY: Oxford University Press.

Golub, G.H. and Ortega, J.M. (1992). *Scientific Computing and Differential Equations: An Introduction to Numerical Methods*, San Diego, CA: Academic Press.

Johnson, L.W. and Riess, R.D. (1982). *Numerical Analysis*, 2nd edn, Englewood Cliffs, NJ: Prentice Hall.

Kahaner, D., Moler, C., and Nash, S. (1989). *Numerical Methods and Software*, Engelwood Cliffs, NJ: Prentice Hall.

Kincaid, D. and Cheney, W. (1991). *Numerical Analysis: Mathematics of Scientific Computing*, Belmont, CA: Brooks/Cole Publishing Company.

Matthews, J.H. and Fink, K.D. (1999). *Numerical Methods Using MATLAB*, 3rd edn, Upper Saddle River, NJ: Prentice Hall.

Ortega, J.M. and Grimshaw, A.S. (1999). *An Introduction to C++ and Numerical Methods*, New York, NY: Oxford University Press.

Press, W.H., Teukolsky, S.A., Vetterling, W.T., and Flannery, B.P. (2002). *Numerical Methods in C++: The Art of Scientific Computing*, Cambridge: Cambridge University Press.

Stoer, J. and Bulirsch, R. (1993). *Introduction to Numerical Analysis*, 2nd edn, New York: Springer-Verlag.

Van Loan, C.F. (1997). *Introduction to Scientific Computing: A Matrix–Vector Approach Using MATLAB*, Upper Saddle River, NJ: Prentice Hall.

Books that emphasize computational science are:

DeVries, P.L. (1994). *A First Course in Computational Physics*, New York, NY: John Wiley & Sons.

Garcia, A.L. (2000). *Numerical Methods for Physics*, 2nd edn, Upper Saddle River, NJ: Prentice Hall.

Giordano, N.J. (1997), *Computational Physics*, Upper Saddle River, NJ: Prentice Hall.

Harrison, P. (2001). *Computational Methods in Physics, Chemistry and Biology: An Introduction*, Chichester: John Wiley & Sons.

Kinzel, W. and Reents, G. (1998). *Physics by Computer: Programming Physical Problems Using Mathematica® and C*, Berlin, Heidelberg, New York: Springer-Verlag.

Koonin, S.E. and Meridith, D.C. (1990). *Computational Physics: Fortran Version*, Redwood City, CA: Addison Wesley Longman.

Landau, R.H. (1997). *Computational Physics: Problem Solving with Computers*, New York, NY: John Wiley & Sons.

Pang, T. (1997). *An Introduction to Computational Physics*, Cambridge: Cambridge University Press.

Thijssen, J.M. (1999). *Computational Physics*, Cambridge: Cambridge University Press.

Vesely, F.J. (1994). *Computational Physics: An Introduction*, New York, NY: Plenum Press.

Yang, D. (2001). *C++ and Object-Oriented Numeric Computing for Scientists and Engineers*, New York, NY: Springer-Verlag.

Data structures and computer science algorithms are examined in:

Ammeraal, L. (1997). *Algorithms and Data Structures in C++*, Chichester: John Wiley & Sons.

Preiss, B.R. (1999). *Data Structures and Algorithms with Object-Oriented Design Patterns in C++*, New York, NY: John Wiley & Sons.

Sedgewick, R. (1998). *Algorithms in C++*, 3rd edn., Reading, MA: Addison Wesley Longman.

Weiss, M.A. (1999). *Data Structures and Algorithms in C++*, Reading, MA: Addison Wesley Longman.

Books, beside extensive product manuals, on software engineering, particularly the UML process and/or Rational Rose, include:

Booch, G., Rumbaugh, J., and Jacobson, I. (1999). *The Unified Modeling Language User Guide*, Reading, MA: Addison Wesley Longman.

Fowler, M. and Scott, K. (1997). *UML Distilled: Applying the Standard Object Modeling Language*, Reading, MA: Addison Wesley Longman.

Jacobson, I., Booch, G., and Rumbaugh, J. (1999). *The Unified Software Development Process*, Reading, MA: Addison Wesley Longman.

Krutchen, P. (1999). *The Rational Unified Process: An Introduction*, Reading, MA: Addison Wesley Longman.

Lee, R.C. and Tepfenhart, W.M. (1997). *UML and C++: A Practical Guide to Object-Oriented Development*, Upper Saddle River, NJ: Prentice Hall.

Pont, M.J. (1996). *Software Engineering with C++ and CASE Tools*, Harlow: Addison Wesley Longman.

Pooley, R. and Stevens, P. (1999). *Using UML Software Engineering with Objects and Components*, Harlow, Harlow: Addison Wesley Longman.

Quatrani, T. (1998). *Visual Modeling with Rational Rose and UML*, Reading, MA: Addison Wesley Longman.

Rumbaugh, J., Michael, B., Permelani, W. Eddy, F., and Lorensen, W. (1991). *Object-Oriented Modeling and Design*, Engelwood Cliffs, NJ: Prentice-Hall.

Sommerville, I. (1996). *Software Engineering*, 5th edn, Harlow: Addison Wesley Longman.

Tkach, D., Fang W., and So, A. (1996). *Visual Modeling Technique: Object Technology using Visual Programming*, Menlo Park, CA: Addison Wesley Longman.

Tkach, D. and Puttick, R. (1996). *Object Technology in Application Development*, Menlo Park, CA: Addison Wesley Longman.

Chapter 21: *The main reference for this chapter is the remarkable paper, which contains a considerable amount of additional material:*

Veldhuizen, T.L. (1999). C++ Templates as Partial Evaluation, in *1999 ACM SIGPLAN Workshop on Partial Evaluation and Semantics-Based Program Manipulation (PEPM '99)*.

More detail on the material in this paper can be found in Barton and Nackman (1994) listed above as well as the references:

Veldhuizen, T.L. (1995). Expression Templates, *C++ Report*, **7**, 26–31

Veldhuizen, T.L. (1995). Using C++ Template Metaprograms, *C++ Report*, **7**, 36–43

Veldhuizen, T.L. (1998). Arrays in Blitz++, in *Proceedings of the 2nd International Scientific Computing in Object-Oriented Parallel Environments (ISCOPE'98)*, Berlin, Heidelberg, New York, Tokyo: Springer-Verlag.

Veldhuizen, T.L. (2000). Techniques for Scientific C++, in *Indiana University Computer Science Technical Report #542 Version 0.4.*

A very simple and readable discussion of memory blocking is given in the online publication:

Maxwell, M.T. and Cameron, K.W. (2002). Optimizing Application Performance: A Case Study using LMBench, *ACM Crossroads Student Magazine*, **8** (5).

Chapter 22: *Simple comparisons of boundary conditions for finite difference implementations of parabolic equations can be found in:*

Yevick, D., Friese, T., and Schmidt, F. (2001). A Comparison of Transparent Boundary Conditions for the Fresnel Equation,. *J. Comp. Phys.*, **168**, 433–444.

Yevick, D., Yu, J., and Schmidt, F. (1997). Analytic Studies of Absorbing and Impedance-Matched Boundary Layers, *Photon. Technol. Lett.*, **9**, 73–75.

Two comprehensive textbooks on finite difference methods are:

Mitchell, A.R. and Griffiths, D.F. (1987). *The Finite Difference Method in Partial Differential Equations*, Chichester: John Wiley & Sons.

Samarskii, A.A. (2001). *The Theory of Difference Schemes*, New York, NY: Marcel Dekker.

Chapter 23: *The Monte-Carlo and importance sampling techniques are discussed in:*

Rubenstein, R.Y. (1981). *Simulation and the Monte-Carlo Method*, New York, NY: John Wiley & Sons.

Three key articles on the multicanonical method are:

Berg, B. and Neuhaus, T. (1991). Multicanonical Algorithms for First-Order Phase Transitions, *Phys. Lett. B*, **267**, 249–253.

Gubernatis, J. and Hatano, N. (2000). The Multicanonical Monte-Carlo Method, *Computing in Science and Engineering*, 95–102.

Okamoto, Y. and Hansmann, U.H.E. (1995). Thermodynamics of Helix-Coil Transitions Studied by Multicanonical Algorithms. *J. Phys. Chem.*, **99**, 11276–11287.

The implementation of the multicanonical method in the estimation of probability distribution functions is discussed further in the paper:

Yevick, D. (2003). Multicanonical Evaluation of Joint Probability Density Functions in Communication System Modeling, *IEEE Photon. Technol. Lett.*, **15**, 1540–1542.

Appendix A: *While a commercial MATLAB installation is accompanied by extensive paper and on-line documentation, numerous books present introductions to MATLAB that are specifically oriented to different scientific areas. Three of these are:*

Biran, A. and Breiner, R. (2002). *Matlab 6 for Engineers*, Upper Saddle River, NJ: Prentice Hall.

King, J. (2001). *MATLAB 6 for Engineers: Hands-on Tutorial*, Flourtown, PA: R.T. Edwards.

Pärt-Enander, E. and Sjöberg, A., (2001). *Användarhandledning för Matlab 6*, Stockholm: Elandars Gotab.

Appendix D: *A more extensive version of the material in this chapter can be found in the web document:*

B. Gobbo, (1999) *Calling Fortran Routines from a C++ Program in Unix*, wwwcompass.cern.ch/compass/software/offline/userinfo/fandc/fandc.html

Appendix E: *Different coding groups generally implement slightly different C++ programming conventions. The discussion here includes numerous IBM Visual Age® product conventions as well as personal preferences but otherwise follows the document:*

Naval Command, Control and Ocean Surveillance Center RDT & E Division (NRaD), now Space and Naval Warfare Systems Center, San Diego and AHNTECH, Inc., San Diego CA, (1995). C++ Language Coding Standard. In cadwes.colorado.edu/~billo/standards/nrad/.

Index